Informing an Effective Response to Climate Change

America's Climate Choices:
Panel on Informing Effective Decisions and Actions Related to Climate Change

Board on Atmospheric Sciences and Climate

Division on Earth and Life Studies

NATIONAL RESEARCH COUNCIL
OF THE NATIONAL ACADEMIES

THE NATIONAL ACADEMIES PRESS
Washington, D.C.
www.nap.edu

THE NATIONAL ACADEMIES PRESS · 500 Fifth Street, N.W. · Washington, DC 20001

NOTICE: The project that is the subject of this report was approved by the Governing Board of the National Research Council, whose members are drawn from the councils of the National Academy of Sciences, the National Academy of Engineering, and the Institute of Medicine. The members of the committee responsible for the report were chosen for their special competences and with regard for appropriate balance.

This study was supported by the National Oceanic and Atmospheric Administration under contract number DG133R08CQ0062. Any opinions, findings, conclusions, or recommendations expressed in this material are those of the author(s) and do not necessarily reflect the views of the sponsor.

International Standard Book Number-13: 978-0-309-14594-7 (Book)
International Standard Book Number-10: 0-309-14594-5 (Book)
International Standard Book Number-13: 978-0-309-14595-4 (PDF)
International Standard Book Number-10: 0-309-14595-3 (PDF)
Library of Congress Control Number: 2010940140

Copies of this report are available from the program office:

Board on Atmospheric Sciences and Climate
500 Fifth Street, N.W.
Washington, DC 20001
(202) 334-3512

Additional copies of this report are available from the National Academies Press, 500 Fifth Street, N.W., Lockbox 285, Washington, DC 20055; (800) 624-6242 or (202) 334-3313 (in the Washington metropolitan area); Internet, http://www.nap.edu

Cover images:

Far left: © City of Palo Alto

Middle left: courtesy of National Aeronautics and Space Administration

Middle right: Photo by Scott Bauer courtesy of National Wildlife Refuge

Far right: courtesy of National Oceanic and Atmospheric Administration

Printed in the United States of America

THE NATIONAL ACADEMIES
Advisers to the Nation on Science, Engineering, and Medicine

The **National Academy of Sciences** is a private, nonprofit, self-perpetuating society of distinguished scholars engaged in scientific and engineering research, dedicated to the furtherance of science and technology and to their use for the general welfare. Upon the authority of the charter granted to it by the Congress in 1863, the Academy has a mandate that requires it to advise the federal government on scientific and technical matters. Dr. Ralph J. Cicerone is president of the National Academy of Sciences.

The **National Academy of Engineering** was established in 1964, under the charter of the National Academy of Sciences, as a parallel organization of outstanding engineers. It is autonomous in its administration and in the selection of its members, sharing with the National Academy of Sciences the responsibility for advising the federal government. The National Academy of Engineering also sponsors engineering programs aimed at meeting national needs, encourages education and research, and recognizes the superior achievements of engineers. Dr. Charles M. Vest is president of the National Academy of Engineering.

The **Institute of Medicine** was established in 1970 by the National Academy of Sciences to secure the services of eminent members of appropriate professions in the examination of policy matters pertaining to the health of the public. The Institute acts under the responsibility given to the National Academy of Sciences by its congressional charter to be an adviser to the federal government and, upon its own initiative, to identify issues of medical care, research, and education. Dr. Harvey V. Fineberg is president of the Institute of Medicine.

The **National Research Council** was organized by the National Academy of Sciences in 1916 to associate the broad community of science and technology with the Academy's purposes of furthering knowledge and advising the federal government. Functioning in accordance with general policies determined by the Academy, the Council has become the principal operating agency of both the National Academy of Sciences and the National Academy of Engineering in providing services to the government, the public, and the scientific and engineering communities. The Council is administered jointly by both Academies and the Institute of Medicine. Dr. Ralph J. Cicerone and Dr. Charles M. Vest are chair and vice chair, respectively, of the National Research Council.

www.national-academies.org

AMERICA'S CLIMATE CHOICES:
PANEL ON INFORMING EFFECTIVE DECISIONS AND
ACTIONS RELATED TO CLIMATE CHANGE

DIANA LIVERMAN (Co-Chair), University of Arizona, Tucson and Oxford University, United Kingdom.

PETER RAVEN (Co-Chair), Missouri Botanical Garden, St. Louis

DANIEL BARSTOW, Challenger Center for Space Science Education, Alexandria, Virginia

ROSINA M. BIERBAUM, University of Michigan, Ann Arbor

DANIEL W. BROMLEY, University of Wisconsin-Madison

ANTHONY LEISEROWITZ, Yale University, New Haven, Connecticut

ROBERT J. LEMPERT, The RAND Corporation, Santa Monica, California

JIM LOPEZ, * Department of Housing and Urban Development

EDWARD L. MILES, University of Washington, Seattle

BERRIEN MOORE, III, Climate Central, Princeton, New Jersey

MARK D. NEWTON, Dell, Inc., Round Rock, Texas

VENKATACHALAM RAMASWAMY, National Oceanic and Atmospheric Administration, Princeton, New Jersey

RICHARD RICHELS, Electric Power Research Institute, Inc., Washington, D.C.

DOUGLAS P. SCOTT, Illinois Environmental Protection Agency, Springfield

KATHLEEN J. TIERNEY, University of Colorado at Boulder

CHRIS WALKER, The Carbon Trust LLC, New York, New York

SHARI T. WILSON, Maryland Department of the Environment, Baltimore

NRC Staff

MARTHA McCONNELL, Study Director

LAUREN M. BROWN, Research Associate

RICARDO PAYNE, Program Assistant

DAVID REIDMILLER, Christine Mirzayan Science and Technology Fellow

---

*Asterisks denote members who resigned during the study process.

Foreword: About America's Climate Choices

Convened by the National Research Council in response to a request from Congress (P.L. 110-161), *America's Climate Choices* is a suite of five coordinated activities designed to study the serious and sweeping issues associated with global climate change, including the science and technology challenges involved, and to provide advice on the most effective steps and most promising strategies that can be taken to respond.

The Committee on America's Climate Choices is responsible for providing overall direction, coordination, and integration of the *America's Climate Choices* suite of activities and ensuring that these activities provide well-supported, action-oriented, and useful advice to the nation. The committee convened a *Summit on America's Climate Choices* on March 30-31, 2009, to help frame the study and provide an opportunity for high-level input on key issues. The committee is also charged with writing a final report that builds on four panel reports and other sources to answer the following four overarching questions:

- What short-term actions can be taken to respond effectively to climate change?
- What promising long-term strategies, investments, and opportunities could be pursued to respond to climate change?
- What are the major scientific and technological advances needed to better understand and respond to climate change?
- What are the major impediments (e.g., practical, institutional, economic, ethical, intergenerational, etc.) to responding effectively to climate change, and what can be done to overcome these impediments?

The Panel on Limiting the Magnitude of Future Climate Change was charged to describe, analyze, and assess strategies for reducing the net future human influence on climate. The panel's report focuses on actions to reduce domestic greenhouse gas emissions and other human drivers of climate change, such as changes in land use, but also considers the international dimensions of limiting climate change.

The Panel on Adapting to the Impacts of Climate Change was charged to describe, analyze, and assess actions and strategies to reduce vulnerability; increase adaptive capacity; improve resiliency; and promote successful adaptation to climate change in different regions, sectors, systems, and populations. The panel's report draws on a wide range of sources and case studies to identify lessons learned from past experiences, promising current approaches, and potential new directions.

The Panel on Advancing the Science of Climate Change was charged to provide a concise overview of past, present, and future climate change, including its causes and its impacts, and to recommend steps to advance our current understanding, including new observations, research programs, next-generation models, and the physical and human assets needed to support these and other activities. The panel's report focuses on the scientific advances needed both to improve our understanding of the intergrated-climate system and to devise more effective responses to climate change.

The Panel on Informing Effective Decisions and Actions Related to Climate Change was charged to describe and assess different activities, products, strategies, and tools for informing decision makers about climate change and helping them plan and execute effective, integrated responses. This report describes the different types of climate change-related decisions and actions being taken at various levels and in different sectors and regions; and it develops a framework, tools, and practical advice for ensuring that the best available technical knowledge about climate change is used to inform these decisions and actions.

America's Climate Choices builds on an extensive foundation of previous and ongoing work, including National Research Council reports, assessments from other national and international organizations, the current scientific literature, climate action plans by various entities, and other sources. More than a dozen boards and standing committees of the National Research Council were involved in developing the study, and many additional groups and individuals provided additional input during the study process. Outside viewpoints were also obtained via public events and workshops (including the Summit), invited presentations at committee and panel meetings, and comments received through the study website, *http://americasclimatechoices.org*.

Collectively, the *America's Climate Choices* suite of activities involve more than 90 volunteers from a range of communities including academia, various levels of government, business and industry, other nongovernmental organizations, and the international community. Responsibility for the final content of each report rests solely with the authoring panel and the National Research Council. However, the development of each report inluded input from and interactions with members of all five study groups; the membership of each group is listed in Appendix E.

Preface

How can America make more informed decisions about climate change? This was the question asked of the Panel on Informing Effective Decisions and Actions Related to Climate Change. We were challenged to identify the opportunities and challenges associated with informing effective decisions and actions, including the different activities, products, strategies, and tools for informing decision makers about climate change and helping them plan and execute effective, integrated responses. We were asked to describe the different types of climate change-related decisions and actions being taken at various levels and in different sectors and regions and to review frameworks and tools for ensuring that the best available technical knowledge about climate change is used to inform these decisions and actions.

Our first challenge was to decide how to set the limits of our panel report given the broad statement of task, the limited time, and the potential for overlap with the work of the three other *America's Climate Choices* panels. We also took into account input received during the public discussion of the study, especially suggestions about the significance of looking at decision makers beyond the Federal government and about the importance of communication and education. We soon recognized that an informed and effective national response to climate change requires that the widest possible range of decisions makers—public and private, national and local—have access to up-to-date and reliable information about current and future climate change, the impacts of such changes, the vulnerability to these changes, and the response strategies for reducing emissions and implementing adaptation. We also acknowledged the importance of information that is needed to assess whether the decisions or responses are successful or should be revised.

We began our work with reflections about America's ability to face grand and complex challenges in the past, where a record of success and learning from experience provided us with an optimistic start to thinking about informing climate choices. We then examined the decisions and actions that have already been taken in relation to climate, who was making the decisions, and what tools and information they were using or lacking. Responding to our task statement we then turned to an assessment of frameworks and tools for making climate-related decisions and identified two key types of information services that are needed in making decisions about climate change: (1) information about climate, climate impacts, and adaptation, and (2) information about greenhouse gas emissions and their management. We recognized that

America needs good international information for effective decisions and can play an important role in maintaining international observational and research activities. Finally we decided to assess what is known about public understanding of climate and the ways in which climate knowledge is communicated and incorporated in formal and informal education systems. We tried as much as possible to maintain a "user" perspective: Is the right information available and accessible for the different types of decisions that people are making? Where is there potential for confusion? How can information services be designed so as to allow monitoring and assessment of climate and climate policy so that we can understand what is happening, evaluate the effectiveness of policies, and make adjustments to increase the effectiveness of decisions?

We were fortunate that our panel included representatives from many different groups, including federal, state, and local government, universities, the private sector, physical and social scientists, and several individuals who have long experience of decision making about climate in a variety of different roles, including international and non-governmental organizations. We believe that this diverse panel reflects the range of actors involved in decision making about climate, but we also invited several people to meet with the panel to share their insights and answer our questions. We are grateful to these speakers: Eric Barron, Mary Nicholls, Ted Nordhaus, Jeff Seabright, Alex Perera, Amanda Staudt, Mark Way, Andrew Castaldi, Michael Liffmann, Brad Udall, Louie Tupas, and Chet Koblinsky. We also relied on a number of previous National Research Council reports that focused on decision making about climate and many of our findings and recommendations echo, reemphasize, and build on those of previous panels and committees. Several members of other panels and the main committee were helpful in defining areas of overlap and providing information in their areas of expertise, especially Kathy Jacobs, Jim Geringer, and Tom Karl. We are especially grateful to Adam Bumpus for his assistance with Chapters 2 and 6.

The study was conducted during a period when climate issues and climate policy were being debated and developed at all levels of government, and our time frame for the report included pivotal negotiations at the international level in Copenhagen, several climate-related bills in Congress, proposals for new approaches to climate services in Federal agencies, state actions to limit emissions and set up greenhouse gas trading, swings in public support for climate policy, and some major private sector actors moving to incorporate climate risks into their investments and decisions. This posed challenges for the panel, as it sometimes seemed as though the world was racing ahead of our cautious deliberations. This is one reason why we have avoided, for the most part, focusing on specific recommendations, and have chosen to emphasize the frameworks, information, and criteria that can be used to inform and evaluate decisions, whatever those decisions may be.

We extend our gratitude to the staff at the National Research Council. We are especially grateful to our study director, Martha McConnell, who was able to maintain a positive attitude and enthusiasm in spite of our time limitations, the challenges in bounding our task and in liaising with other panels, and coordinating a large committee of talented but very busy members. Martha was assisted in all aspects of her work by Lauren Brown. David Reidmiller, Joe Casola, Paul Stern, and Michael Craghan also assisted at times during the study process. We also want to acknowledge the director of the Board on Atmospheric Sciences and Climate, Chris Elfring, who engaged often with our panel to provide advice and coordination with other panels.

Diana Liverman (Co-Chair)

Peter Raven (Co-Chair)

Acknowledgments

This report has been reviewed in draft form by individuals chosen for their diverse perspectives and technical expertise, in accordance with procedures approved by the National Research Council's Report Review Committee. The purpose of this independent review is to provide candid and critical comments that will assist the institution in making its published report as sound as possible and to ensure that the report meets institutional standards for objectivity, evidence, and responsiveness to the study charge. The review comments and draft manuscript remain confidential to protect the integrity of the deliberative process. We wish to thank the following individuals for their participation in their review of this report:

CHRIS WEST, UK Climate Impacts of Programme, Oxford
ROSS ANDERSON, High Roads for Human Rights, Salt Lake City, Utah
NED FARQUHAR, Department of the Interior, Bureau of Land and Minerals Management, Washington, D.C.
CHARLES O. HOLLIDAY JR., Bank of America, Washington, D.C.
KIRSTEN DOW, University of South Carolina, Columbia
KAI LEE, The David and Lucile Packard Foundation, Los Altos, California
PETER GOLDMARK, Environmental Defense Fund, New York, New York
BONNIE VAN DORN, Association of Science-Technology Centers, Washington, D.C.
MICHELE BETSILL, Colorado State University, Fort Collins
F. STUART CHAPIN III, University of Alaska, Fairbanks
STEPHEN SCHNEIDER, Stanford University, Palo Alto, California
CHARLES REDMAN, Arizona State University, Tempe
ELTON SHERWIN, Ridgewood Capital, Palo Alto, California
HADI DOWLATABADI, University of British Columbia, Vancouver
CARLO C. JAEGER, Potsdam Institute for Climate Impact Research, Germany

Although the reviewers listed above have provided many constructive comments and suggestions, they were not asked to endorse the conclusions or recommendations nor did they see the final draft of the report before its release. The review of this report was overseen by M. Granger Morgan, Carnegie Mellon University, and Robert A. Frosch, Harvard University. Appointed by the National Research Council, they were responsible for making certain that an independent examination of this report was carried out in accordance with institutional procedures and that all review comments were carefully considered. Responsibility for the final content of this report rests entirely with the authoring committee and the institution.

Institutional oversight for this project was provided by:

BOARD ON ATMOSPHERIC SCIENCES AND CLIMATE

ANTONIO J. BUSALACCHI, JR. (Chair), University of Maryland, College Park
ROSINA M. BIERBAUM, University of Michigan, Ann Arbor
RICHARD CARBONE, National Center for Atmospheric Research, Boulder, Colorado
WALTER F. DABBERDT, Vaisala, Inc., Boulder, Colorado
KIRSTIN DOW, University of South Carolina, Columbia
GREG S. FORBES, The Weather Channel, Inc., Atlanta, Georgia
ISAAC HELD, National Oceanic and Atmospheric Administration, Princeton, New Jersey
ARTHUR LEE, Chevron Corporation, San Ramon, California
RAYMOND T. PIERREHUMBERT, University of Chicago, Illinois
KIMBERLY PRATHER, Scripps Institution of Oceanography, La Jolla, California
KIRK R. SMITH, University of California, Berkeley
JOHN T. SNOW, University of Oklahoma, Norman
THOMAS H. VONDER HAAR, Colorado State University/CIRA, Fort Collins
XUBIN ZENG, University of Arizona, Tucson

Ex Officio Members

GERALD A. MEEHL, National Center for Atmospheric Research, Boulder, Colorado

NRC Staff

CHRIS ELFRING, Director
EDWARD DUNLEA, Senior Program Officer
LAURIE GELLER, Senior Program Officer
IAN KRAUCUNAS, Senior Program Officer
MARTHA McCONNELL, Program Officer
MAGGIE WALSER, Associate Program Officer
TOBY WARDEN, Associate Program Officer
JOE CASOLA, Post-Doctoral Researcher
RITA GASKINS, Administrative Coordinator

KATIE WELLER, Research Associate
LAUREN M. BROWN, Research Associate
ROB GREENWAY, Program Associate
SHELLY FREELAND, Senior Program Assistant
AMANDA PURCELL, Senior Program Assistant
JANEISE STURDIVANT, Program Assistant
RICARDO PAYNE, Program Assistant
SHUBHA BANSKOTA, Financial Associate

Contents

SUMMARY   1

1   INTRODUCTION   19
     The Challenge of Climate Change, 20
     Decision Makers, Their Information Needs, and the Challenge of
        Resource Allocation, 25
     Barriers to Effective Decision Making, 29
     The Scope and Purpose of This Report, 42
     Organization of This Report, 44

2   MANY DIFFERENT DECISION MAKERS ARE MAKING CHOICES TO RESPOND
     TO CLIMATE CHANGE   47
     States, Cities, and Local Governments, 49
     Business Sector, 59
     Non-governmental Organizations, 72
     Federal Government, 73
     The Law and Climate Change, 84
     Conclusions and Recommendations, 86

3   DECISION FRAMEWORKS FOR EFFECTIVE RESPONSES TO CLIMATE CHANGE   91
     Why Responding to Climate Change Needs a Decision Framework, 93
     Other Ways of Making Decisions, 97
     Fundamental Elements of a Risk Management Framework, 103
     The Utility of Adaptive Governance in Decision Making About Climate Change, 117
     Conclusions and Recommendations, 121

4   RESOURCES FOR EFFECTIVE CLIMATE DECISIONS   123
     Whose Decisions? Which Resources?, 125
     Decision Support Tools: Their Characteristics and Uses, 130
     Understanding Impacts and Informing Adaptation Decisions, 150
     Understanding the Value of Information for Resource Allocation, 153
     Assessments as Tools for Climate-Related Decision Making, 161
     Conclusions and Recommendations, 164

5   CLIMATE SERVICES: INFORMING AMERICA ABOUT CLIMATE VARIABILITY AND
    CHANGE, IMPACTS, AND RESPONSE OPTIONS   167
    The Need for Climate Services, 170
    Potential Functions of Climate Services, 178
    Recommendations from Prior Reports, 185
    Institutional Considerations, 189
    Goals for Climate Services Operation, 197
    Metrics for Evaluating Performance of Climate Services, 201
    Conclusions and Recommendations, 201

6   INFORMING GREENHOUSE GAS MANAGEMENT   205
    Greenhouse Gas Accounting Systems, 205
    Information on Emissions and Energy Use and the Public Response to
        Climate Change, 222
    Institutional Options for Informed Greenhouse Gas Management, 229
    Competition and Equity Considerations, 230
    Conclusions and Recommendations, 231

7   INTERNATIONAL INFORMATION NEEDS   235
    Economic and Market Couplings, 235
    Shared Resources and Ecological Services, 237
    Human Health, 239
    Humanitarian Reasons, 239
    National Security, 241
    Ways Forward, 242
    Conclusions and Recommendations, 249

8   EDUCATION AND COMMUNICATION   251
    K-12, Higher Education, and Informal Science Education, 252
    The General Public, 258
    Communication and Education for Decision Makers, 274
    Conclusions and Recommendations, 276

REFERENCES   283

APPENDIXES

A   Panel on Informing Effective Decisions and Actions Related to Climate Change: Statement of Task, 301
B   The American Experience with Complex Decisions: Past Examples, 303
C   Comparison of CO_2 Emissions for States Versus National, United States, in 1999 and 2000, 311
D   State Greenhouse Gas Emission Targets and Baselines, 313
E   America's Climate Choices: Membership Lists, 315
F   Panel on Informing Effective Decisions and Actions Related to Climate Change: Biographical Sketches, 319

Summary

Global climate change is a significant long-term challenge for the United States. Across the nation, individuals, businesses, and federal, state, and local governments are already consciously making decisions to respond to climate change. To make informed decisions, people need a basic understanding of the causes, likelihood, and severity of the impacts of climate change and the range, cost, and efficacy of different options to limit or adapt to it.

Individuals are choosing whether to make their homes and transportation more energy efficient, or support climate and energy policies. Private companies are reducing their carbon footprints, and some are planning for climate impacts. Humanitarian and environmental non-governmental organizations (NGO's) are deciding how to guide their members and respond to climate change. Resource managers are deciding how to manage water, forests, and coastal ecosystems to reduce the risks of climate change. Cities and states are starting to limit emissions and develop adaptation plans—today, more than 50 percent of Americans live in a jurisdiction that has enacted a greenhouse gas emissions (GHG) reduction goal.

This growing number of people and organizations responding to climate change has not only increased the demand for information but also provides the basis for an effective national capacity to respond to climate change. These efforts can be thought of as a set of policy experiments which can inform future action by other federal and non-federal actors. Three key lessons are drawn from these experiences:

1. A broad range of tailored information and tools is needed for the diversity of decision makers and to engage new constituencies.
2. Most decision makers will need to make climate choices in the context of other responsibilities, competing priorities, and resource constraints.
3. There is a critical need to coordinate a national response that builds on existing efforts, learns from successes and failures, reduces burdens on any one region or sector, and ensures the credibility and comprehensiveness of information and policy.

The Panel on Informing Effective Decisions and Actions Related to Climate Change, a part of the congressionally requested study on *America's Climate Choices* (ACC), was charged to describe and assess climate change-related activities, decisions, and actions at various levels and in different sectors and to examine the available decision frameworks and tools to inform these decisions and actions. The panel focused this charge by asking the following questions:

1. Who is making decisions and taking action on climate change in the United States? What are their needs for information and decision support, and what are the barriers to good decisions?
2. What decision making frameworks and methods are being used, and which are the most effective?
3. How might climate and greenhouse gas information systems and services support more effective decisions and actions?
4. What is known about the most effective ways to communicate about climate change, especially with the public and through formal and informal education?

The panel coordinated with the other ACC panels and notes that many of the findings of the companion reports are consistent with the independent findings of our panel. For example, the reports *Limiting the Magnitude of Future Climate Change* (NRC, 2010d) and *Adapting to the Impacts of Climate Change* (NRC, 2010a) highlight the importance and leadership of local and state governments and the private sector in reducing GHG emissions and adapting to climate impacts. The reports *Advancing the Science of Climate Change* (NRC, 2010b) and *Adapting to the Impacts of Climate Change* (NRC, 2010a) recommend a risk management approach to respond to climate change. Collectively, these reports conclude that there is strong, credible, scientific evidence that climate change is happening and is caused largely by human activities, and provide a number of options to limit emissions and adapt to the impacts.

An effective national response to climate change will require informed decision making based on reliable, understandable, and timely climate-related information tailored to user needs. For example, state and local authorities need improved information and tools to plan to both reduce emissions and adapt to the impacts of climate change, and a better understanding of how the public views climate change. Private firms who plan to disclose climate risks need standardized methods of reporting and better information about how climate impacts, policy, and consumer concerns are changing. Educators and organizations seeking to communicate about climate change need more accessible and reliable information about climate, guidelines for effective communication, and information about helpful networks.

Good information systems and services are essential to effectively and iteratively manage climate risks. They help decision makers evaluate whether particular policies and actions are achieving their goals or should be modified and underpin the effective communication of climate change choices to Congress, students, and the public.

Transparency, accountability, and fairness in the measurement, reporting, and verification of data on climate change, risks and vulnerabilities, sources of GHG emissions, and climate policy is a priority.

Although the findings and recommendations of this report are mainly directed at the federal government—especially the federal role in the design of information systems and services to support and evaluate responses to climate change—it is also relevant to decision makers in state, local, and tribal governments, and in the private and nongovernmental sectors who are making decisions about climate change.

COORDINATE A COMPREHENSIVE, NATIONWIDE RESPONSE TO CLIMATE CHANGE

Today, decisions and actions related to climate change are being informed by a loose confederation of networks and other institutions created to help guide climate choices (Figure S.1). In the panel's judgment, the federal government has the responsibility and opportunity to lead and coordinate the response to climate change, not only to protect the nation's national security, resources, and health, but also to provide a policy framework that promotes effective responses at all levels of American society. Although actions taken to date offer many lessons, a patchwork of regional, state, and local policies has emerged, prompting some state and business leaders to call for the development of a more predictable and coherent policy environment at the federal level. Federal policy can benefit from comprehensive information about the actual effectiveness of emission reduction and adaptation actions across the nation. A clearinghouse could provide careful reporting and verification of what climate-related decisions are being implemented.

Even the federal response is difficult to evaluate because the number of agencies beginning to respond to climate change has expanded far beyond the core research functions of the U.S. Global Change Research Program (USGCRP)—for example, to agencies with responsibility for infrastructure, security, and housing—and because of the lack of clear, accessible, and coordinated information on federal responsibilities and policies. Many federal agencies have not yet incorporated or "mainstreamed" climate change into their own agency planning processes. Effective and visible incorporation of climate concerns as central to the ongoing activities of the federal agencies would be a major step forward. This explicit demonstration of leadership could help galvanize and maintain the development of responses in the private sector, states, regions, and localities. The panel concludes that there is an urgent need to improve the coordination of climate information, decisions, assessment, and programs across federal agencies to ensure an effective response to climate change across the nation.

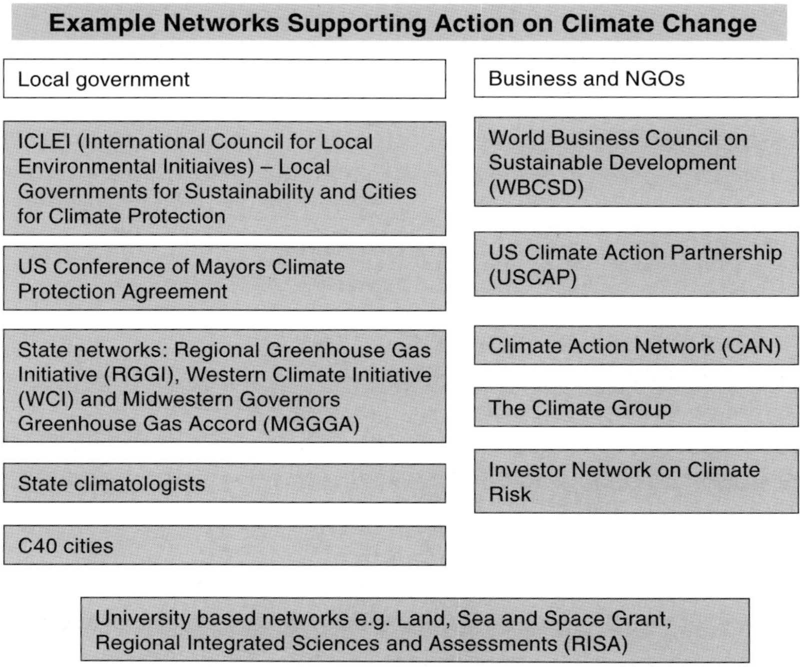

Example Networks Supporting Action on Climate Change

Local government	Business and NGOs
ICLEI (International Council for Local Environmental Initiaives) – Local Governments for Sustainability and Cities for Climate Protection	World Business Council on Sustainable Development (WBCSD)
US Conference of Mayors Climate Protection Agreement	US Climate Action Partnership (USCAP)
State networks: Regional Greenhouse Gas Initiative (RGGI), Western Climate Initiative (WCI) and Midwestern Governors Greenhouse Gas Accord (MGGGA)	Climate Action Network (CAN)
	The Climate Group
State climatologists	Investor Network on Climate Risk
C40 cities	

University based networks e.g. Land, Sea and Space Grant, Regional Integrated Sciences and Assessments (RISA)

FIGURE S.1 Example networks supporting action on climate change.

Recommendation 1:

To improve the response to climate change, the federal government should

a) Improve federal coordination and policy evaluation by establishing clear leadership, responsibilities, and coordination at the federal level for climate-related decisions, information systems, and services.

The roadmap for federal coordination might include leadership and action through executive orders, the Office of Science and Technology Policy, an expanded USGCRP, a new Council on Climate Change, the reorganization of existing agencies, or even the establishment of new organizations, regional centers, or departments within the government.

b) Establish information and reporting systems that allow for regular evaluation and assessment of the effectiveness of both government and non-gov-

ernmental responses to climate change, including a regular report to Congress or the President as suggested in our companion reports.

This could include aggregating and disseminating "best practices" with a web-based clearinghouse and creating ongoing assessments to enable regular exchange of information and plans among relevant federal agencies, regional researchers, decision makers, NGOs, and concerned citizens.

Recommendation 2:

To maximize the effectiveness of responses to climate change across the nation, the federal government should

a) **Assess, evaluate, and learn from the different approaches to climate-related decision making used by non-federal levels of government and the private sector;**

b) **Enhance non-federal activities that have proven effective in reducing greenhouse gas emissions and adapting to the projected impacts of climate change through incentives, policy frameworks, and information systems; and**

c) **Ensure that proposed federal policies do not unnecessarily preempt effective measures that have already been taken by states, regions, and the private sector.**

The potential of the aggregated emission reductions from non-federal actors is considerable and, if successful, would ease the task and lower the costs for the federal government. The federal government can enhance and complement these responses through carefully designed integrative policies; however, there is also a risk that federal action could preempt or discourage decisions by other actors, and miss opportunities to credit those non-federal actors who have taken early action.

ADOPT AN ITERATIVE RISK MANAGEMENT APPROACH TO CLIMATE CHANGE

Many climate-related decisions must address and incorporate uncertainty, the expectation of surprises, and factors that underlie the need to improve long-term decision making and crisis responses. Decision makers will differ in their assessment of the degree of risk that is unacceptable. These are not issues that are unique to climate choices. Decision makers in government and the private sector, as well as individuals, frequently make decisions with only partial or uncertain information and update these decisions as conditions change or more information becomes available. These include decisions with long-term implications such as saving for retirement, buying insurance, investing in infrastructure, or launching new products. The range of possibilities depends on future conditions and shifts such as a recession or technology breakthrough and the availability of information. Most people recognize the need to act despite uncertainty and that it is impossible to eliminate all risk. Effective management can benefit from a systematic and iterative framework for decision making (see Figure S.2).

Examples of the effective use of iterative risk management for climate choices discussed in this report include the UK Climate Impacts Programme, the NYC and Chicago Adaptation Plans, Tulsa flood management, Swiss Re insurance, the National Ad-

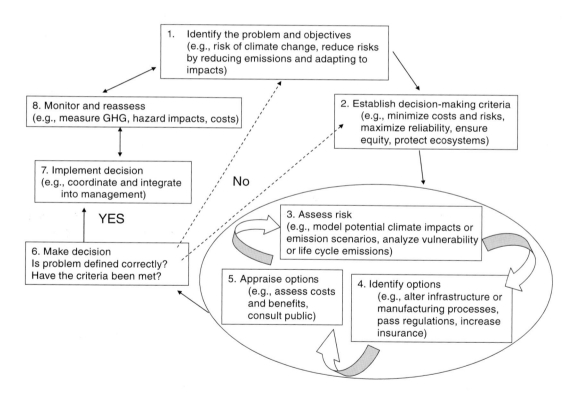

FIGURE S.2 An iterative risk management and adaptive governance approach for climate change at multiple levels of government and public and private sectors in which risks and benefits are identified and assessed and responses are implemented, evaluated, and revisited in sustained efforts to develop more effective policies or to respond to emerging problems and opportunities. SOURCE: Adapted from Willows and Connell (2003).

aptation Plans of Action (United Nations Framework Convention on Climate Change), and water management in Southern California.

Recommendation 3:

Decision makers in both public and private sectors should implement an iterative risk management strategy to manage climate decisions and to identify potential climate damages, co-benefits, considerations of equity, societal attitudes to climate risk, and the availability of potential response options. Decisions and policies should be revised in light of new information, experience, and stakeholder input, and use the best available information and assessment base to underpin the risk management framework.

There are important areas in which iterative risk management is already being used to manage climate risks. For example, the Federal government uses the Federal Crop Insurance Corporation (FCIC) and National Flood Insurance Program (NFIP) to share and reduce the risks of current weather variability for farmers and homeowners. However, the insurance programs do not take into account climate change, its impact on likely losses, and the fiscal implications. In the private sector, some firms already report on their management of environmental impacts to government and shareholders, but reporting can be inconsistent and many firms still do not take into account climate risks (e.g., responsibility for emissions, policy uncertainty, climate impacts) in their planning and disclosure.

Recommendation 4:

The federal government should review and revise federal risk insurance programs (such as FCIC and NFIP) to take into account the long-term fiscal and coverage implications of climate change. The panel endorses the steps that have already been taken by federal financial and insurance regulators, such as the Securities Exchange Commission, to facilitate the transparency and coordination of financial disclosure requirements for climate change risks.

IMPROVE THE RANGE AND ACCESSIBILITY OF TOOLS TO SUPPORT CLIMATE CHOICES

Tools and methods for making decisions about climate change range from basic graphs to complex computer-based tools such as Earth system models, impact models, economic (including cost-effectiveness and cost-benefit) models, and integrated assessment models. There are several important challenges in the use of these tools and methods for evaluating the potential outcomes of different decisions:

- A mismatch exists between the global- or national-scale climate and energy models and the needs of local or sectoral decision makers.
- There is a lack of agreement over approaches to economics, uncertainties, and subjective judgments in the development of tools.
- Users can misunderstand the assumptions and limits of tools and methods and require technical training and stakeholder engagement.
- As recommended in the 2007 National Research Council report *Analysis of Global Change Assessments* (NRC, 2007a), assessments that synthesize information and evaluate progress toward goals are an important decision support tool and need a clear mandate and goals, adequate funding, engagement of users, strong leadership, interdisciplinary integration, careful treatment of uncertainties, independent review, nested approaches, and development of relevant tools and communication strategies.

Recommendation 5:

a) **The federal government should support research and the development and diffusion of decision support tools and include clear guidance as to their uses and limitations for different types and scales of decision making about climate change.**

b) **The federal government should support training for researchers on how to communicate climate change information and uncertainties to a variety of audiences using a broad range of methods and media.**

CREATE INFORMATION SYSTEMS AND SERVICES TO SUPPORT LIMITING EMISSIONS, ADAPTATION, AND EVALUATING THE EFFECTIVENESS OF DECISIONS AND ACTIONS

There is a growing demand for better information on climate change, including climate variability, observed climate changes, potential impacts, trends in greenhouse gas emissions, and options for limiting emissions or adaptation. Some of these demands are a result of new regulatory or reporting structures (e.g., state and regional GHG trading schemes, Environmental Protection Agency (EPA) requirements, and city and corporate emission reduction commitments); growing concerns that climate change is affecting local water resources, ecosystems, or human health; and the assessment obligations of the U.S. Global Change Research Act.

These information needs can be met by a range of services at scales from the local to the international, including systems that cover both climate risks and GHG management. The federal government has a critically important role because it provides and supports large infrastructure for data collection and analysis (such as satellites, climate models, or *in situ* monitoring systems), it can make information easily accessible to diverse populations, and it can set standards for information quality. Non-federal governments and the private sector also have important roles to play by sharing results from the actions they take.

Our main insights and recommendations focus on four critical sets of information:

1. Climate services,
2. Greenhouse gas information systems,
3. Consumer information relating to greenhouse gas emissions, and
4. Information about the international context.

Climate Services

Although a long-term goal might be to establish a single federal "climate service" that could provide information on both climate change and GHGs, the panel decided to discuss climate information and GHG information separately and make distinct recommendations. The federal system is changing rapidly, including agency announcements of climate services initiatives and the establishment of national and regional GHG regulations and registries. Proposals have been made for a multilevel network as well as a national assessment process that would include many federal agencies and regional centers and take advantage of expertise within, for example, the National Weather Service, Cooperative Extension programs, the Regional Integrated Sciences and Assessments (RISA), Regional Climate Centers (RCC), Sea Grant and Land Grant programs, the private sector, and universities. Many of these already provide important models of how to interact with stakeholders and provide climate information relevant to local and larger-scale decisions; climate services should build on, enhance, and avoid unnecessary damage to these efforts.

Key functions to meet national needs for state-of-the-art information on climate change, its impacts, and response options to reduce risk may be overlooked if the system is based only on existing federal capabilities (see Box S.1). In addition, there are benefits in integrating federal activities—such as climate observations or modeling—

BOX S.1
Summary of Core Climate Service Functions

1. A user-centered focus that responds to the decision making needs of government and other actors at national, regional, and local scales;
2. Research on user needs, response options, effective information delivery mechanisms, and processes for sustained interaction with multiple stakeholders;
3. Enhanced observations and analyses designed specifically to provide timely, credible, authoritative, relevant, and regionally useful information on climate change and vulnerability, and effectiveness of responses;
4. Trustworthy and timely climate modeling and research to support federal decision making about limiting emissions and adaptation;
5. A central and accessible web portal of information that includes a system for sharing response strategies and access to decision support tools;
6. Capacity building and training for linking knowledge to action across the nation; and
7. An international information component.

A detailed discussion of core climate service functions is presented in Chapter 5.

---

with regionally based, bottom-up research, vulnerability analyses, and adaptation options.

The panel does not recommend a specific institutional home or structure for climate services, but it is our judgment that no single government agency or centralized unit can perform all the functions required. Coordination of agency roles and regional activities is a necessity for effective climate services, and efforts should be made to build upon existing relationships of trust between stakeholders and climate information providers (such as those developed at regional centers and regional agency offices).

To inform and be effective, climate services need a clear set of principles to guide products and activities. This includes leadership and institutional support at the highest level; credible, timely, and clearly communicated science (regional, natural, and social); equitable access to information and input from users; adequate and independent budget; and ongoing evaluation of effectiveness to adapt services to new information.

Recommendation 6:

The nation needs to establish a coordinated system of climate services that involves multiple agencies and regional expertise, is responsive to user needs,

has rigorous scientific underpinnings (in climate research, vulnerability analysis, decision support, and communication), performs operational activities (timely delivery of relevant information and assessments), can be used for ongoing evaluation of climate change and climate decisions, and has an easily accessible information portal that facilitates coordination of data among agencies and a dialogue between information users and providers.

Greenhouse Gas Information Systems

The importance of monitoring, reporting, and verification of emissions has emerged as a key issue in climate negotiations and climate policy. High quality, harmonized information on emissions from multiple sources and at multiple scales is needed to detect trends, verify emissions reduction claims, develop policies to manage greenhouse gases, and inform citizens. Both public and private organizations report information on emissions, often using standards and methods geared toward a specific application (e.g., regulation, carbon trading, and international treaty reporting). The resulting plethora of GHG information systems has created confusion for consumers, businesses, and policy makers and threatens to undermine the legitimacy of responses. Harmonization of different approaches is essential to ensure that GHG emission reporting is transparent, accountable, and fair (Box S.2). The federal government should provide enhanced GHG information and management systems, perhaps as part of climate services. This could assist decision makers in the public and private sectors, with advice on policy, emissions reporting, and practical steps toward GHG emissions reductions.

Recommendation 7:

The nation should establish a federally supported system for greenhouse gas monitoring, reporting, verification, and management that builds on existing expertise in the EPA and the DOE but could have some independence. The system should include the establishment of a unified (or regionally and nationally harmonized) greenhouse gas emission accounting protocol and registry. Such an information system should be supported and verified through high quality scientific research and monitoring systems and designed to support evaluations of policies implemented to limit greenhouse gas emissions.

BOX S.2
Summary of Elements of an Effective Greenhouse Gas Accounting System

1. Accounting principles to allow accurate, transparent, relevant, consistent, and accessible information;
2. A strong scientific basis in research on GHG science, monitoring, and the design of accounting systems;
3. A national accounting system and standards to report the full range of GHG emissions using consistent methods, boundaries, baselines, and acceptable thresholds;
4. Information available at the zip code and firm level;
5. High-quality verification schemes, including for carbon offsets, agricultural land use, and forests;
6. Methods to facilitate GHG management in supply chains and to control emissions at the most effective stage in the production-consumption chain;
7. A national greenhouse gas registry to track emissions from specific entities, support a variety of policy choices, and link to international systems that might benefit American firms and citizens;
8. Ongoing evaluation and feedback with users to support adaptive management and to adapt to new science and monitoring technologies.

A detailed discussion of GHG management systems is presented in Chapter 6.

Consumer Information Relating to Greenhouse Gas Emissions

Consumers and firms can play an important role in the national response to climate change by choosing to reduce their energy use or purchase low-carbon products. A significant proportion of consumers may respond to smart billing and meters that provide feedback, information on energy efficiency (e.g., product ratings and appliance labels), and carbon calculators or GHG labeling, especially when legitimated by federal or industry-wide standards.

Accurate and available information about GHG emissions can also—when coupled with incentives, regulation, and technology—foster changes in behavior. Existing federal efforts could be expanded to promote reliable information and advice for consumers, best practices, and services to enterprises on reporting, measuring, and practical steps to limiting emissions.

Recommendation 8:

The federal government should review and promote credible and easily understood standards and labels for energy efficiency and carbon/greenhouse gas information that can build public trust, enable effective consumer choice, identify business best practices, and adapt to new science and new emission reduction goals as needed. The federal government should also consider the establishment of a carbon or greenhouse gas advisory service targeted at the public and small and medium enterprises. Core functions could include information provision, assessment of user needs and national progress in limiting emissions, carbon auditing guidelines and reporting standards, carbon calculators, and support for research.

International Information

Information from other countries is essential to U.S. choices about responding to climate change for reasons that include (1) the economic and market couplings of the United States with the rest of the world, such as in agriculture; (2) shared water and other natural resources; (3) disease spread and human health; (4) humanitarian relief efforts; and (5) human and national security. The United States needs to be an active participant in improved acquisition and sharing of global data and increased monitoring, understanding, and surveillance of climate change and variability, greenhouse gases, forests, land use and biogeochemical cycles, vulnerabilities, and the effectiveness of emissions reduction and adaptation responses.

A wide range of users, including farmers, businesses, humanitarian and conservation NGOs, transboundary resource managers, and security agencies, can benefit from international information about climate change and climate change policies. Many federal agencies support the collection, analysis, and dissemination of international information and the United States needs to be a leader in establishing international consistency between systems, standards for monitoring greenhouse gas emissions and other critical earth system variables, and supporting economic and social data that inform both domestic and international decision making about the impacts and responses to climate variability and change.

Recommendation 9:

The federal government should support the collection and analysis of international information, including (a) climate observations, model forecasts, and

projections; (b) the state and trends in biophysical and socioeconomic systems; (c) research on international climate policies, response options, and their effectiveness; and (d) climate impacts and policies in other countries of relevance to U.S. decision makers.

IMPROVE THE COMMUNICATION, EDUCATION, AND UNDERSTANDING OF CLIMATE CHOICES

Communicating about climate change and climate choices is challenging for a variety of reasons, including the invisibility of GHGs; the time lag between GHG emissions and climate impacts; the complexity and uncertainty of projections; the different ways in which people approach and frame climate change in the context of other priorities and concerns; the impact of the media, special interests, and advocacy in polarizing debate; and the difficulties scientists have establishing bridges to the public and policy makers.

The climate-related decisions that society will confront over the coming decades will require an informed and engaged public and an education system that provides students with the knowledge they need to make informed choices about responses to climate change. Today's students will become tomorrow's decision makers as business leaders, farmers, government officials, and citizens. Our report finds that much more could be done to improve climate literacy, increase public understanding of climate science and choices, and inform decision makers about climate change, including an urgent need for research on effective methods of climate change education and communication. Table S.1 summarizes some simple guidelines for effective climate change communication.

Public Understanding of Climate Change

Although nearly all Americans have now heard of climate change, many have yet to understand the full implications, the options for a national response, and the opportunities and risks that lie in the solutions. Public beliefs and attitudes can shift from year to year in response to media coverage and other events. For example, high unemployment, heavy snowfalls in the eastern United States, and criticisms of climate scientists were concurrent with a decline in public concern in early 2010. Majorities of Americans, however, are still concerned about climate change, want their elected officials at all levels to take more action, and support policies such as renewables, regulation, and incentives to reduce GHG emissions. Likewise, many Americans are interested in mak-

TABLE S.1 Guidelines for Effective Climate Change Communication

Principle	Example
Know your audience	There are different audiences among the public. Learn what people (mis)understand before you deliver information and tailor information for each group.
Understand social identities and affiliations	Effective communicators often share an identity and values with the audience (e.g., a fellow CEO or mayor, parent, co-worker, religious belief, or outdoor enthusiast).
Get the audience's attention	Use appropriate framing (e.g., climate as an energy, environmental, security, or economic issue) to make the information more relevant to different groups.
Use the best available, peer-reviewed science	Use recent and locally relevant research results. Be prepared to respond to the latest debates about the science.
Translate scientific understanding and data into concrete experience	Use imagery, analogies, and personal experiences including observations of changes in people's local environments. Make the link between global and local changes. Discuss longer time scales, but link to present choices.
Address scientific and climate uncertainties	Specify what is known with high confidence and what is less certain. Set climate choices in the context of other important decisions made despite uncertainty (e.g., financial, insurance, security, etc.). Discuss how uncertainty may be a reason for action or inaction.
Avoid scientific jargon and use everyday words	Degrees F rather than degrees C. "Human caused" rather than "anthropogenic." "Self-reinforcing" rather than "positive feedback." "Range of possibilities" rather than "uncertainty." "Likelihood" or "chance" rather than "probability." "Billion tons" rather than "gigatons."
Maintain respectful discourse	Climate change decisions involve diverse perspectives and values.
Provide choices and solutions	Present the full range of options (including the choice of business as usual) and encourage discussion of alternative choices.
Encourage participation	Do not overuse slides and one-way lecture delivery. Leave time for discussion or use small groups. Let people discuss and draw their own conclusions from the facts.
Use popular communication channels	Understand how to use new social media and the internet.
Evaluate communications	Assess the effectiveness of communications, identify lessons learned, and adapt.

ing individual changes to save energy and reduce their own impact, but they confront barriers such as up-front capital costs and lack of knowledge about what actions to take. Americans also express a clear desire for more information about climate change, including how it might affect their local communities. Responses to climate change do vary considerably among different segments of the American public, and communication efforts must recognize and address the diversity of views and framings of the climate issue.

Formal and Informal Education

Although many schools, museums, arts, and professional organizations have begun to include climate change in their curriculum and outreach programs, the panel concluded that the United States could make considerably more progress in national, state, and local climate education standards, climate curriculum development, teacher professional development, production of supportive print and web materials, and making educational institutions themselves more sustainable. Recent efforts include a Climate Literacy Framework, but many federal activities are relatively small and new and the panel found little information on measurable outcomes for these programs or for climate education more broadly. A nationally coordinated climate change education network would help support, integrate, and synergize these diverse efforts by conducting research on effective methods, sharing best practices and educational resources; building collaborative partnerships; and leveraging existing education, communication, and training networks across the country. While there are risks of confusion and contradictions in the provision of information from multiple sources, respectful debate about how to interpret information is in itself educational and can inspire interest in science and public policy, as well as individual actions. Educational efforts should include the human dimensions of climate change and climate change solutions—not just the natural science of climate change.

Communicating with Decision Makers

Given the complexities of climate change science and policy, decision makers also benefit from regular communication of new scientific insights and response options. Federal agencies can play a role in providing brief, evidence-based, and readable summaries to Congress and other stakeholders. Research is urgently needed to identify the climate change information, timing, and formats different decision makers need and the information systems that can best support their decision making.

A nationally coordinated effort is needed to assess the state of formal and informal climate change education and communication in the United States, identify knowledge gaps and opportunities, and evaluate the advantages and disadvantages of different national organizational structures and approaches to promote climate change education and communication. This requires coordination between relevant organizations involved in education and increased federal funding for research on education and communication.

Recommendation 10:

The federal government should establish a national task force that includes formal and informal educators, government agencies, policy makers, business leaders, and scientists, among others, to set national goals and objectives and to develop a coordinated strategy to improve climate change education and communication.

The informational needs of American society to respond to climate change range from basic awareness and understanding of the problem itself to highly technical information used only by specialists in specific fields. Communicators at all levels of government and across all sectors of society will thus need to provide a wide range of different information types for different audiences, from individual households to the nation as a whole. When information is tailored to user needs, communicated clearly, and accompanied by decision support tools that enable the exploration of alternatives and encourage flexible responses, decision makers can develop more informed, credible, and effective responses to climate change. The federal government can and should play a leading role in setting national goals, objectives, and strategies and coordinating effective information systems to support America's climate choices.

CHAPTER ONE

Introduction

Global climate change has become one of the nation's most significant long-term policy challenges, and addressing this challenge will require an array of often complex decisions by many different sectors of society and levels of government. Each decision will take on a distinct character, will involve a different mix of participants, and will be made in the context of many other policy issues. The options for responding to climate change involve a broad range of strategies, including (1) limiting greenhouse gas (GHG) emissions to slow the rate and limit the extent of climate change, (2) taking adaptation actions to reduce potential damages from climate change impacts, (3) expanding research and development to provide better low-carbon options for the national and global economy, and (4) improving scientific understanding about climate change and its impacts to enable better informed decision making.

Just as the participants and issues will vary, the needs for information and institutional support will differ across different groups and levels of decision making. For example, the general public would benefit from a better basic understanding of climate change and how it interacts with important values about economic growth, national security, quality of life, health, human rights, and the natural landscape. The general public also needs better understanding of how to think about the various risks of climate change and the responses to it (including the risks of not responding). Likewise, farmers and transportation planners want climate change forecasts at local and regional scales, including projections of the likelihood, severity, timing, and location of specific climate impacts. Decision makers in business and government require economic cost-benefit analyses and information to judge how best to allocate finite resources and make tradeoffs between competing values. People need information, which is often derived by trial and error, to help clarify over time particular aspects of each climate related problem, the emerging options available to respond to the problem, the plausible range of outcomes, and the types of institutions required for supporting effective action in the face of uncertainty.

Decision makers—public and private, national and local—need access to up-to-date and reliable information about current and future climate changes, the impacts of such changes, the vulnerability to these changes, and the response strategies for reducing emissions and implementing adaptation. Also important is the information that is needed to assess whether the decisions or responses are successful or should be revised in the light of experience and new knowledge. After considerable discussion of the task, and the relation of its work to the other three *America's Climate Choices* (ACC) panels, the *Informing* panel chose to investigate the following key questions for this report:

- Who is making decisions and taking action on climate change in the United States? What are their needs for information and decision support, and what are the barriers to good decisions?
- What decision making frameworks and methods are being used, and which are the most effective?
- How might climate and greenhouse gas information systems and services support more effective decisions and actions?
- What is known about the most effective ways to communicate about climate change, especially with the public and through formal and informal education?

This report sets out to identify the types of decisions that may need to be made about climate change by governments, the private sector, and society. It examines the ways in which information to support these decisions can be provided more effectively through the development of new, authoritative and accessible information, especially about climate impacts and GHG emissions. Finally, it looks at the development of decision tools that facilitate the use of information and integrate the key values, data, and processes that interact to shape alternative futures.

Although we hope that our findings will be of interest to a wide range of decision makers, our recommendations are directed primarily toward the federal government and its role in informing and coordinating a national response to climate change.

THE CHALLENGE OF CLIMATE CHANGE

The ACC companion reports (*Limiting the Magnitude of Future Climate Change* (NRC, 2010d), *Adapting to the Impacts of Climate Change* (NRC, 2010a), and *Advancing the Science of Climate Change* (NRC, 2010b) provide detailed overviews of the causes, consequences, and range of responses to climate change in the United States and globally. Collectively they communicate a sense of urgency about the risks of climate change and the need to make immediate decisions related to reducing GHG emissions, implementing adaptation strategies, and investing in research.

This ACC panel agrees with the conclusions of the report *Advancing the Science of Climate Change* (NRC, 2010b) that "[c]limate change is occurring, is caused largely

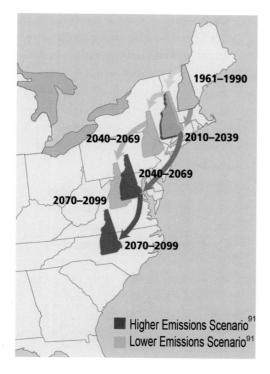

FIGURE 1.1 Yellow arrows track what summers are projected to feel like under a lower emissions scenario, while red arrows track projections for a higher emissions scenario. By late this century, residents of New Hampshire would experience a summer climate more like what occurs today in North Carolina. SOURCE: Frumhoff et al. (2007).

by human activities, and poses significant risks for—and in many cases is already affecting—a broad range of human and natural systems." This is consistent with the analyses of the Intergovernmental Panel on Climate Change (IPCC, 2007b), which found that the global climate is warming, that this warming is very likely due to green-house gases from human activity, and that unless we reduce GHG emissions, the climate will warm by 2°F to 11.5°F (1.1°C to 6.4°C) by the end of the century and will have serious impacts on ecosystems, water resources, low latitude agriculture, coasts, ocean acidification, and increased risks of abrupt or irreversible change (Figure 1.1). The IPCC also recommends an iterative risk management approach[1] that includes adaptation and emissions reduction strategies and that takes into account damages, co-benefits, sustainability, equity, and attitudes toward risk (IPCC, 2007b).

New research and data have confirmed and updated the trends and analyses of the IPCC and have suggested further reasons for concern. Climate data analyses show that the earth has continued to warm, sea ice and glaciers are shrinking, and regional

---

[1] An iterative risk management framework defines risk as the impact of some adverse event multiplied by the probability of its occurrence (see IPCC, 2007b).

changes in the United States are occurring, including increases in winter temperatures and intense drought in the Southwest (Karl et al., 2009). Greenhouse gas emissions have continued to increase, with carbon dioxide (CO_2) concentrations reaching to 385 ppm in 2008, the highest level in 2 million years. Emissions from fossil fuel use and cement production have increased by an average 3 percent per year since 2000 with a growing proportion of emissions driven by economic development in Asia (LeQuéré et al., 2009). There are also indications that the capability of land and oceans that naturally take up or absorb carbon dioxide is weakening, contributing to higher levels of this GHG in the atmosphere (House et al., 2008; LeQuéré et al., 2009).

These trends are occurring despite efforts in some parts of the world to limit emissions and are moving toward the higher end of the emission scenarios used by IPCC and thus toward faster and more intense climate changes. Several recent modeling studies suggest that the delays in limiting emissions and the difficulties in turning things around even with immediate deployment of low-carbon technologies and forest protection mean that there is a very high chance of exceeding 450 ppm of carbon dioxide equivalent[2] (CO_2e) GHG concentrations with consequently higher risks of higher temperatures (for more discussion see Calvin et al., 2009; NRC, 2010d). These results suggest that delays in acting now may make it more difficult and more expensive for decisions makers to respond later.

Although the extent of future climate change and the exact nature and severity of impacts remain uncertain, continuing to emit GHGs at the current rate is expected to create long-term or irreversible changes in earth systems and a variety of undesirable consequences that will require profound adaptations on the part of both human and natural systems (IPCC, 2001, 2007b; NRC, 2010b; Solomon et al., 2009). Responding effectively to these risks requires effective long-term planning because decisions and actions taken now will have important implications for decades to come. The emissions reduction strategies and adaptation responses that will reduce the magnitude of climate change and reduce its impacts require active collaborations across science, technology, industry, government, and the public.

The earth and climate systems, like economic and social systems, exhibit complex and chaotic behaviors that can be unpredictable and are difficult to model. Not only is the climate a chaotic system, but humans are pushing it into poorly understood patterns and processes where there are chances of rapid climate change and surprise. The uncertainties regarding the details of future climate change depend on which decisions

---

[2] For GHG emissions inventories and mitigation, the common practice is to compare and aggregate emissions by using global warming potentials (GWPs). Emissions are converted to a carbon dioxide equivalent (CO_2e) basis using GWPs as published by the IPCC.

society makes about future energy and resource use, the complexity associated with the interactions between natural and human caused climate change and with other environmental changes, the difficulties of modeling climate at the regional scale, and the incomplete understanding of processes (e.g., carbon cycle and ice sheet dynamics) and of the vulnerability and adaptive capacity of human and natural systems at the local scale.

In Chapters 5 and 6, the panel addresses the need for monitoring, reporting, verification, and information systems that can help manage these uncertainties. Decisions need to be based on scenarios that cover the range of possible developments in socioeconomic and environmental systems. Uncertainties in understanding climate change and response options—as in other areas of economic, technological, social, military, and environmental policy—are an important reason for action that can help reduce risks. Uncertainty is not a reason for inaction. Rather, with knowledge of uncertainties we can anticipate a range of possibilities, some of which may be so severe that we should act now to reduce the chances of their occurrence. Effective decisions require the best available information, including information about the level and nature of uncertainty, so that policy makers and others can make careful judgments about what to do.

Communicating uncertainty often poses a problem for those trying to generate support for measures that might reduce the risks associated with climate change, especially in explaining the science and the choices to the public (Figure 1.2). An effective American response to climate change requires a solid base of information and a strong set of institutions that can evaluate the risks, costs, and opportunities presented by climate change, can make the best possible decisions about how to respond, and can communicate the decisions and the rationale behind them clearly to the relevant audiences.

Decision makers and stakeholders will need sector specific information to respond to climate change and may assign different values both to the impacts of climate change and to the costs and benefits of policy actions to limit or adapt to these impacts. Hence, a fundamental part of climate change policy must be deciding how to allocate finite resources among the diverse options available for limiting emissions, adaptation, or research. For that, decision makers need a clear understanding of and accurate information on the costs, risks, tradeoffs, and potential benefits of each option for various segments of society. This is not unique to the problem of climate change; most important decisions are made without perfect clarity. The choice of options is seldom either-or, but rather it is a judgment about what constitutes the right mix and also

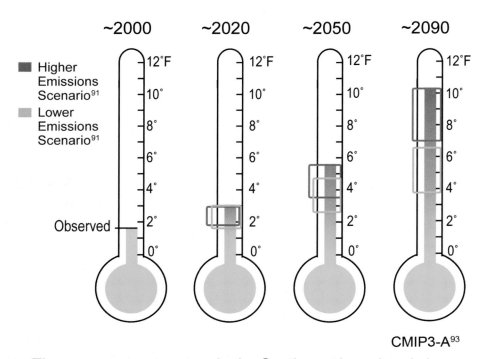

CMIP3-A[93]

The average temperature in the Southwest has already increased roughly 1.5°F compared to a 1960-1979 baseline period. By the end of the century, average annual temperature is projected to rise approximately 4°F to 10°F above the historical baseline, averaged over the Southwest region. The brackets on the thermometers represent the likely range of model projections, though lower or higher outcomes are possible.

FIGURE 1.2 This simple visualization of how climate change might affect temperatures in the southwestern United States portrays uncertainty in two ways: through a low and high emissions scenario (one where fossil fuel use continues to increase, one where use is limited) and through brackets that show the range of uncertainty for temperatures under each scenario. SOURCE: USGCRP (2009).

how the right combination of options is likely to vary across geographic regions and over time.

DECISION MAKERS, THEIR INFORMATION NEEDS, AND THE CHALLENGE OF RESOURCE ALLOCATION

Table 1.1 provides a range of examples of who may need to make decisions about climate change, ranging across scales of government to the private sector, non-governmental organizations (NGOs), and individuals, and the type of decisions they may wish to make. This table makes it clear that information and decision tools need to be made available to a broad audience and that focusing only on the U.S. Federal government would miss many of the key decisions and responses that will be made across America.

The table shows that decision makers are faced with many different decisions but of course most decision makers have limited financial, human, or political resources and cannot take action on everything so must make choices and set priorities about where to allocate scarce resources. Not surprisingly, many of the fundamental choices regarding climate change policy involve the allocation of resources. In making such allocations, decision makers are concerned with how to establish objective, defensible, return on investment criteria for their constituents, or, in the case of business, their shareholders. Resource allocation takes into account not only near-term priorities but also long-term objectives, which include the economy, non-climate decisions, and the effect on future generations. They need to balance and communicate the costs of not acting with those of taking action; they must decide how much to spend on different types of actions, such as emissions reductions and adaptation, and which sectors and places should receive resources to respond. Because climate change decisions often involve benefits related to health, safety, equity, and environmental concerns, policy makers must decide whether and how to include such non-monetary benefits in the return on investment analysis. The ability to create and implement effective climate policy will likely come down to the availability of resources, and the choices policy makers make will be directly linked to the price assigned to the harms against which one hopes to protect.

As more communities become aware of the need to prepare for the inevitable impacts of climate change, policy makers may be faced with the choice of directing resources to programs designed to limit GHG emissions or to programs that seek to build resilience against future impacts. At the local level, strategies to limit emissions tend to focus on energy (conservation, efficiency, and development), low-carbon technol-

TABLE 1.1 Examples of Decision Makers and the Choices They Make

Who	Example Decisions
Federal government	Whether to participate in international agreements and bilateral/multilateral assistance programs relating to climate change
	Whether to regulate GHG emissions and, if so, what policy mechanisms (e.g., cap-and-trade, carbon taxes, standards, etc.) to use, how these mechanisms are designed, and what agencies and institutions will administer them
	How and where to reduce GHG emissions from federal activities
	How to adapt to climate change on federal lands and jurisdictions
	Priorities for funding research, technology development, and observing systems
	Setting standards and guidelines for carbon management (e.g., energy efficiency, information labels, GHG reporting, and carbon disclosure for investors), coastal protection, water allocation, etc.
	How to ensure the security of food, water, and health for the U.S. population, how to respond to potential national security risks of climate change, and how and whether to respond to human security concerns in other regions of the world
	What is the best way to educate and communicate about climate change to the public
State, tribal, and local government	How to control GHG emissions, especially from utilities, transport, and buildings, and whether to join regional trading initiatives, and how to encourage citizens to reduce their emissions
	Setting renewable portfolio and energy efficiency standards
	How to incorporate climate change into land use planning, infrastructure projects, and disaster planning
	How to amend the building code to reduce GHG emissions and to address the impacts of climate change, including the increased potential for flooding, droughts, high winds, heat waves, and disruption of utility services, as well as the need for buildings to be inhabitable without energy
	How and whether to limit emissions from state and local government operations
	How to facilitate adaptation through policy decisions about insurance cover, environmental protection, land use, etc.
	Potential information campaigns and educational guidelines
Private sector	How to reduce GHG emissions from operations and supply chains, and whether to participate in regional and global carbon markets and offsetting

TABLE 1.1 Continued

Who	Example Decisions
	How to develop good information for consumers about carbon in products and other sustainable practices
	Whether and how to influence government and international policy through best practice, lobbying, business networks, etc.
	Whether and how to insure climate risks
	How to adapt to climate risks and respond to climate impacts in a globalized market
	Whether to invest in businesses and technologies that are vulnerable to climate risks or that are not limiting their emissions
	Whether to start up a new business focused on solutions to climate change
	How to respond to pressure from NGOs, shareholders, and investors concerning climate change
	How and what to communicate about climate change (especially from media and cultural sector)
	Funding research and development
Non-governmental organizations (e.g., trade, religious, environmental, humanitarian, foundations)	How to reduce their own GHG emissions and influence the emissions of their members or the public
	Where and how to facilitate adaptation
	Whether and how to influence government, the private sector, and the public through information, communication, action, networks, and lobbying
	Funding research and responses to climate change
Individuals	How seriously to judge the threat of climate change and how to weigh current costs against future benefits
	How to prepare by adapting homes, lifestyles, and landscapes to climate change
	What actions to take to reduce their emissions in household energy use, travel, and purchase of household goods and food
	Should their investments (including pensions) be in portfolios with low climate risk or in climate responsible businesses
	Whether and how to try and inform or influence others (families, employers, educators, politicians, and neighbors) or hold them accountable for actions on climate change

ogy, and transportation. Adaptation strategies tend to focus on infrastructure (roads, bridges, ports, and coastal development), water (conservation, supply, and management), disaster preparation, and public health and safety concerns. Not surprisingly, many local policy makers are searching for initiatives that address both limiting emissions and adaptation, such as land use planning, distributive energy systems, open space preservation, and green space development.

Some of these strategies are already functions of local government, including comprehensive planning, building and energy codes, neighborhood outreach, equipment purchasing, and infrastructure planning and development. Others are new to local government, including involvement in markets for GHGs, carbon taxes, or ways to pilot the development of new technologies and energy sources, where the economics of resource allocations are less clear.

One of the most significant resource allocation questions is how to address the challenge of the nation's infrastructure and the barriers it poses to both emissions reduction and adaptation. Power generation, transport, protected areas, water resources, and urban development are the result of major infrastructural investments that have locked the nation into pathways and patterns of high GHG emissions and vulnerabilities to the impacts of climate change. Infrastructure may be the nation's greatest barrier and, as discussed below, its most powerful opportunity to limit emissions and adapt to climate change. Decision makers must consider whether to replace this infrastructure now or over longer periods of replacement and reinvestment, and whether to prevent the building of new infrastructure that increases climate risks.

As climate change policy is addressed, decision makers must also decide how quickly to implement new policy. This choice is especially challenging given both the general need for more and better scaled information on climate change and its impacts, and also the uncertainties associated with new energy development. Decision makers are wary of making wrong choices, such as picking the wrong technology or building the wrong "infrastructure of the future." On the other hand, many policy makers have embraced the idea that taking action to respond to climate change is urgent and that they play a vital role in catalyzing change through successful policy action, even providing inspiration for other decision makers to take action. Thus, some policy makers are innovating new decision making processes that embrace failure as an element of future success and evaluate the benefits of being "first movers" in the development of new technologies, models of action, and policy.

Economic information such as costs and benefits of different actions, return on investments and avoided damages, and distributional and competitive effects on local economies, firms, and households is essential to such resource allocation decisions.

However, such information is fraught with assumptions about how to cost and value different actions now and into the future, including those that have non-monetary effects on areas such as ecosystems and health. For example, a local decision maker may worry about how to balance the costs and risks of regulating local industrial emissions with the emerging possible impacts of climate change on local tourism or water supplies and costs of adapting these sectors. But the decision maker is also faced with the problem of competing priorities where they may feel that other urgent issues—such as poverty, housing, and crime—demand the bulk of available financial and human resources, leaving little for responding to climate change. We discuss some of these challenges, and some tools that may help with such complex economic decisions, in Chapter 4.

Decision makers are increasingly aware of the multidisciplinary nature of climate change policy even as they work to make the most of available resources. They need to create policy and stakeholder teams that stretch beyond traditional notions of jurisdiction; they are also seeking ways to leverage resources not only across disciplines but also across physical and temporal scales to maximize strategies and investments. For example, they seek to take advantage of economies of scale to improve purchasing power and the marginal costs of new technologies.

Decision makers also confront the choice of how to best integrate policies and initiatives across multiple geographic and temporal scales. An example might be a local neighborhood development initiative to help a community become more energy efficient, walkable, and environmentally friendly. This initiative would benefit greatly if integrated with a larger regional plan that involves building new or more efficient public transit along a nearby corridor, and an energy plan to construct a connected energy district. Thus, decision makers today need information and support to help them make the major infrastructure investment choices that will be effective across a wide range of possible future conditions.

BARRIERS TO EFFECTIVE DECISION MAKING

Framing of Climate Change Affects Decision Making and Responses

Decision making about climate change is often conducted not only under conditions of scientific uncertainty but also by people who may be unfamiliar with the details and weight of scientific evidence. Under these conditions, human judgment is greatly influenced by a number of factors, including the "framing" of the problem itself (Ferree et al., 2002; Gamson and Modigliani, 1989; Nisbet and Mooney, 2007; Tversky and

Kahneman, 1981)."Frames" often take the form of a relatively small set of interpretive stories or contextual clues that guide attention, highlight certain problem features (and not others), and influence subsequent decision making. The use of different frames can lead to dramatically different choices. Nisbet (2009) argues that people "rely on frames to make sense of and discuss an issue; journalists use frames to craft interesting and appealing news reports; policy makers apply frames to define policy options and reach decisions; and experts employ frames to simplify technical details and make them persuasive." The way an issue is framed often affects the way in which people use information and choose information sources and can constrain the range of decisions and choices they see as available to themselves and others.

Climate change itself has been framed in many different ways, each of which leads decision makers to think differently about how to respond (see Table 1.2). For example, one of the dominant sources of conflict in international climate negotiations derives from three alternative framings of the source of GHG emissions (national, per capita, and historical) and, therefore, who is responsible for reducing emissions. Using a national frame, China is now the world's largest emitter of CO_2e, and the United States is second. Russia is the third largest emitting country and India is now the fourth largest national emitter of CO_2e (Table 1.2).

The national frame alone suggests that China, Russia, the United States, and India must reduce their emissions immediately if the world is to restrain climate change.[3] The per capita frame (dividing national emissions by the number of people in each country), however, tells a different story and leads to very different conclusions. The United States is by far the largest emitter in the world by per capita. By contrast, the average Chinese and Indian emit significantly less. Using this frame, Chinese, Indian, and other developing country negotiators argue that it is unfair for developed countries like the United States, who continue to emit far more carbon per person, to demand emission reductions from countries that are struggling to lift the living standards of billions currently living in poverty. These two contrasting frames lie at the heart of the current international climate debate and lead to different ways of assigning responsibility for action.

Finally, the historical frame identifies the primary sources of the already high concentrations of GHGs in the atmosphere today and further complicates the story. Cumulatively, since 1751, at the level of individual countries, the United States has been by far the largest emitter of carbon, while the USSR is the second largest emitter (based

---

[3] Carbon Dioxide Information Analysis Center, Oak Ridge National Laboratory; see *http://cdiac.ornl. gov/trends/emis/meth_reg.html*.

TABLE 1.2 Three Frames for Carbon Emissions (in Million Metric Tons)—National, Per Capita, and Historical—Lead to Different Conclusions about Responsibility for Climate Change

	National	Per Capita	Historical[a]
China	1,664,589	1.27	27,650,786
United States	1,568,806	5.18	88,592,149
Russia[b]	426,728	2.99	6,383,064
USSR[c]			30,762,582
India	411,914	0.37	8,260,796
Japan	352,748	2.8	13,649,731
Germany	219,570	2.67	20,483,021
United Kingdom	155,051	2.56	15,439,857
Canada	148,549	4.55	6,746,257
South Korea[d]	129,613	2.68	2,636,425
Italy	129,313	2.19	5,118,897

[a] Since 1899
[b] Since 1992
[c] 1830-1991
[d] Since 1945
SOURCE: data from CDIAC, Potsdam Institute for Climate Impact Research, Boden et al. (2009)

on data from 1830-1991).[4] China is the third largest emitter in the historical frame. Arguably then, the United States and China bear particular historical responsibility for the already high concentrations of GHGs in the atmosphere. These three frames—national, per capita, and historical—thus each lead to different conclusions about "who" is primarily responsible for climate change and therefore who should lead in reducing emissions. In addition, Chakravarty et al. (2009) suggest a framework for allocating emissions at the individual level. Each of these frames is supported by data that can be presented in several alternative ways and where the accuracy of the information may be questioned.

Other ways of framing the climate change issue, from environmental, economic, energy, health, or ethical perspectives, also draw on particular data and analyses at a variety of scales. For example, environmental framings often draw on information about how climate change will affect species in specific localities, and economic framings draw on information about the costs of responding in different sectors and regions.

---

[4] Carbon dioxide data for Russia is 1992-2006 (CDIAC, *http://cdiac.ornl.gov/*)

In the media and in public discourse, too, climate change is often framed in very different ways. Sometimes it is framed as an "environmental problem"—understood through the lens of human impacts on the natural environment, ecosystems, and particular species (e.g., polar bears, coral reefs, and tropical rainforests). Another common frame portrays climate change as a "Pandora's box"—a long list of potentially disastrous impacts (e.g., sea level rise, drought, floods, heat waves, infectious diseases, famines, and water shortages; Nisbet, 2009).

Climate change also has been described within a "political frame" as a Democratic versus Republican, liberal versus conservative issue, or a debate between specific political leaders (e.g., Al Gore versus George W. Bush; Dunlap and McCright, 2008). Another common frame is climate change as an "economic problem"—one where either the impacts (sea level rise, drought, floods, heat waves, etc.) or proposed policies to address it (carbon taxes, cap-and-trade systems, etc.) are potential threats to local, regional, national, and international economic growth and development.

More recently, new frames have emerged, including climate change as a "moral and ethical problem," often asserted through specific religious beliefs and teachings; a "national security problem" with major geopolitical implications; a "public health problem" with serious potential consequences for human well-being; a "human rights problem"; or as an "economic opportunity" for the development of green jobs and new competitive industries (CNA Corporation, 2007; Leiserowitz, 2007; Nisbet, 2009).

Scientific uncertainty itself is used as a frame that has often been deployed strategically by groups and special interests seeking to cast doubt on the reality of human caused climate change (McCright and Dunlap, 2000, 2003). Meanwhile, environmentalists sometimes attempt to amplify other scientific uncertainties to motivate action (e.g., the possibility of tipping points of abrupt and catastrophic climate change; Nisbet, 2009). Each of these frames calls attention to different features of the issue, resonates in different ways with different audiences, and implies a need for different kinds of policy responses (Maibach et al., 2009).

Some people now argue that climate change should be reframed as a clean energy issue. Climate change remains a relatively low national priority (Leiserowitz et al., 2009), but energy and energy independence are important to the public and policy makers across political party lines. Solving the nation's energy challenges will require many of the same policies and investments needed to reduce GHG emissions, such as improved energy efficiency, conservation, and the development of new renewable sources of energy. Some argue, therefore, that a more effective way to address climate change is to focus not on emission targets and timetables (a "pollution frame") but on national investments to develop the new "clean energy" economy of the 21st century

(Nordhaus and Shellenberger, 2007). Finally, others argue that climate change itself should be placed within the broader context of sustainable development, which attempts to integrate and harmonize economic prosperity, environmental protection, human development, and security (Kates et al., 2005).

Each of these frames can now be found circulating in scientific reports, media stories, political debates, and public discourse. Meanwhile, decision makers from different sectors of society strategically select, ignore, amplify, and downplay these various frames as a way to either raise or lower public concerns about the issue and support or oppose particular policy options (Leiserowitz, 2006, 2007). Framing is "an unavoidable reality of the communication process" (Nisbet, 2009); thus, efforts to inform climate change decision making and action will invariably involve some element of framing—highlighting certain features of the issue and ignoring or discounting others. Likewise, different individuals and groups within American society respond to each of these frames in very different ways.

National efforts to inform climate change decision making and action by diverse individuals and institutions across the United States must recognize that framing matters and different audiences are likely to respond to various frames in different ways. Stakeholder groups often do not take multiple perspectives into account. For example, decision makers may frame emissions reductions and adaptation as unnecessary regulation and government interference, without taking into account the risks associated with inaction in the face of long-term climate change. This kind of overly narrow problem framing can limit the choice of action or negatively affect the quality of the decisions that are made.

Resource allocation constraints and conflicting frames are only two of the barriers to effective decision making about climate change. At the most fundamental level, efforts to inform effective decisions in the climate change arena face the same major barriers, mainly political and economic, as other major social problems, such as health care. Attempts to address social problems invariably provoke some degree of political disagreement, and the search for solutions to major problems almost always require major investments, which may lead to additional controversy because such investments may necessitate tradeoffs. Those who want to inform and make decisions must wrestle with uncertainties concerning information, the efficacy of proposed solutions, the possibility of unanticipated consequences resulting from decisions, the challenge of implementing the solution, and sustaining the action over time. There are also a number of barriers that influence many decisions about climate change that are specific to certain types of decision makers but that can be overcome through particular strategies (Table 1.3).

TABLE 1.3 Some Barriers to Effective Climate Decisions and Actions and Roads around Those Barriers

Barrier to Action	Overcoming the Barriers
Different framings of climate change	Emphasize process as well as information products, establish and support processes for stakeholder engagement and knowledge transfer, understand how different framings influence decision making, and learn how to best communicate climate change
Economic and resource allocation constraints	Increase resources available to respond to climate change, provide good information on the implications of alternative allocations, consider long-term issues such as those of infrastructure
Political context	Provide decision-relevant information to inform political initiatives for limiting climate change and adapting to it, provide good science to address misinformation and disinformation, educate the public regarding climate choices, and consider responsibility to future generations and people in other nations
Information gaps	Conduct research and observations to improve information, increase access to reliable information at appropriate scales, and communicate information clearly
Institutional and organizational constraints	Identify institutional barriers to decision making including rules, cultures, and organizational structures; promote interagency and public-private partnerships, establish consistent standards and targets, seek institutional stability, and maintain and enhance boundary organizations
Lack of insight into effective decision processes	Establish principles of effective decision making including stakeholder engagement, linking information producers and users, and adaptive management

Barriers Associated with the Way Decisions Are Made

One key barrier to decision making, especially by government, is a misunderstanding of how the most effective decisions are made. For example, those seeking to aid in the decision making process often assume that the provision of sound scientific findings, assembled in the right format, delivered to "end users" (that is, decision makers) will automatically improve the quality of decisions and actions that are subsequently taken. Or they assume that the best way to inform decisions is to conduct research to reduce uncertainties in scientific projections. Different decision makers require different levels of certainty (Slocum et al., 2003) and types of information which may depend on personal values, their institutional rules, and handling of potential risk. This way of thinking contrasts with what is currently known about factors that affect the decision making process. Prior National Research Council reports, including *Public Participation in Environmental Assessment and Decision Making* (NRC, 2008b) and *Informing Decisions in a Changing Climate* (NRC, 2009a), provide extensive discussions of barriers to effective and informed decision making. For example, the literature on public engagement in decision making for the environment points out the importance of early and continuous stakeholder engagement in decision related activities. It also emphasizes the need to conduct scientific investigations in ways that address the concerns of decision makers, which can be quite different from those of the scientists who generate the information. The reports review experience with decision support in coping with climate variability and change, and stress the importance of developing appropriate frameworks for supporting climate related decisions. They find that sound decisions require both *good information* and *well-structured processes* for developing, providing, and using that information and concluded that efforts to inform decisions are more likely to be judged effective when they follow the six principles described in Box 1.1 (NRC, 2009a).

Processes that feature ongoing, two-way communication between information producers and decision makers provide the best way to identify decision makers' needs and ensure that useful information is produced and that its intended users are prepared to receive it. These engagement processes can lead to the development of social networks consisting of information producers, users, and boundary organizations that perform key communication functions for particular constituencies, that is, groups of information users with similar needs. Much of the guidance offered in the reports is counterintuitive, where the notion that the needs of decision makers is competing with science-related needs or ideas that scientists may have about "end-user" information requirements. Of rising importance in the years to come is *joint production of knowledge* in accord with users' needs. This perspective is elaborated in the report *Informing Decisions in a Changing Climate* (NRC, 2009a).

Effective decision making can be hampered by insufficient attention to the development of appropriate decision support processes. It is not only the products used to facilitate decision making that are important, such as scientific forecasts, scenarios, maps, cost-benefit analyses, and epidemiological data, but also the decision making process itself. For example, if the processes by which the information has been developed appear biased, decision makers may be reluctant to use the information (NRC, 2008a). Early engagement with stakeholders allows trust to build between information producers and users (Moser, 2005; NRC, 1996, 2008c). Early engagement also helps to ensure that the decision support information will be as complete and responsive to

---

BOX 1.1
Principles of Effective Decision Support

1. **Begin with users' needs.** Decision support activities should be driven by users' needs, not by scientific research priorities. These needs are not always known in advance, and they should be identified collaboratively and iteratively in ongoing communication among knowledge producers and decision makers. The latter can usefully be thought of as constituencies—groups and networks of decision makers that face the same or similar climate related events or choices and therefore have similar information needs.

2. **Give priority to processes over products.** To get the right products, start with the right process. Decision support is not merely about producing the right kinds of information products. Without attention to process, products are likely to be judged inappropriate by intended users—although excessive attention to process without delivery of useful products can also be ineffective. To identify, produce, and provide decision support, processes of interaction between decision support providers and decision makers are essential.

3. **Link information producers and users.** Decision support systems require networks and institutions linking information producers with decision makers. The cultures and incentives of science and practice are different, for good reason, and, in order to build productive and durable relationships, those differences need to be respected. Some ways to accomplish this rely on networks and intermediaries, such as boundary organizations.

4. **Build connections across disciplines and organizations.** Decision support services and products must account for the multidisciplinary character of the needed information, the many organizations that share decision arenas, and the wider societal context in which decisions are made.

5. **Seek institutional stability.** Decision support activities need stable support. This can be achieved through formal institutionalization, less formal but long-lasting network building, norms for routinizing decision making, and mandates, along with committed funding and personnel. Stable institutions are needed to ensure consistent and effective decision making, to foster trust in decision making processes, and to sustain policy initiatives.

6. **Design for learning.** Decision support systems should be structured to enable flexibility, adaptability, and learning from experience.

SOURCE: NRC (2009a).

---

actual need as possible and that any additional research will be undertaken. If unintended social, political, and economic consequences begin to manifest themselves as the decision process proceeds, further obstacles may arise. As policies are implemented, new facts will arise (e.g., different costs and benefits than assumed, or changing social or economic trends), and course corrections will become necessary (NRC,

2009a). Information about climate change and its consequences is continually changing and decisions may not be right the first time; decisions will have to be revisited
from time to time within constituency networks (see Chapter 3).

Political Challenges Can Delay Making Informed and Effective Decisions

Like other major social policy issues that the nation faces, such as immigration and
border security, health care, and defense, climate change is a politically charged topic,
in part because the costs and tradeoffs of policy strategies threaten different political
and economic interests. Political discourse plays a significant role in problem formulation and framing in the climate change arena. Advocates for different strategies for
responding to climate change seek to mobilize support for particular policy options
and close off the exploration of others by seeking to control the framing of the issue
(Johnston and Noakes, 2005; Snow et al., 1986). For example, the tobacco industry
fought for decades to frame tobacco use as an individual lifestyle choice, rather than
an addiction, in an effort to forestall stricter regulation of its product. In opposition,
the anti-tobacco movement argued for the regulation of tobacco as an addictive
substance and a health hazard that is harmful even to those exposed to second-hand
smoke. Ultimately, decision makers rejected the arguments framing tobacco use as
an individual choice and tied decisions to scientific evidence about the harm that it
causes.

The dominance of particular interest groups in any decision making process can
result in delay or even lack of consideration of certain policy options. Although a
number of important federal activities are under way, as of this writing consideration
of federal climate change legislation has been postponed in Congress, and decisions
about international action on climate change under the United Nations Framework
Convention on Climate Change have also been delayed following disagreements
and confusion at the Copenhagen negotiations. The reasons behind these delays and
disagreements are complex and include differences of opinion about the urgency of
responding to climate change. We discuss the drivers and attitudes toward climate
policy opposition and why attitudes toward climate change are shifting in Chapter 8.
This apparent stalemate at the federal and international levels stands in marked
contrast to the multiple actions described in Chapter 2 and elsewhere in this report,
which have been undertaken by regions, states, local communities, non-governmental
organizations, and the private sector.

Federal inaction may constitute a significant barrier to effective decision making for
many reasons. Without guidance at the national level on emissions reductions and

adaptive actions, various sub-national laws and regulations are emerging. A policy vacuum is emerging in areas where only the federal government can act, such as setting nationwide GHG emissions reduction targets; establishing mechanisms to regulate carbon; providing "policy certainty" for corporations seeking to make long-term investments (see Chapter 3 for more discussion); generating substantial revenues that can be directed toward climate change responses; and making large-scale investments in research and development.

Because political stalemates of this type are common, scholars of the policy process have developed the idea of "policy windows" (Kingdon, 1995), a metaphor that emphasizes the idea that opportunities for action may appear under unusual circumstances and then disappear later, as politics returns to "business as usual." When policy windows open, choices that had formerly been off the table or impossible to undertake may be given a second look, and decisions that previously seemed impossible may be made. Scholarship in this area emphasizes the importance of "policy entrepreneurs" (Kingdon, 1995; Roberts and King, 1991) that are prepared and able to take advantage of policy windows to temporarily disrupt systems of political dominance. For example, disaster events are well known for their ability to open policy windows, if only temporarily (Birkland, 1997). In addition, new scientific findings such as those contained in IPCC reports, as well as meetings and conventions that require a response on the part of policy makers, can help create policy windows.

The broader social, economic, and political context should be taken into account in understanding why some initiatives move smoothly through the policy process while others do not, how and why policy windows open in some cases but not in others, and why some policy entrepreneurs succeed while others fail. In the case of climate change, this means recognizing the impact on decision support activities of the current national and global macroeconomic climate, public opinion regarding national policy issues, and other significant political issues such as health care.

Politics and Science

Response to climate change can also be hampered by interest groups that spread conflicting information about climate change science, promoting confusion among decision makers and the public. For example, some groups state that the climate is not changing, assert that the science is controversial or highly uncertain, and deny scientific facts offered by others knowledgeable about the field (Jacques et al., 2008; McCright and Dunlap 2000, 2003; Rowland, 2000). Other groups may overstate the case, focusing only on the more dramatic scenarios or implying that the science is more certain than suggested by the literature.

Claims that climate change science is controversial and that climate change is not in fact occurring continue to be made and are one reason that people put off or decide against acting on climate change. As we discuss in Chapter 8, there are segments of the American public who are unconvinced of the risks of climate change and the need to act, but the majority of Americans are concerned and would like to see action. Some people are certainly confused or overwhelmed by the debate about climate change, especially by the way the science is portrayed in the media, and by the impact of special interest funding. As policy debates go forward, it will be important for decision makers to examine the scientific and policy bona fides of those claiming expertise and providing information about climate change and for government to provide the best possible assessments of the science with clarity and careful evidence. As Justice Brandeis noted, "sunlight is the best disinfectant," and transparency is essential in matters of policy. Decision makers must also recognize that the nation has been here before and draw appropriate lessons from history (see Appendix B) where efforts to deny and undermine scientific findings, such as with asbestos, tobacco, and other scientific issues that have been presented as controversial, notwithstanding a preponderance of scientific evidence to the contrary. Advice on issues has also shifted, for example, on mammograms for middle-aged women.

Institutional Barriers to Effective Decision Making

A full discussion of the institutional and organizational factors that complicate decision making with respect to climate change is beyond the scope of this report. As is the case with policy making more generally, institutional and organizational inertia can stand in the way of sound climate decision making. The purpose of institutions is to establish, implement, and sustain norms and codes of conduct within policy arenas. By design, institutions are slow to change; without such stability, social life would lack structure and predictability. Stability is also important because users need trusted sources of information over time. However, when institutions are incapable of adapting to new circumstances and information, or cannot do so in a timely manner, it impedes effective responses to climate change. A recent NRC report, *Restructuring the Federal Climate Change Science Program* (NRC, 2009d), evaluates the institutional challenges for the U.S. Global Change Research Program.

The ability of any organization to take on adaptive measures will depend on the extent to which it sees the immediacy and importance of the need to act in the context of its other interests and responsibilities. The following factors guide that vision and ability of an organization to act:

- Knowledge, understanding, and experience with climate change and its effects;
- Compatibility of its mission with climate change issues;
- Jurisdiction or domain, including mechanism of interagency coordination;
- Capacity (human, financial); and
- Capability (politics, organizational culture).

Legislation, regulation, and other types of external pressure (e.g., crises, elections, social movements, and media attention) are typically required to stimulate institutional and organizational change. Boundary organizations have sometimes been able to overcome these barriers by communicating and collaborating among organizations such as scientific agencies, research centers, local government agencies, and corporations (Cash, 2001; Cash et al., 2003; Fennell and Alexander, 1987).

Institutional barriers also arise because of insufficient coordination among federal agencies whose activities are relevant to climate change research, emission reductions, and adaptation strategies. Organizations at federal, state, and local levels and in the private sector that are not currently involved in climate change programs (e.g., the U.S. Census Bureau) have information they can contribute to support climate related decision making. Many agencies do not consider climate change as part of their authority, particularly at the sub-federal level. Climate change polices usually reside in environment and energy agencies and are often limited to emission reduction strategies rather than adaptation (see *Adapting to the Impacts of Climate Change* NRC, 2010a).

Lack of clarity regarding institutional and organizational roles, responsibilities, and authority also hampers decision making. Decision makers characteristically act within bounds, and as a result they need clear guidance in areas such as targets, mandates, and other aspects of limiting and adapting to climate change. However, at the present time, such guidance is unavailable in many areas. As discussed in Chapter 6, the nation currently lacks consistent standards for how to report emissions reductions, as well as sound ways of monitoring emissions and verifying compliance with whatever standards might be developed. Without these and other types of information, there can be no national strategy for managing GHGs. Focusing on another policy vacuum, federal disaster related legislation offers virtually no guidance on what states and local communities should do in response to the threat of climate change, largely because climate related concerns have not yet been incorporated into such legislation.

As in other policy arenas, some of these kinds of ambiguities can be clarified through amendments to existing legislation, rulemaking activities, and the development and adoption of standards. Others will likely be settled only through a protracted process of court decisions, new legislation, and cross agency negotiation. This is the case, for example, for the ruling that the Environmental Protection Agency has the authority to regulate GHGs. Many decisions on carbon regulation hinge upon the outcome of that case, decisions that cannot be made without further policy clarification.

Informational Barriers

Effective decision making is based on sound information. Sectors throughout our national economy (water, agriculture, fisheries, financial, health, and energy) need up-to-date reliable information tailored to their specific needs. Some of the major informational barriers to effective decisions (discussed at length in the remainder of this report) include the lack of detailed, timely, and consistent information on GHG emissions and the activities that produce them, uncertainties in how climate will change at the regional scale and what it means for sectors, landscapes, livelihoods, human needs, and the need to link information about climate change and responses in the United States to what is happening internationally.

Some decision makers may understandably resist using information when uncertainty is high (Slocum et al., 2003), while others may find even uncertain projections useful. Related problems develop when different model projections predict different outcomes or when emission reductions are inconsistently reported or reported more than once. Decision makers are then faced with the challenge of determining which projections and emission reports appear to be most credible. For example, information on hurricane landfall projections is used by government responders to decide how to allocate emergency resources; by local officials to decide whether to issue evacuation orders; by businesses choosing whether to close and lose business, or remain open and risk damage or injury; and by residents deciding whether and when to evacuate.

Information needs to be provided in a timely fashion if decision makers are to make the best possible decisions. For example, a utility company making decisions with respect to long-term investments may find decadal and multi-decadal climate projections quite useful. In contrast, year-to-year projections may be more appropriate for decision makers in the agricultural sector. In addition, decision makers are often required to act rapidly, even if appropriate information is not available, because waiting for more definitive information may mean losing resources and momentum. Other decisions are tied to budget or election cycles, which mean that major projects must be designed and planned years before they will be implemented. Ultimately, decision makers will judge when the available information is adequate for decision making purposes, and this judgment will depend on personalities and the particular circumstances.

Another critical barrier to information use is that of accessibility. If information is not easily accessible then people may make decisions without it. Because information about climate change is available from multiple sources, including different agencies within the federal government, decision makers may waste time and become confused through trying to find relevant information, especially through the internet, and frustrated in attempts to find information specific to their needs and location. If information providers fail to understand the information needs of their users they will miss opportunities to increase the effectiveness of decisions. The constant stream of information about climate change from the media, NGOs, and government can also lead to issue fatigue and failure to pay attention to important new information.

THE SCOPE AND PURPOSE OF THIS REPORT

In 2008, Congress directed the National Academy of Sciences to "investigate and study the serious and sweeping issues relating to global climate change and make recommendations regarding what steps must be taken and what strategies must be adopted in response to global climate change." This report, *Informing an Effective Response to Climate Change*, is part of the resulting *America's Climate Choices* suite of activities (see Foreword). This panel was asked to consider what can be done to inform effective decisions and actions related to climate change. More specifically, the panel was asked to describe and assess different activities, products, strategies, and tools for informing decision makers about climate change and helping them to plan and execute effective, integrated responses (see statement of task in Appendix A). Companion reports provide information and advice on *Limiting the Magnitude of Future Climate Change* (NRC, 2010d), *Adapting to the Impacts of Climate Change* (NRC, 2010a), and *Advancing the Science of Climate Change* (NRC, 2010b).

The panel recognizes that climate change is but one among many important issues policy makers face. However, climate change touches all aspects of our nation's economy, prosperity, human health and safety, and security (Figure 1.3). An effective response requires actions across a wide range and scale of public and private agencies and organizations, as well as by individual citizens. Thus, the panel devoted considerable attention to understanding the information needs of different entities and the

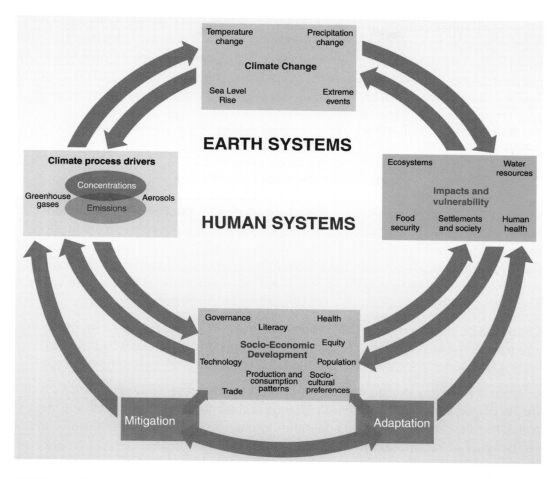

FIGURE 1.3 Complexities of the connections between climate change and many aspects of our economy, prosperity, society, and security. SOURCE: IPCC (2007b).

institutions needed to provide comprehensive information that might inform their attitudes, decisions, and actions.

Climate change presents a technical, social, and political challenge that is in some ways similar to, although in other ways quite unique from, many challenges the United States has faced before. The United States has the proven ability to revolutionize technology and the nation's infrastructure, mobilize around a common purpose, work with other nations to combat common threats, and solve major environmental problems at far less cost than originally expected (see Appendix B). Previous generations have successfully addressed problems of similarly daunting complexity, uncertainty, and scale.

The task assigned to this panel had considerable potential for overlap with topics dealt with in other panels because information and decision making constitute a core component of reducing emissions (the *Limiting* panel), adaptation (the *Adapting* panel), and scientific research (the *Science* panel). Although we tried to negotiate boundaries and overlaps with other panels, we were asked to prepare an independent panel report. We had extensive discussions about our task description and hope that the topics we have covered are those of greatest immediate help in decision making.

ORGANIZATION OF THIS REPORT

Chapter 2 describes who is making decisions about climate change in the United States and finds that many other actors beyond the federal government are making decisions that are a significant contribution to the overall national responses. Chapter 2 analyzes different types of decision makers, federal agencies, the courts, state government, cities, companies, and environmental NGOs and some of the decisions they are making. An initial assessment is made of some of their information needs.

Chapter 3 builds on previous NRC reports and the IPCC, proposing iterative risk management as the best approach to informed decisions, and discusses why this framework is best suited for a variety of decision makers in responding to climate change.

Chapter 4 evaluates specific decision support tools and other resources used for a variety of decisions related to climate change at the international, national, and state scales. Tools used in business, for adaptation choices, for limiting GHG emissions, and for the value of information are evaluated.

Chapter 5 addresses the question of what type of climate services are needed for an effective response to climate change in the United States and outlines the justification, functional components, institutional consideration, and principles for climate services. This chapter was written during a period of considerable debate about the needs for and management of an official National Climate Service. The chapter addresses some of the issues under debate and on the provision of information on climate, climate change, and climate impacts (emissions are addressed in Chapter 6).

Chapter 6 examines a variety of strategies and institutions that are needed to ensure the tracking of emissions and provide government, the private sector, and individuals with reliable information about the GHG implications of their decisions, practices, and lifestyles.

Chapter 7 discusses issues surrounding the need for international information on adaptation strategies and GHG emissions. Informed decisions and effective responses to climate change within the United States require that the nation contribute to international information gathering and that U.S. decision makers, including farmers and other businesses, need to receive reliable and usable information about what is happening with climate change impacts and responses elsewhere in the world.

Chapter 8 discusses what is known about communication and education about climate change, with some recommendations for improvements. During the early stages of the *America's Climate Choices* study there were several opportunities for public input. One of the main issues of concern was that of public understanding of climate change and the ways in which information and education might provide the public with improved insights and strategies for responding to climate change. Chapter 8 is a response to this issue. Because other panels focused on training and capacity building, we did not undertake a comprehensive analysis of training beyond that within the education system.

CHAPTER TWO

Many Different Decision Makers Are Making Choices to Respond to Climate Change

Our companion *America's Climate Choices* (ACC) panel reports have made a number of recommendations regarding climate-related decisions that might be taken, including actions to reduce greenhouse gas emissions, to adapt to the impacts of climate change, and to implement adaptable response strategies and policies. Specific actions to implement these recommendations will require new information, mechanisms, and institutions for providing that information for decision makers. This report describes what types of information will be required and how it might best be delivered. One can get an initial sense of these information requirements by examining actions decision makers are already taking.

Many different people and organizations—from individuals and small companies to communities, state governments, federal agencies, and multinational corporations—have already made decisions and taken actions to respond to climate change (see Table 2.1). Some have decided to reduce their greenhouse gas emissions or to plan for adapting to the impacts of climate change; some have decided not to act; and others have decided to inform themselves and others about the science, costs, and benefits of climate change and potential responses. Understanding the nature, effectiveness, and interactions of all these decisions is an important step in maximizing the effectiveness of America's response to climate change.

This chapter identifies major groups of decision makers, their motivations, and the kind of actions they take, and identifies what information is missing that may prevent a sustained response. Emphasis is placed on non-federal decision makers, many of whom have acted in advance of federal policy and whose decisions, actions, and successes may be affected by any new choices made at the federal level. The reach of some of these non-federal responses has been considerable. For example, decisions on automobile emissions standards made by the state of California have rippled throughout the U.S. economy. When groups of states set up carbon trading markets, or when major corporations (such as Dupont or Walmart) decide to reduce emissions in their operations and supply chains, the effects can reverberate nationwide or even globally.

TABLE 2.1 A Typology of Decision Makers Responding to Climate Change

	International	National	Regional	Local
Government	Intergovernmental organizations (e.g., World Bank, UNFCCC), international networks of local governments (e.g., ICLEI, WCI)	Federal agencies, Executive, Congress, Judiciary	Tribal and state governments (agencies, executives, legislatures and judiciaries), regional offices of federal agencies, interstate networks and agencies (e.g., RGGI)	City, county, and other local government
Private sector	Multinational corporations, international business networks (e.g., WBCSD)	Corporate HQs, national business networks	Regional corporate offices, companies and business associations	Local businesses and associations
Non-profit organizations	International environmental and humanitarian organizations and networks (e.g., CAN, Oxfam)	Environmental and other NGOs	Regional offices of NGOs, regional organizations	Local NGOs
Citizens	International citizens networks	Voters, citizen and consumer networks	Voters, citizen networks	Individual as voters, consumers, agents

NOTE: CAN, Climate Action Network; ICLEI, Local Governments for Sustainability; NGO, non-governmental organization; RGGI, Regional Greenhouse Gas Initiative; UNFCCC, United Nations Framework Convention on Climate Change; WBCSD, World Business Council for Sustainable Development; WCI, Western Climate Initiative.

The non-federal actors making climate related decisions often have a variety of goals they are trying to achieve. For instance, a state, city, tribe, or business might be pursuing goals such as

- Reducing their greenhouse gas emissions;
- Reducing vulnerability to climate change now and in the future;
- Reducing energy costs and exposure to the volatility of energy costs;
- Making it easier to respond to future federal or other regulations that would significantly reduce greenhouse gases (or we might phrase this as responding more easily to a future low-carbon economy);
- Establishing economic leadership, promoting economic development in green technology sectors;
- Fostering a reputation as an environmental leader; and
- Aiming to use a leadership role to catalyze
  o Emission reduction efforts by others,
  o Regional, national, and global investments in low-carbon technologies, and
  o Diffusion of best practice measures to adapt to climate change.

Ideally, states, cities, and businesses would have information available that would help them evaluate and improve their progress in pursuing each of these goals. Some decisions and actions related to climate change are being informed by a loose confederation of networks and other institutions created to help guide climate choices (Figure 2.1). In general, however, little information is currently available to plan and track any of these actions.

STATES, CITIES, AND LOCAL GOVERNMENTS

State and local governments have unique advantages for implementing policies that are intimately linked with the economy and society because of the familiarity with local circumstances and stakeholders. This section illustrates some of the climate actions being taken at the state and local levels.

Climate Change Actions by U.S. States

Actions by U.S. states to respond to climate change can be significant on the global scale. If the 50 states were treated as nations and compared to other national jurisdictions, they would represent 35 of the world's top emitters (Marland et al., 2003; Peterson and Rose 2006; see Appendix C). Many states have taken a lead on climate

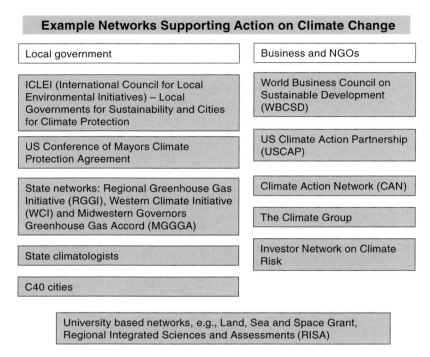

FIGURE 2.1 Example networks supporting action on climate change.

change by passing statewide greenhouse gas emission reduction targets and developing climate action plans guided by advisory boards and commissions (Pew Center, 2009; Rabe, 2008; Table 2.2). Twenty-one states have set emissions targets, albeit with varying stringencies and timelines and using different actions to achieve reduction goals (Pew Center, 2009; Appendix D).

California was the first state to set enforceable statewide greenhouse gas emissions targets (WCI, 2008; Box 2.1). Many other states, however, have coordinated their actions in regional initiatives that include cap–and-trade systems with overall emission reduction targets (Figure 2.2). The Regional Greenhouse Gas Initiative (RGGI) in the northeastern states, for example, capped CO_2 emissions for large fossil fuel electric generation plants starting in 2009 with the goal of stabilizing emissions by 2014 and then reducing them by 10 percent by 2018. The Western Climate Initiative (WCI) calls for reducing 90 percent of the region's greenhouse gas emissions by 15 percent below 2005 levels by 2020. The Midwestern Greenhouse Gas Reduction Accord proposes an economy-wide program to reduce emissions 20 percent below 2005 levels by 2020. These trading schemes have encouraged the inclusion of Canadian and Mexican provincial and state-level governments, as well as allowing some use of international

TABLE **2.2** Summary of State Climate Plans and Greenhouse Gas Reduction Commitments (as of May 2010)

	Completed Climate Change Action Plan	Climate Change Commissions and Advisory Board	Emission Reduction Targets	Actively Participates in Regional Initiatives	State Adaptation Plan
Number of states	36 (4 of which are in progress)	23	20	32 (including observers)	15
Examples	Climate action plan to assist state decision makers identify cost-effective GHG reductions appropriate to their state	Advisory boards to implement climate action plans	Targets and timelines range includes: CA:1990 levels by 2020 (mandatory) VA:30% below BAU by 2025 VT:25% below 1990 by 2012	MGGRA:6 (plus one Canadian Province) WCI:7 (plus 4 Canadian provinces; and 6 U.S., 2 Canadian, and 6 Mexican observers) RGGI:10	Adaptation plans may form part of the Climate Action Plan, although adaptation is not as well developed as mitigation at state level

NOTE: MGGRA, Midwest Greenhouse Gas Reduction Accord; WCI, Western Climate Initiative; RGGI, Regional Greenhouse Gas Initiative.
SOURCES: Pew Center (2009a); EPA (2008).

BOX 2.1
California and Climate Change

In 2006, California passed the California Global Warming Solutions Act (AB-32), committing the state to reduce greenhouse gas emissions to 80 percent below 1990 levels by 2050, and capping emissions at 1990 levels by 2020—an overall reduction of 30 percent from "business as usual" emissions projections. The act is unusually prescriptive in setting out specific policy goals and benchmarks within the legislation. In addition to the emission reduction goals, the act mandates reporting of emissions from large industrial sources; allows a cap-and-trade system; allows adoption of regulations to control greenhouse gases associated with landfills, fuels, ports, and consumer products; creation of a process to credit voluntary reductions; and requirements to evaluate environmental justice and technology options. Meeting the 2020 emissions target will require changes in the energy, transportation, agriculture, and waste sectors. Other climate-related legislation under development in California ranges from promoting alternative fuels to banning incandescent light bulbs.

Explanations for California's extensive actions on climate change include (Corfee-Morlot, 2009; Franco et al., 2008; Mazmanian et al., 2008)

1. Relatively high awareness and support for climate change policy among California residents compared to the rest of the nation;
2. A significant perception of risks to the state from the impacts of climate change due largely to the results of impact studies showing, among other things, disappearance of California snowpack, threats to water resources, and increased fire risks (see figure below);
3. Political ambition and ethical commitment of leaders who see climate as a high-profile issue and California as a major influence on national policy;
4. The economic and technological advantages of being an "early mover" in a green economy;
5. Willingness to reach out and learn from international experiences; and
6. A long history of efforts to manage air pollution and promote energy efficiency.

Information critical to the state's decisions to act on climate change included research on climate change impacts in the state, economic cost-benefit analyses, and documented successes in other countries.

Although it is too soon to assess the success of AB-32 in reducing greenhouse gas emissions (and there is some opposition to repeal AB-32), California has already limited emissions growth through earlier regulations and investments. For example, per capita electricity consumption has leveled off to less than 8,000 kWh/person since 1980, while the U.S. average has risen to more than 12,000 kWh/person (Kammen and Pacca, 2004).

Residents evacuate due to wildfires in Southern California. SOURCE: Dan Steinberg, AP Photo.

FIGURE 2.2 Regional cap-and-trade initiatives in the United States. SOURCE: Pew Center (2009).

offsets (Blakes, 2009). In addition, California is collaborating with the United Kingdom on emission reductions (California State, 2006).

States have also taken a lead in promoting legislation for other climate-oriented initiatives, such as mandating increases in the percentage of renewables in the energy supply mix, promoting energy efficiency, instituting net-metering and green energy pricing, and setting energy efficiency standards for resources and appliances. Other actions include setting new vehicle emissions standards, promoting alternative fuels such as biodiesel or ethanol, providing incentives to buy low-carbon fuels and vehicles, and creating agricultural plans to promote biomass storage. In addition, some states are funding research and development on clean coal technology and on carbon capture and storage (Pew Center, 2009). Many states are also collecting information and are concerned about changes beyond their boundaries because of the impacts on their own economies and livelihoods. For example, Arizona and Washington pay attention to what may be happening in Mexico and Canada, respectively, and information providers such as the regional assessment centers in the Pacific Northwest (e.g., Climate Impacts Group) and the Southwest (e.g., Climate Assessment of the Southwest) both have important non-U.S. and border components. The reasons for state action on climate change include perceptions of vulnerability, judgments that the costs of inaction are greater than the costs of response, and opportunities for long-term economic advantage (Box 2.1). Decision makers have also responded to climate change for short-term leadership advantages and because they feel a strong ethical obligation to act on climate change.

Through these actions, U.S. states have acknowledged their decision-making responsibilities for responding to climate change as well as their financial responsibilities for the cost of future climate change impacts. Their steps to limit the magnitude of climate change may accelerate or guide national policy (Peterson and Rose, 2006). Some argue that effective greenhouse gas emission reductions in the United States will only succeed with bottom-up approaches, which can then be complemented by top-down rule (Victor et al., 2005; Wiener et al., 2006).

States (and other jurisdictions such as cities and counties) that have taken action on climate change are concerned that federal actions may undermine or overlook their investments and policies relating to climate change. National policies can be designed to avoid undermining state action. For example, if state emissions reductions are greater than those established under a national cap-and-trade system, the federal emission allowances for state reductions could be retired rather than freed up to be used in other states.

State decision makers have information needs that include detailed data on greenhouse gas emissions down to the local level, regional projections of how climate change may affect their jurisdiction, and information on the distribution of costs and benefits of climate impacts, emission reductions, and adaptation options. They must also understand how climate change and policy will affect other states and countries, especially where they are part of regional carbon markets, where they share vulnerable resources, such as water, or where state economies are closely tied to national and international markets.

Climate Change Actions by City and Local Governments

As locations with a dense amount of buildings, cities are significant producers of greenhouse gas emissions, accounting for as much as 70 percent of global fossil fuel emissions. Cities across North America began to adopt targets and timetables for reducing their greenhouse gas emissions in early 1990s, in large part because of the emergence of transnational municipal networks[1] that focused specifically on greenhouse gas reduction (Rabe, 2009; Schroeder and Bulkeley, 2009). For example,

---

[1] Such as ICLEI–Local Governments for Sustainability, formerly known as the International Council for Local Environmental Initiatives (ICLEI) Cities for Climate Protection and the International Solar Cities Initiative.

Los Angeles has committed to reduce emissions 35 percent below 1990 levels by 2030 and New York has a major program for both emission reductions, 30 percent below 2005 levels by 2030, and adaptation (see *ACC: Adapting to the Impacts of Climate Change*, NRC, 2010b). The World Mayors Council on Climate Change also galvanized political action, committing to reduce greenhouse gases by 80 percent below 1990 levels by 2050 in industrialized countries through further collaboration with transnational initiatives (ICLEI, 2007). Domestically, the U.S. Conference of Mayors' Climate Protection Agreement is a large network of local governments working on climate change with 1,017 mayors.

City-level action on climate change has been driven by opportunities to save money through energy efficiency, creating jobs and generating tax revenues through the development of green technologies, demonstrating leadership, and reducing vulnerability, especially to sea level rise and water shortages. A recent report looked at a sample of U.S. cities and found an extensive array of actions to address climate change, including retrofitting government-owned buildings to become more energy efficient, converting fleets to hybrids, and planning for the long-term impacts of climate change (Figure 2.3; CDP, 2008).[2] These cities are moving quickly to adopt the emissions reporting standard developed by ICLEI–Local Governments for Sustainability, the Climate Registry, the California Climate Action Registry, and the California Air Resources Board, illustrating the potential for rapid diffusion of useful information and decision tools. However, considerable opportunities remain for cities to take further action on climate change, but their reach is often limited by what they are actually able to control; emissions (from transport, for example) and adaptation (of water systems, for example) are often influenced by other federal, state, and local jurisdictions.

Counties are also taking action on climate change and may become increasingly important should the Environmental Protection Agency (EPA) choose to regulate carbon emissions as air pollutants. Among counties taking early action on climate change are King County, Washington (see Box 2.2); Fairfax and Arlington counties, Virginia; Nassau County, New York; Miami-Dade, Florida; and Cook County, Illinois; with most pledging to reduce emissions 80 percent by 2050 and with a request to the federal government to raise fuel economy standards to 35 mpg.

As with many cities, county action has been based on information about potential

---

[2] The 18 cities examined were Annapolis, MD; Arlington, VA; Atlanta, GA; Burlington, VT; Chicago, IL; Denver, CO; Edina, MN; Fairfield, IA; Haverford, PA; Las Vegas, NV; New Orleans, LA; New York, NY; North Little Rock, AR; Park City, UT; Portland, OR; Rohnert Park, CA; Washougal, WA; and West Palm Beach, FL. Information here based on press releases until final report is released.

FIGURE 2.3 Green roofs such as this one on Chicago City Hall are aimed at conserving energy. SOURCE: CCAP (2008).

regional vulnerabilities to climate change and relies on detailed inventories of both county government greenhouse gas emissions as well as those across the county.

The Impact of State and Local Government Action

Many states, cities, and local governments have taken both political and practical action to limit the magnitude of and adapt to climate change, showing leadership nationwide. As a result of these actions, 53 percent of Americans now live in a jurisdiction that has enacted a greenhouse gas emissions cap (Lutsey and Sperling, 2008). Few data exist on the aggregated impacts of city or state action on climate change, but it appears that actual emissions reductions are likely to be influenced most strongly by key competencies, governance structures, and legal frameworks at the local and national levels (Bulkeley and Betsill, 2003; Schroeder and Bulkeley, 2009). There is also some evidence that municipalities seem to give a higher priority to limiting climate change than adapting to its impacts (Granberg and Elander, 2007; Hanak et al., 2008).

BOX 2.2
Case Study of Local Government and City Action: King County, Washington

King County, Washington, which includes the city of Seattle, provides an example of the vital role of scientific information and political leadership in responding to climate change. As early as 1988, Ron Simms, then a young member of the King County Council, tried and failed to get climate change onto the county agenda. By 2003, as chief executive of King County, Simms was able to create a county inventory of greenhouse gas emissions and to appoint a county program manager for climate change. In 2005, Simms decided to begin planning for adaptation, even though he was ahead of both public opinion and that of the King County Council and its staff.

In partnership with the University of Washington's Climate Impacts Group (CIG), Simms organized a county-wide workshop on projected scenarios of climate change and its likely impacts on the county, state, and region for 20, 50, and 100 years into the future. The workshop drew more than 700 participants and focused on water, agriculture, forest ecosystems, the coastal zone, and fish and shellfish. Background materials, including a climate impacts white paper, sector specific fact sheets, and a climate change primer for non-specialists were provided to help guide discussions.[a] Participants sought to identify resource and information needs for the sectors being affected by climate change and approaches for improving sector specific adaptation plans, including risk assessments and responses (see Box 2.3 for a sector specific example).

Following the workshop, demands for information and assistance grew quickly, and the groundswell of interest within the county fueled opportunities for launching policy innovations. When a white paper by the University of Washington CIG[b] showed that 10 climate models chosen to project future climate in the Pacific Northwest converged in their predictions up to the year 2050, Simms charged his staff

---

Through their relationships and networks, state and local governments have "reached up" to affect national, and even international, climate change issues (Engel, 2006) and can be seen as innovators through "policy entrepreneurship," playing a role as first movers on climate change (as they have for other environmental issues), functioning as policy innovators, and testing policies that can be then used at federal levels (Litz, 2008).

In general, more research is needed to understand gaps between words and action at both the state and the city levels, to examine concrete changes in policy and progress that result in emissions savings, and to investigate the possibility for scaling up and out from the city level.

to use the projections as a benchmark for designing adaptation policies. The projections were used to inform investment decisions relating to flood prevention and wastewater treatment, as well as plans to build, repair, or replace societal infrastructure. By 2007, King County had published and committed itself to a Climate Plan as well as to emissions reductions (King County, 2007). King County and the University of Washington also wrote a guidebook on planning for climate change for local, regional, and state governments, and by June 2009, more than 1,000 hard copies and 1,000 electronic copies of the guide had been distributed to U.S. cities and states, and 17 more to cities overseas.

The King County model was scaled up to assist in multi--governmental regional planning processes for issues such as water management. Water suppliers for the cities of Everett, Seattle, and Tacoma, for example, used the climate change projections to evaluate impacts to their own systems and to adjust operations and management plans. The governor and state legislature have asked the University of Washington CIG to perform new assessments every 4 years; they have also required state agencies to create interdepartmental teams for climate policy assessment, planning, and implementation.

The King County example demonstrates the importance of political leadership in stepping ahead on issues such as climate change and the value of local government partnering with universities to provide regionally relevant climate projections and impact information from a credible source.

---

[a]See *http://cses.washington.edu/cig/outreach/workshops/kc2005.shtml*; *http://www.kingcounty.gov/exec/globalwarming/environmental/2005-climate-change-conference.aspx*.

[b]"Scenarios of Future Climate for the Pacific Northwest" discussed how climate model runs from the IPCC could be downscaled to the Pacific Northwest region at 1/8 degree resolution, explored the uncertainties, and compared the predictions to observations for the 20th century.

BUSINESS SECTOR

The business sector can play a powerful role in responding to climate change because it produces a large share of greenhouse gas emissions and possesses substantial financial, technological, political, and organizational resources (Levy and Newell, 2005; Newell, 2000, 2008). Approximately 60 percent of U.S. greenhouse gas emissions in 2007 resulted from electricity generation from fossil fuels and from fossil fuel–based transport, both associated with some of the countries largest corporations, including oil companies, utilities, appliance manufacturers, and the auto industry.[3] Whether emissions are assigned to sectors (e.g., energy, agriculture, or industrial processes), electrical utilities, business supply chains, or to consumers, it is clear that corporate America is intricately linked to greenhouse gas emissions and thus to actions and policies to reduce them.

---

[3] See *http://www.epa.gov/climatechange/emissions/downloads09/ExecutiveSummary.pdf*.

BOX 2.3
Information Needs of a Hydroelectric Dam Manager

A hydroelectric dam manager is faced with the challenge of balancing environmental protection goals, energy production, water supply, agriculture, stream navigation, and recreation. Managers are often making long-term plans regarding dam maintenance and infrastructure, which requires an understanding of climate variability, and a range of climate impacts and associated adaptation methods, on multiple time and small space scales. Weather and climate influence stream flow, demands for energy and water, watershed, and ecosystems. Specifically, the dam manager needs to understand how projected global- or continental-scale changes in temperature and precipitation relate to local changes in watersheds. Decisions that need to be made may not apply to traditional planning parameters, such as the "100-year flood." Widespread monitoring of ecosystem changes and the downstream effects of infrastructure as well as improved modeling of projected climate changes are crucial for the dam manager to make effective decisions. Long-term decisions are difficult to make given the uncertainty about future conditions. Particularly with infrastructure, using climate change planning scenarios and robust decision making can help to identify strategies and decrease vulnerability to a range of potential stresses, versus optimizing a single goal (Groves and Lempert, 2007; Groves et al., 2008).

The business sector is also vulnerable to climate change impacts. Although the insurance and agricultural industries are most clearly affected, a wide range of other sectors face potential damages (or opportunities), depending on how higher temperatures, rainfall extremes, changes in water availability, and sea level rise affect their operations, supply chains, or demand for products. Businesses also risk the possibility of litigation if their activities can be linked to high emissions and damages (Allen, 2003).

Although many American corporations were initially skeptical of or resistant to action on climate change, a large number of companies, including some of the most economically powerful and highest greenhouse gas emitters, are now publicly disclosing their emission profiles, voluntarily reporting and aiming to manage them, and even actively lobbying for a stronger regulatory landscape in which to make new green investments (Hoffman, 2005; Okereke, 2007; O'Riordan, 2000; USCAP, 2009).[4] The business response to climate change is enhanced when there is confidence in the continuity of policies such as carbon prices and incentives to invest in low-carbon technologies (Stern et al., 2006).

---

[4] The Business Environmental Leadership Council, which represents 45 U.S. companies (most in the Fortune 500) with combined revenues of more than $2 trillion, publicly supports mandatory regulation on climate change in the United States and internationally (Pew Center, 2008).

Business action on climate change can be motivated by concerns for profit and comparative advantage, and also driven by wider societal pressures and concern for the environment. Okereke (2007) suggests that companies implementing climate activities with a genuine sense of societal and ethical responsibility are more likely to make deeper emissions cuts than those whose actions are based purely on economic rationality (even if this does favor some emissions reductions). Market mechanisms to address climate change, such as the clean development mechanism (CDM) or emissions trading, can also motivate corporate action because of the potential profits to be made in the creation, trade, and service industry surrounding emissions reductions and credits (Bumpus and Liverman, 2008). Jones and Levy (2007) suggest that emissions trading can be used by corporations to provide flexibility and defer expensive shifts in resource allocation toward low-emission energy sources. In addition, many businesses are facing pressure from their shareholders to address climate risks and recently the Securities and Exchange Commission (SEC) announced that companies must provide information to investors about the business risks associated with climate change.

Business action on climate change has been fostered and to some extent coordinated by business networks (see Figure 2.1), such as the World Business Council for Sustainable Development (WBCSD), the Climate Group, and the U.S. Climate Action Partnership (USCAP), which promote politically feasible and profitable ways to reduce greenhouse gases. These include the promotion of market mechanisms for emissions reductions (such as emissions trading and offsets), incentives for clean technology investment, and learning-by-doing approaches.

Whereas actions by companies in the fossil fuel, utility, and transport sectors have the largest potential to reduce the risks of climate change, large corporate entities such as Walmart or GE have considerable power to influence consumers through their product offerings or to demand lower emissions from companies in their supply chains (CDP, 2009b). In the financial sector, banks, investors, risk managers, and corporate leaders are driving carbon-conscious investment strategies (see Box 2.4). Other businesses are providing innovative solutions to climate change, from developing low-carbon technologies to creating products such as crops and buildings that are better adapted to a warmer world. Many of these business decisions are long-term strategic investments.

BOX 2.4
HSBC: An Example of Business Response

HSBC—a global banking and financial services organization with 10,000 offices in 83 countries and territories—recognized early that climate change represents a threat to its business as well as an opportunity for decreasing greenhouse gas emissions. HSBC has been carbon neutral since 2005,[a] making it the first major bank and first FTSE[b] 100 Company to do so (Cogan, 2008). Emissions have been lowered by reducing energy consumption, sourcing renewable energy where possible, and offsetting remaining emissions. HSBC's climate strategy includes the following elements:

- The appointment of former World Bank chief economist Nicholas Stern as Special Adviser on Economic Development and Climate Change to the Group Chairman.
- Partnerships to share experience, gain knowledge, and participate in third-party initiatives. Through such partnerships, HSBC is educating 100,000 HSBC employees about climate change, supporting the work of conservation organizations to limit the impacts of climate change on cities, people, forests, and water, and encouraging low carbon consumption and the development of environmentally friendly products and services.
- Establishment of the Climate Change Center of Excellence, which evaluates the implications of climate change for the HSBC Group, acts as a central source of climate knowledge, and supports the implementation of the firm's carbon finance strategy.

HSBC's decision to become carbon neutral involved a steep learning curve, but it resulted in an additional business opportunity where it can provide advice and potentially source credible carbon offsets for its corporate clients. It should be noted that HSBC's carbon neutrality program is only for their direct operations; it does not yet cover the carbon footprint of those firms to whom it lends.

---

[a] http://www.hsbc.com/1/2/sustainability/protecting-the-environment.
[b] FTSE is an independent company jointly owned by The Financial Times and the London Stock Exchange. (SOURCE: http://www.ftse.com/index.jsp)

The Impact of Business Decisions and Actions

It is difficult to quantify the potential contributions of businesses to limiting the magnitude of climate change because of the wide range of corporate responses and the inconsistency, incompleteness, and quality of reporting. Voluntary disclosure of carbon management strategies and emissions to groups such as the Carbon Disclosure Project (CDP) has increased over time, and information is now available for 475 major companies. Although the total emissions of these companies are known (the Global 500 reporting companies are responsible for 11.5 percent of total global emissions; see CDP, 2009a), it is difficult to assess emission reduction successes to date because the data are aggregated in different ways, even within the same corporate entity (Kolk et al., 2008; see Chapter 6). Even experienced analysts find it difficult to assess individual company greenhouse gas profiles, and even harder to examine changes over time due to inconsistencies in data and reporting. Attempts at international standardization, such as those promulgated by the Climate Disclosure Standards Board (CDSB, 2009), aim to ameliorate this problem.

Table 2.3 provides a sample of actions and commitments by major U.S. firms that are aimed at voluntary emission reductions of 10 percent or more. By 2005, as many as 60 U.S. companies had set emission reduction targets, ranging from as little as 1 percent per year for 4 years to a 25 percent reduction from 1990 levels by 2011 (Hoffman, 2005). Jones and Levy (2007) suggest that, although business has been energetic in its reaction to climate change, the results have been ambiguous and dwarfed by the creation of non-climate friendly products and services and by increases in unit sales that offset the energy used to produce a unit. As a result, external evaluations may come to different conclusions when examining firms' achievements.

The impact of business decisions to respond to climate change should be measured not only in terms of their greenhouse gas emissions reductions but also in terms of their choices about adaptation, energy efficiency, and technology investments and innovations. Although there are some good case studies of business adaptation to climate change—especially in the insurance sector—there are few measures of how well business is adapting to climate change in the United States. Many firms do report improvements in energy efficiency and other energy-saving measures but there is no focused assessment of corporate investment in low-carbon technologies or other relevant technologies such as carbon sequestration.

Major financial and investment firms are now starting to develop indices to evaluate how well firms are managing risks associated with climate change. For example, the Global Framework for Climate Risk Disclosure (CRDI, 2007) proposed by a consortium, including major pension schemes and socially responsible investors, that firms disclose emissions, physical risks of climate change, and efforts to manage greenhouse gases and climate impacts through their financial reports, the CDP, or the Global Reporting Initiative. Investors have explicitly asked for the disclosure of information on direct and indirect emissions since 1990 using the Greenhouse Gas Protocol (see Chapter 6) and information on how climate and weather generally affect their business and its operations, including supply chains. A 2007 study (Gardiner et al., 2007) found that many companies did not have or disclose this information. Only 47 percent

TABLE 2.3 Examples of Greenhouse Gas Reduction Commitments by Business Sectors in the United States

Sector	Company	Commitment and tangible GHG reductions so far	Reduction type	Commitment to future GHG reduction targets
Oil and Gas	BP	Achieved 10% reduction from 1990 levels in operational GHG emissions	Absolute	-
	Chevron Corp	Increased energy efficency by 27% of output	Intensity	Currently setting annual GHG reduction plan and targets
Energy, Power Generation and Utilities	Exelon Corporation	Reduced emissions by 38% between 2001 and 2008	Absolute	Aims to reduce >15MM lCO2e by 2020
	Public Service Enterprise Group (PSEG), Newark, NJ	Reducing US GHG emissions by 31% per kilowatt from 2000 to 2008	Intensity	-
	American Electric Power	4% reduction of US GHG emissions from 2001 to 2006	Absolute	6% Reduction from 2001 by 2010
Cement/ Industrials	St. Lawrence Cement	Global GHG reductions 16% per ton of cementious product from 2000 to 2006	Intensity	Global GHG reduction of 20% per ton of cementious product by 2012
	Alcoa	-	-	Reduce total US GHG by 4% from 2008 to 2013

Chemicals	3M	Reduce total US GHG by 60% between 2002 and 2007	Absolute	-
	Johnson & Johnson	Cut GHGs by 12.7% from 1990 to 2007 (while sales increased 400%)	Absolute	-
	SC Johnson	Total US GHG reduction of 17% from 2000 to 2005	Absolute	Total US GHG reduction of 8% from 2005 to 2010
Auto Industry	Ford	Reduced global facility GHGs by 39% between 2000 and 2007	Absolute	Agreed to cut emissions by 60-80% by 2050
	General Motors	Reduced US GHG emissions by 23% from 2000 to 2005	Absolute	Reduce CO_2 emissions by 40% from 2000 to 2010
	Mack Trucks	32% GHG reduction per unit from 2003 to 2007	Intensity	12% GHG reduction per unit by 2012
Pharmaceuticals	Baxter International Inc.	27% GHG reduction per unit of production from 2000 to 2005	Intensity	Reduce total US GHG emissions by 5% from 2005 to 2012
	Pfizer Inc.	Reduction of global GHG by 43% per $ million revenue 2002-2008	Intensity	Reduce total US GHG emissions by 20% from 2007 to 2012
	Roche and US Affiliates	Total US GHG reduction 11% from 2001 to 2006	Absolute	Total US GHG reduction of 15% by 2010

continued

TABLE 2.3 Continued

Sector	Company	Commitment and tangible GHG reductions so far	Reduction type	Commitment to future GHG reduction targets
Technology	Sun Microsystems Inc.	Total US GHG reductions of 23% from 2002 to 2007	Absolute	Total global GHG reductions of 20% from 2007 to 2012
	United Technologies	Reducing global GHG emissions by 46% per revenue dollar	Intensity	Total global GHG reductions by 12% from 2006 to 2010
	Dell Inc.	-	-	Reduce global GHG emissions by 15% per dollar and maintain net global zero emissions to 2010
Banking	HSBC North America	First carbon neutral bank (global)	Absolute (with offsets)	Reduce total US GHGs by 10% from 2005 to 2010
	Citigroup Inc.	-	-	Total global GHG emissions by 10% from 2005 to 2011

SOURCES: CDP (2008); EPA (2009a); GM (2009); PRNewsWire (2009).

of S&P companies responded to a survey compared to 72 percent internationally in the FT500 and many did not want their responses to be public. Few companies were thinking about impacts of climate change and adaptation, and overall only 25 percent were providing the information that investors were looking for. Prototype climate risk rating indices aim to include information on overall carbon footprints, ability to manage climate risk exposure, the rate of change in carbon and climate risk, the ability to benefit from climate-driven opportunities, and risk and costs of climate regulations.[5] Ceres and Price Waterhouse have proposed information systems that would link risks of water scarcity to municipal bond investments. In February 2008, Citigroup, JP Morgan, Chase, and Morgan Stanley launched the Carbon Principles, a voluntary framework aimed at addressing climate risks associated with financing carbon-intensive projects in the U.S. power sector. Bank of America, Credit Suisse, and Wells Fargo also endorsed the principles later that year. In December 2008, a second group of global financial institutions—including Credit Agricole, HSBC, Munich Re, Standard Chartered, and Swiss Re—announced their adoption of the Climate Principles, a set of commitments on climate business strategies developed by The Climate Group[6] that include developing financial products and services to help clients manage climate risk and opportunity, incorporating climate change issues into research activities, and considering "practical ways to assess the carbon and climate risks" of lending and investment activities.

The choice to consider climate risks in investment decisions has important implications for both private and public information systems and raises the bar for all providers of such information including the federal government to ensure the accuracy, legitimacy, and transparency of information that is being used by the business community and other actors. We discuss how to improve such information in Chapters 5 and 6 of this report.

The panel judges that climate policy at all levels of government can benefit from understanding the motivations, drivers, and barriers to corporate actions. Accurate accounting of business response to climate change can legitimate the actions of responsible firms and expose lack of real action by others. Federal policy needs to acknowledge the variable nature of corporate action on climate change while realizing that, once clear federal policy guidance is provided, business itself has the ability to provide incentives for structural shifts and innovation. For the business community, the decisions and actions of government at all levels are extremely important in enabling and framing the business response to climate change. For some business groups the

---

[5] An example is Riskmetrics, Carbon Beta (*http://www.riskmetrics.com/sites/default/files/Carbon_Beta.pdf*).

[6] See *http://www.theclimategroup.org/programs/the-climate-principles/*.

TABLE 2.4 Summary of Action on Climate Change from Environmental NGOs

U.S. Environmental NGO	Climate Policy Objectives	Research
Conservation International[a]	"Tackling climate change by conserving nature" / "Climate community and biodiversity conservation" Need for immediate adaptation	In-house primary scientific and economic research on REDD and adaptation planning and forecasts
Environmental Defense Fund	Supports market-driven responses to climate change in line with science (80% cut by 2050). Lobbies for strong inclusion of business as agents of solutions	Uses in-house primary science to define policy goals and influence governmental policy
Greenpeace USA	Independent global campaigning organization; strong advocacy position, nonviolent direct action, lobbying, diplomacy	Position based mainly on "bearing witness" to environmental destruction; commissions science reports and has some in-house capabilities[c] carries out research with others such as WRI
National Audubon Society	Supports mandatory measures to reduce GHGs in the context of supporting ecosystems for birds and other wildlife. Direct lobbying in Congress through public policy office in DC.	Less research, more focused on direct policy lobbying for federal protection of lands

Partnerships	Public Education	Position on Carbon Markets	Action on Own Operational Emissions
Business, government, scientists, other NGOs Many corporate (mainly U.S.); global NGOs including field operations and projects on REDD in developing countries	Limited; has carbon calculator	Supports both market and non-market action on REDD	
Large-scale corporate engagement to push market-driven policies and prove strategic (not philanthropic) gains for business.[b] Linked with NRDC for market policies	Science for public education (especially on climate change); learning displays and exhibits	Believes "economic incentives = environmental gains"; "weds markets and social goals"; supports cap and trade; international REDD	
Partners with and supports grassroots movements on climate change, selectively with corporate entities taking leadership on issues	Website, direct action, high-profile initiatives to highlight current problems and possible solutions	Pro-deep emissions cuts from industry; unclear exactly on market position	Not clear
Limited; mainly networked through Audubon's own community based centers	Aims to educate members to be more energy efficient (with cost savings) and to lobby government (at different levels) for GHG reductions.	No policy stance	Not clear

continued

TABLE 2.4 Continued

U.S. Environmental NGO	Climate Policy Objectives	Research
Natural Resources Defense Council (NRDC)	Uses wide support base and over 300 lawyers plus scientists to be effective action group to pushing for new state and federal laws to cap pollution, to partnering with business and industry to boost energy efficiency and championing the widespread use of clean energy[d]	Strong ability for litigation and courtroom action backed by in-house science and economic analysis
The Nature Conservancy (TNC)	Largest eNGO by revenue. Leading conservation organization to protect ecologically important lands. Lobbying through information from project activities.	Project activities on deforestation and helping adaptation in project areas.
Sierra Club	Oldest grassroots environmental organization in the U.S. Main position is to lobby and work with Congress to create federal legislation; then to engage at international level[e]	Strong roots in environmentalism, less in-house research but bases on scientific analysis of problems, e.g., uses science to refute use of coal-fired power stations

[a] See *http://www.conservation.org/learn/climate/Pages/strategy.aspx* (CI 2008).

[b] See *http://www.edf.org/page.cfm?tagID=1475; http://www.edf.org/page.cfm?tagID=1746*.

[c] See *http://www.greenpeace.org/international/about/greenpeace-science-unit-2*.

Partnerships	Public Education	Position on Carbon Markets	Action on Own Operational Emissions
Partners with business to create green markets, new clean energy choices (e.g., Center for Market Innovation). Pro-new models for green business. Focus on building green, sustainable communities; new energy investment.	Some public education through website; mainly through illustrative court cases and focus on policy solutions	Supports "cap-and-Invest" for U.S. climate policy: sale of pollution allowances to bring in new government revenue, allowing funds to be redirected toward infrastructure improvements and incentives for innovation.	Not clear on organizational emissions reductions
Partners with large corporate entities, IGOs such as World Bank. Part of U.S. Climate Action Partnership (USCAP), alongside major business interests	Some through membership and website. Most through project implementation and lobbying. Has carbon footprint calculator for people and organizations	Actively supports mandatory reductions and cap-and-trade system. Fully supports markets for carbon reductions in REDD. Part of World Bank REDD initiatives	Not clear (although has on-site carbon footprint calculator and provides carbon offsets)
Campaigning and forming partnerships with other eNGOs. Less direct corporate partnerships.	Large public education aspect considering Sierra Club outings, website, and social media (Twitter etc).	Not clear policy on carbon markets	Not clear

[d] See *http://www.nrdc.org/about/priorities.asp; http://www.nrdc.org/globalWarming/cap2.0/default.asp; http://www.marketinnovation.org/.*

[e] See *http://www.sierraclub.org/carbon/*

uneven policy landscape that has emerged from the decisions of state and local government has become a barrier to effective action and is one reason why some business leaders and networks are calling for clear federal action to provide a more predictable and coherent policy environment. While some businesses would like federal action to support a signal in the form of a carbon price, others see the federal role in terms of support for research and development of low-carbon technologies or setting basic standards and guidelines for reporting emissions, efficiency, and climate risk. For companies with significant international presence, the relationship between U.S. policy and that of other countries and international regimes is also of great interest, especially, for example, the linkages of U.S. to international carbon trading schemes and reporting standards.

NON-GOVERNMENTAL ORGANIZATIONS

Environmental and other non-governmental organizations (NGOs)[7] in the United States have been active in the climate change arena for several decades. Their actions have focused more on informing policy makers and the public than on limiting their own direct emissions (which are small) but have also included partnering in the implementation of responses to climate change. Table 2.4 illustrates some of the contributions made by U.S. environmental NGOs to the climate change agenda, such as generating peer-reviewed articles, establishing pilot projects on climate adaptation, or educating the media and the public on climate issues. Some NGOs provide instructions on how to reduce emissions through energy efficiency or offer offsets to their members. Others work closely with both government and private-sector partners in negotiating policies and standards for carbon offsets, forest protection, renewables, and ecosystem protection. Although NGOs may need to be embedded in the policy networks of countries to have substantial effects on issues (e.g., Hall and Taplin, 2007) and necessarily have vested interests and potential accountability problems (Jepson, 2005), they have a strong role in shaping the agenda from various different competencies.

Reasons for NGO action on climate change vary but include the alignment of climate

---

[7] The NGO community encompasses a wide range of groups. Within the international climate policy sphere, for example, the United Nations Framework Convention on Climate Change (UNFCCC) recognizes several categories of NGOs, including business, environmental organizations, labor unions, and research institutions. Most U.S. environmental groups have developed a climate change program, and many humanitarian organizations have also decided to address climate change, especially the challenges of adaptation in the developing world. The NGO community also includes several organizations that are skeptical about climate change.

change responses with their core missions (e.g., ecosystem conservation or vulnerability of the poor), opportunities to attract new members and funds, concern with long-term sustainability, environmental advocacy, and education regarding the impacts caused by climate change (e.g., on human rights). In some cases, concerns have been raised about the source and legitimacy of scientific information provided by NGOs, about political bias and cooptation by government or the private sector, and about the transparency and governance of some organizations. But overall, NGOs have an important role in communicating climate change information and influencing opinions and, ultimately, the choices made as people and decision makers move to respond to climate change.

NGOs rely extensively on information provided by government agencies and on international scientific assessments provided by groups such as the Intergovernmental Panel on Climate Change (IPCC). They also play an important role in informing the public about climate change.

FEDERAL GOVERNMENT

Last but not least, the Federal government has a wide range of agencies with responsibility for making decisions on climate change (Table 2.5). As the scope of the response to climate change expands beyond science and research to implementation of solutions, even more federal agencies will need to be involved. This section describes the climate decision-related activities of the federal government and how they have changed over time.

History of the Federal Role in Climate Change Decision Making

Federal involvement in climate change issues began in earnest in 1978 when the National Climate Program Act was passed to support both climate research and policy recommendations to "assist the Nation and the world to understand and respond to natural and human-induced climate process and their implications" (P.L. 95-367). The resulting interagency program yielded high-quality scientific research, useful information, and increased international participation, but it fell short of its mandate to integrate the growing understanding of climate variability and impacts into national planning efforts (NRC, 1986a). Although the need for management strategies to deal with socioeconomic consequences of climate variation was recognized (NRC, 1985), the executive guidance and funding to address it was absent (GAO, 1990).

By the mid-1980s, global change science had matured into an interdisciplinary field of research (Correll, 1990; CRS, 1990), but little progress had been made in improving

TABLE 2.5 Examples of Federal Departments and Agencies Who Are Affected by or Involved in Decisions about Climate Change

Federal Department or Agency[a]	Example Climate-Related Decisions
Agency for International Development	Famine early warning, disaster relief, capacity-building programs in the developing world
Agriculture	Adapting agriculture and natural resources through Cooperative Extension advice and education, Federal Crop Insurance, forest management (USFS), resource conservation (NRCS). Research on carbon, water, ecosystems and atmospheric chemistry (ARS), greenhouse gas monitoring, methane management
Commerce	Information provision through National Weather Service; NOAA climate research and information through labs, cooperative institutes and grants, climate observations and monitoring; Post regulation of carbon emissions: may need to consider trades implications of carbon content of imported goods
Congress and Senate	Development and review of major policy including climate change and energy legislation
Council on Environmental Quality	Development and coordination of environmental and energy policy. Oversight of October 2009 Executive Order that requires Federal agencies to set a 2020 greenhouse gas emissions reduction target within 90 days
Defense	Management of defense lands, U.S. Army Corps of Engineers work on water resources management and flood protection. Climate risks to national security
Energy	Information provision through Energy Information Administration; research through national labs; emissions monitoring; R&D financing for next-generation technologies; climate modeling
Environmental Protection Agency	Research on impacts of climate change; regulation of air and water quality; coordination of voluntary emission reduction programs

TABLE 2.5 Continued

Federal Department or Agency[a]	Example Climate-Related Decisions
Health and Human Services	Disease control (CDC), medical and health research (NIH)
Homeland Security	Federal emergency management
Housing and Urban Development	Public investment and regulation of housing, parks, and urban planning
Interior	Climate impacts and adaptation include land management (BLM), Indian Affairs (BIA), fish and wildlife (FWS), National Parks (NPS), and water (Reclamation, USGS). USGS research on water, land, and ecosystems
National Aeronautics and Space Administration (NASA)	Earth observations and monitoring, research on earth system including atmospheric composition, carbon, water, land use, and climate variability
National Science Foundation	Research into atmospheric composition, climate, carbon, water, ecosystems, human dimensions, and education
Office of Science and Technology Policy	Advice on science and technology policy, assessments
Office of Energy and Climate Change Policy	Coordinate administration policy on energy and climate change
Securities and Exchange Commission (SEC)	Consider carbon and climate disclosure requirements for publically traded companies
State	International treaties
Transportation	Research on impacts of climate on infrastructure, energy efficiency, emissions; financing alternative mass transit

[a] Departments of Education, Justice, Labor, Treasury, and Veterans Affairs do not yet have major programs focused on climate change (but may do so in the future).

communication between scientists and public decision makers beyond the Department of Energy's *Carbon Dioxide Research: State-of-the-Art Report Series* (DOE, 1985). In 1990, the U.S. Global Change Research Act (P.L. 101-606) was enacted to "advance scientific understanding of global change and provide usable information on which to base policy decisions related to global change." However, the U.S. Global Change Research Program (USGCRP) established by the act was structured to focus on scientific research and monitoring to reduce uncertainty about the causes and consequences of climate change, not on policy issues (NRC, 1990).

In 1994, Congress added a new program element: scientific assessments of global climate change and its impacts on the environment and on various socioeconomic sectors. Although these assessments were to be produced at least every 4 years, only two comprehensive assessments (2001 and 2009) of the potential consequences of climate change for the United States have been published. The 2001 report was the first to integrate key findings from regional and sectoral analyses and to begin a national process of research, analysis, and discussion about the coming changes in climate, their impacts, and what Americans could do to adapt (NAST, 2001). It was intended to initiate an ongoing course of interaction and reporting that would be improved over time (NRC, 2007c). However, opposition to the report both within and beyond government delayed further assessment until 2009.

As the USGCRP evolved over time, an important area of research was added. In 2002, the interagency program was broadened to include research that could yield results within a few years by improving either decision-making capabilities or public understanding. Although the program remains focused on fundamental research on the physical climate system (NRC, 2009a,b), examples of agency efforts aimed at supporting decision making include

- NOAA's National Integrated Drought Information System, which assists local and regional planners in developing drought-resilient communities and the Regional Integrated Sciences and Assessments (RISA) program, which provides climate information to regional stakeholders;
- NASA's remote sensing imagery, which shows decision makers the rapid shrinking of ice sheets and other climate effects;
- NSF's research centers on decision making under uncertainty;
- DOE's integrated assessments;
- EPA's research on climate change impacts; and
- the International Research Institute for Climate and Society, which focuses on enhancing society's ability to understand, anticipate, and manage the impacts of climate fluctuations.

Informing Congress

Congress has a critical role to play in *America's Climate Choices* in framing the climate change issue, providing opportunity for public debate, establishing policy, and requesting information. Over the last two decades Congress has helped set an agenda for climate research and policy and has frequently requested information about a wide array of scientific, technical, and economic information related to climate change and to our nation's response strategies. For example, Congressional hearings have produced a number of calls for improved information about climate change that illustrate the range of people who need and use such information (Box 2.5).

At the national level, there is a wealth of available information on an array of issues, not just climate change, that are all competing for the attention of policy makers in Congress. What are the major sources of that information and analysis, and what is the quality of the information? In general, the mechanisms for Congressional information and analysis may be divided into two categories: active information gathering (or Congressional "pull" of information), and passive information receipt (or public "push" of information).

Active information gathering and analysis is initiated by members or committees (Congressional "pull") and may include the following:

- formal hearings;
- directed studies or investigations by the Congressional agencies (Congressional Research Service [CRS], Congressional Budget Office [CBO], or Government Accountability Office [GAO]);
- inquiries to, or studies by, individual federal agencies (such as NOAA, DOI, EPA, DOD, USDA, DOE, etc.);
- directed studies by Congressionally chartered non-profit organizations;
- directed studies performed by congressional staff, and legislative fellows; and
- personal study by members and staff.

Passive information gathering comes to members or committees uninvited (public "push") and may include the following:

- briefings,
- think tank reports,
- news media reports (TV, internet news sources, newspaper, etc.),
- publications of various kinds (books and periodicals),
- constituent or citizen input to individual members through letters (Box 2.6),
- policy positions advocated by important lobbyists and political donors,

- national reports (e.g., USGCRP *Global Climate Change Impacts* report), and
- internationally coordinated reports (e.g., IPCC).

Within each of the information and analysis categories, the degree of accuracy and objectivity varies widely. The CRS and CBO are agencies of Congress that are directly accountable to Congress for the quality of their information and analysis. They have produced many helpful reports on climate issues. For example, in 2009 the CRS produced reports on cap-and-trade (R49809), ocean acidification (R41043), climate policy (RL34513), climate science (RL34266), the carbon cycle (RL34059), and biochar (R41086). These reports rely heavily on the scientific literature and information from

---

BOX 2.5
Example of Congressional Testimony Calling for Climate Information

"Decision-makers at all levels of government and in the private sector need reliable and timely information to understand the possible impacts and corresponding vulnerabilities that are posed by climate change so they can plan and respond accordingly."
  Sarah Bittleman, Office of Oregon Governor on behalf of Western Governors Association, May 3, 2007

"Decision makers need information tailored to specific local fisheries and ecosystems"
  Scott Doney, Woods Hole Oceanographic Institution, May 7, 2007

"To make best use of available information in a changing climate, water management will need to adopt more flexible tools than those that have sufficed in the past."
  Christopher Milly, USGS, June 6, 2007

"The RCCs [Regional Climate Centers] have been increasingly called upon for information related to future climate conditions. Users are more aware of variations in climate conditions and require information to assist them in managing year-to-year climate variations and adapting to changing climate conditions."
  Arthur T. DeGaetano, Northeast Regional Climate Centers, May 5, 2009

"Given the dynamic nature of managing our water supply system, with our multiple objectives, capricious weather and the need to balance immediate and short term issues with longer term planning horizons, it is critical that we have access to real-time monitoring and forecasting information. Seattle relies on several federal agency monitoring and forecasting services to help inform our decision-making."
  Paul Fleming, Manager, Seattle Public Utilities, May 5, 2009

federal agencies and usually compare the advantages and disadvantages of different policy options. The CBO focuses on cost estimates and has recently produced reports on the cost of reducing emissions, flood insurance, offsets, and climate impacts. The GAO is an independent non-partisan agency that audits federal spending, reports on government program effectiveness, and conducts policy analysis and in the past year has produced reports on geoengineering (GAO-10-546T), emissions trading (GAO-10-377), adaptation (GAO-10-113) and aviation and climate change including recommendations for federal actions. Their reports are based on the scientific literature, interviews, and surveys of agencies, and they often provide testimony directly to congressional committees.

"For the user, we need an accessible go-to entity we can count on to help us sift through the ever-changing science, gather the raw data, benchmark against the experience of others, educate our publics, and work with us in assessing our vulnerabilities."
David Behar, San Francisco Public Utilities Commission and Water Utility Climate Alliance, May 5, 2009

"It is critical to communicate information on climate risks and adaptation strategies to the agricultural community. Farmers and local communities will ultimately be responsible for implementing adaptation strategies. While impact studies have been conducted at universities and research centers across the country, in most cases, this information has not been adequately conveyed to farmers. There is a significant gap between top-down analysis and bottom-up implementation. Additional outreach is needed to convey what information is available to farmers so that they can begin developing adaptation strategies."
Heather S. Cooley and Dr. Juliet Christian-Smith, June 18, 2009

"A key requirement for adapting to climate change is the availability of information detailing what those changes are likely to be. In addition, technical support in how to use such information in decision making on adaptation will be critical."
Stephen Seidel, Pew Center on Global Climate Change, October 22, 2009

"We must develop and deploy smart grid technology in a manner that empowers consumers with greater information, tools and choices about how they use electricity, including access to real-time energy information. And energy information should be made available based on open non-proprietary standards to spur the development of products and services to help consumers save energy and money."
Dan W. Reicher, Google, October 28, 2009

BOX 2.6
A Sampling of Positions Advocated by State and Local Governments to Strengthen the American Clean Energy and Security Act[a]

Increasing Energy Efficiency

- Increase the ability of local governments to effectively implement GHG emission reductions through improvements in building codes by altering the state–local share of funding for municipalities responsible for 100 percent of code enforcement in their jurisdictions;
- Authorize state bonding authorities to support local funds to enable property owners to take out loans for energy efficient upgrades; and
- Expand the building efficiency labeling requirement to include existing buildings as well as new buildings.

Critical Planning

- Increase funding for transportation planning;
- Require climate adaptation plans for coastal cities with populations over 500,000 people, which accounts for 15 percent of the U.S. population;
- Consolidate multiple planning requirements to increase effective implementation; and
- Require, promote, and fund strategic adaptation measures at the federal, state, and local levels of government.

Although lobbying organizations are known as advocacy groups and their information may not be impartial or objective, they nevertheless constitute a considerable body of topical experts who provide information to Congress.

The panel judges that poor coordination between the federal agencies is limiting the effectiveness of the national response to climate change. Future synergies between the Executive branch and Congress may help establish clear agency responsibilities as well as ensure regular evaluation and assessment of federal climate change policies. The report *Limiting the Magnitude of Climate Change* (NRC, 2010a) suggests a "Climate Report of the President," analogous to the Economic Report of the President, as one such mechanism to inform Congress. The Informing panel judges that this type of mechanism may help in evaluating the effectiveness of our actions.

Carbon Reduction Goals

- Establish reduction targets of 20 percent by 2020 and 83 percent by 2050 from 2005 GHG emission levels, and periodically review and adjust, as necessary, the level and timing of reductions required under the cap in light of new science; and
- Support a renewable energy supply standard of at least 20 percent by 2020.

Implementing Cap and Trade

- Any needed moratoriums on regional cap-and-trade programs should be consistent with compliance periods in existing programs and should not begin until a federal program is operational.
- Increase the effectiveness of carbon markets through flexibility and transparency and maintaining standards for eligibility for past emissions reductions and credit future reductions.
- Increase the effectiveness of carbon markets by allocating 15 percent of total allowances from "excess" free allowances to the State Energy and Environment Development Fund.

---

[a] State Voice Group Recommendations to the Senate Concerning Climate Legislation. H.R.2454 American Clean Energy and Security Act of 2009 sets forth provisions concerning clean energy, energy efficiency, reducing global warming pollution, transitioning to a clean energy economy, and providing for agriculture and forestry related offsets. This passed the U.S. House of Representatives on June 26, 2009.

Federal Role in Responding to Climate Change

Many decision makers and stakeholders have urged the federal government to act more aggressively in responding to climate change. Box 2.6 summarizes suggestions from state and local governments regarding possible federal decisions and investments.

The federal government has the authority and resources to act in a wide range of policy areas relating to climate change, including pollution control, energy supply, energy efficiency, transportation, agriculture and forestry, and waste prevention. Federal investment in energy research and development and standards and regulations for energy efficiency (e.g., Corporate Average Fuel Economy [CAFE] standards) have played an indirect role in reducing the carbon intensity of some sectors of the U.S. economy, although it has not stopped increases in U.S. greenhouse gas emissions. The federal government has also played a significant role in the control of other environmentally significant emissions through regulation (e.g., lead and chlorofluorocarbons) and establishment of environmental markets for sulfur dioxide (SO_2).

In 2002, the stated U.S. goal was to reduce the greenhouse gas intensity of the U.S. economy by 18 percent from 2002 to 2012 (U.S. Office of Global Change, 2007) through the following programs:

- *Voluntary Greenhouse Gas reporting program.* This Department of Energy (DOE) registry for voluntary reporting of emissions and reductions was established in 1994. In 2005, 221 organizations reported to DOE.
- *Climate Leaders.* This is an EPA partnership with industry to develop comprehensive climate change strategies. Industries that reduce greenhouse gas emissions receive government recognition.
- *Climate VISION.* This DOE, EPA, Department of Transportation, and U.S. Department of Agriculture (USDA) program was established in 2003 to work with key industrial sectors to voluntarily reduce emissions intensity. In 2007 the program reported an almost 10 percent improvement in intensity and a 1.4 percent reduction in emissions from 2002 to 2006 in 13 power- and energy-intensive sectors.
- ENERGY STAR®. This EPA voluntary labeling program was created in 1996 to identify and promote energy-efficient products to reduce greenhouse emissions in products, businesses, and public buildings.
- *Save Energy Now.* This DOE program created by the Energy Policy Act of 2005 aims to reduce U.S. industrial energy intensity 25 percent by 2017.
- *Voluntary Programs to Reduce High Global Warming Potential Gases.* This EPA program aims to reduce emissions of perfluorocarbons (PFCs), hydrofluorocarbons (HFCs), and sulfur hexafluoride (SF_6). HCFC-22 production dropped from 36 to 17 tons CO_2-eq from 1990 to 2007.
- *Methane voluntary programs.* These are EPA partnerships to reduce emissions in the fossil fuel, waste, and agriculture sectors. Methane emissions dropped by 10 percent between 1990 and 2003.
- *Targeted incentives for agricultural greenhouse gas sequestration.* This is a 2003 USDA program to reduce greenhouse emissions by 12.4 million metric tons of CO_2e by 2012.
- *Tax incentives for greenhouse reductions.* These tax incentives help implement the Energy Policy Act of 2005 by providing energy tax incentives for alternative fuel vehicles, residential solar, and alternative energy.

The success of these programs is difficult to quantify because most are voluntary and not comprehensively monitored. However, new mandatory programs are being implemented. For example, EPA recently issued rules that require mandatory annual reporting of greenhouse gases for facilities that emit more than 25,000 tons CO_2e per year or entities that supply fossil fuels (EPA, 2009b). A new set of federal initiatives was launched in 2009, including a commitment to long-term reductions in greenhouse emissions, tighter automobile mileage standards, and incentives for renewable energy and energy efficiency. The American Recovery and Reinvestment Act of 2009[8] implements several new policies that address, in part, efforts to reduce GHG emissions. Current legislative proposals to introduce a federal cap-and-trade system, a national climate service, and other responses would greatly increase the need for information and the reach of federal policy and actions.

Federal actions responding to climate change include research and monitoring, investments in infrastructure and disaster relief, and the establishment of laws, regulations, and international agreements. All of these decisions require information and decision tools. A GAO survey of agency officials that included questions about information needs for adaptation found that the majority of respondents would find information on state and local climate impacts and vulnerability, regional climate impacts and vulnerability, and best practices very or extremely useful, and many were interested in better tools for accessing and interpreting information and interacting with stakeholders as well as the establishment of the federal climate information service (GAO-10-133, Table 12). At least two-thirds of the 185 respondents said that finding information on local climate impacts, costs and benefits, thresholds, baselines, and certainty was moderately to extremely challenging (GAO-10-133, Table 9).

Perhaps the greatest gap in assessing the federal response to climate change is a comprehensive framework of measures for evaluating the effectiveness of federal choices and actions—whether they be investment in research, efforts to reduce disaster vulnerability, or the costs and benefits of emission reduction and regulations. Evaluating the effectiveness of federal choices and action relies on robust and up-to-date information systems such as monitoring climate impacts and greenhouse gases further addressed in Chapters 5 and 6. For example, there is very little information on the impacts of voluntary energy and emission reduction programs or on the effectiveness of adaptation measures taken to date. The wide range of agencies involved in research, monitoring, and policy formulation on climate change means that coordination is essential in order to avoid duplication, ensure comprehensive and complementary policy, and provide clear contact points for those beyond the federal government. The USGCRP has provided an important coordination role for research and observa-

---

[8] For example, the American Recovery and Reinvestment Act of 2009 provides $61.3 billion for renewable energy programs including $11.3 billion for smart grid investments, $6.3 billion for state and local governments to modernize buildings, $5 billion for low-income weatherization programs, $3.4 billion for clean coal pilot programs exploring carbon capture and sequestration, $500 million to train workers for green jobs, $2 billion for tax credits to consumers buying plug-in hybrid cars, and $400 million for establishing Advanced Research Projects Agency–Energy within the DOE.

tions but does not implement practical responses to climate change that involve a much broader range of agencies and may need coordination at a higher level.

THE LAW AND CLIMATE CHANGE

One often overlooked actor in climate change is the legal system. U.S. courts are also decision makers on climate change. In 2003, for instance, EPA determined that it did not have the authority to regulate CO_2 emissions, a decision that was challenged in court by Massachusetts and 11 other states. In 2005, the U.S. Court of Appeals in the District of Columbia Circuit supported EPA's contention. However, the U.S. Supreme Court agreed to hear the case in 2006, and on April 2, 2007, the court ruled in favor of the state plaintiffs by a 5-to-4 margin.

The courts are also involved in class action litigation for liability concerning climate change–related damages. For example, in February 2008 a suit was filed on behalf of the Alaskan village of Kivalina (Figure 2.4),[9] alleging that Exxon Mobil and other energy companies have contributed to global warming, which will require an Inupiat village to incur hundreds of millions of dollars in expenses to relocate. Some state attorneys general have filed public nuisance lawsuits against power companies and automobile manufacturers, alleging that greenhouse gas emissions from their activities and products contribute to global warming and harm the states' environment, economies, and citizens.[10] After Hurricane Katrina, Mississippi property owners sued oil, coal, and chemical companies, alleging that their activities contributed to climate change and magnified the effects of the hurricane.[11] Enterprising attorneys and interest groups are focusing on securities law as a basis to identify an entity's greenhouse gas emissions and its actions to limit its potential exposure. Other court cases seek to challenge the regulation of emissions on the basis of cost or jurisdiction.

The courts have specific information needs as they evaluate cases involving climate change. Administrative challenges based on the failure of federal agencies to act on climate change may require careful documentation of the impacts of not acting and detailed understanding of agency responsibilities and actions. Litigation for climate change damages engages highly complex questions of attributing climate change and climate impacts to human caused emissions, and then allocating responsibility for those emissions to potential private emitters (or public agencies that control emissions) who can be sued for damages.

---

[9] U.S. Court of Appeals Ninth Circuit, docket number 09-17490.
[10] Case 1:06-cv-00020, filed January 30, 2006.
[11] U.S. Court of Appeals Fifth Circuit, No. 07-60756, October 16, 2009.

FIGURE 2.4 Alaskan Village of Kivalina. SOURCE: Alaska Army National Guard.

According to an American Bar Association paper (Keteltas et al., 2008):

> Even if plaintiffs are able to overcome the political question doctrine that, for the time being, has ended early tort cases, would-be tort plaintiffs must overcome other hurdles, the most significant of which is causation. The plaintiff's challenge in proving specific harms were caused by climate change—measured against traditional standards of admissibility of scientific evidence in a courtroom—is daunting. To comprehend the scope of this challenge, it is worth considering the types of current and prospective injuries alleged in the tort cases filed to date. These injuries range from heat-related deaths and respiratory illness, to erosion, crop damage, inundation of coastal properties, harm to water supplies from salt water, damage to commercial shipping from reduction in water levels, and prospective harm from impacts of more severe weather events. Given that climate change may involve the cumulative effects of more than a century of emissions, tying one defendant's emissions (or even an entire industry's) to a localized climate event that caused any one of these individual harms—or, in other words, establishing specific causation—could well be impossible. Even as national and international bodies improve their cli-

mate change models, none are focused on tying the emissions of an individual corporation, or even an industry, to a specific local event.

Although climate scientists are working on questions of attribution, the ability to link emissions to impacts is an active area of research. There could potentially be opposition to the collection and analysis of climate information by defendants in such cases brought before the courts that could pose a potential barrier to informing effecting decisions in responding to climate change. But as human caused climate change and impacts become clearer, there will be increasing demand for scientific information and scientists as expert witness in the legal system. Avoiding such litigation may be one reason for potential defendants, such as fossil fuel producers, to take action to reduce emissions.

CONCLUSIONS AND RECOMMENDATIONS

A major finding of this chapter is that many different U.S. organizations are making decisions and taking actions in response to the risks of climate change, including a wide range of non-federal actors. Although the federal government shoulders much of the responsibility for making decisions on responding to climate change, state and local governments, business, NGOs, and the courts are all playing an increasingly important role in reducing greenhouse gas emissions, informing citizens, planning for adaptation, and even carrying out scientific research. Their aggregated commitment to greenhouse emission reductions is considerable, with a large number of states responsible for the bulk of U.S. emissions committing to reductions of up to 80 percent by 2050 and many corporations cutting emissions by more than 10 percent over even shorter periods. However, these commitments have not been fully implemented; inconsistent and incomplete reporting makes it difficult to compare the different targets and baselines for emission reductions and to monitor the effectiveness of the response in terms of both emissions and adaptation. Non-federal groups use and demand a wide range of information in deciding to act on climate change, including climate observations and models, impact assessments, cost-benefit analysis, emissions inventories, and simulation of policy options. Access to up-to-date reliable information remains a challenge for a range of decision makers.

The non-federal actors are also a source of innovation and experimentation in the response to climate change. As Gustavsson et al. note: "[t]he sheer magnitude of policy initiatives at the local level, their diversity, and the experimental and practical nature of many local projects, are bound to bring forward genuinely new ideas and solutions that in the end can have an impact on a larger scale" (2009).

In addition to the organizational decision makers from government, business, and NGOs discussed in this chapter, the U.S. public constitutes an important set of individual decisions and choices about how to respond to climate change. People can choose to reduce their own emissions, to begin adapting their lives to climate change, or to influence policy and businesses through their political activities, consumption, and investment choices. Individuals can benefit from the type of information provided to organizations and decision makers but may also have questions and require answers tailored to the ordinary citizen.

Government climate policy can benefit from understanding the motivations and barriers to corporate actions. Private companies in several sectors have taken significant steps to reduce their carbon footprints in pursuit of future business goals, and some business sectors, such as the insurance industry, are planning for the impacts of climate change. The finance sector is also starting to consider the risks of climate change and options to reduce those risks through investment strategies and ratings. Accurate accounting of business response to climate change can legitimize the actions of responsible firms and expose the lack of action by others. However, emission reduction commitments have not necessarily been fully implemented; few data exist on aggregated impacts or on the possibility for scaling up local and state action. Inconsistent and incomplete reporting makes it difficult to compare different targets and baselines for emission reductions to monitor the effectiveness of the response.

The federal government can take advantage of such innovation while also providing leadership to accelerate responses in the face of growing risks, to encourage those who have not acted, to provide consistency, to link to international efforts, and to prevent a patchwork of regulations. The panel finds that evaluating the effectiveness of federal choices and action relies on robust and up-to-date information systems such as monitoring climate impacts and greenhouse gases further addressed in Chapters 5 and 6. Future actions by the federal government can take advantage of the best of existing decisions and structures at the sub-national level, include representatives of these groups in governance arrangements, and evaluate the impact of federal policy on actions of state and local governments, businesses, and civil society. The federal government can also facilitate "knowledge to action" by providing coordination, information, and guidance about climate change impacts (Chapter 5) and greenhouse gas management (Chapter 6). An appropriate federal action can spur greater overall emissions reductions while preserving room for states to continue to act as policy innovators, drivers, and implementers.

Although this report is targeted primarily at government, many of our findings and recommendations are relevant to individual information needs and behavior, including, for example, our discussion in Chapter 5 of the need for accessible federal information portals with information on climate impacts, or our discussion in Chapter 6 of consumer information about greenhouse gas emissions and energy use.

As we argue in Chapter 6, the increased use of standardized methodologies for emissions inventories could improve estimates of emissions reductions. To the extent that these actions prove successful, they should ease the task and lower the costs of the federal government.

Recommendation 1:

To improve the response to climate change, the federal government should

a) **Improve federal coordination and policy evaluation by establishing clear leadership, responsibilities, and coordination at the federal level for climate-related decisions, information systems, and services.**

The roadmap for federal coordination might include leadership and action through executive orders, the Office of Science and Technology Policy, an expanded USGCRP, a new Council on Climate Change, the reorganization of existing agencies, or even the establishment of new organizations, regional centers, or departments within the government.

b) **Establish information and reporting systems that allow for regular evaluation and assessment of the effectiveness of both government and non-governmental responses to climate change, including a regular report to Congress or the President as suggested in our companion reports.**

This could include aggregating and disseminating "best practices" with a web-based clearinghouse, creating ongoing assessments to enable regular exchange of information, and plans among relevant federal agencies, regional researchers, decision makers, NGOs, and concerned citizens.

Recommendation 2:

To maximize the effectiveness of responses to climate change across the nation, the federal government should

a) **Assess, evaluate, and learn from the different approaches to climate related decision making used by non-federal levels of government and the private sector;**

b   Enhance non-federal activities that have proven effective in reducing green-house gas emissions and adapting to the projected impacts of climate change through incentives, policy frameworks, and information systems; and

c) Ensure that proposed federal policies do not unnecessarily preempt effective measures that have already been taken by states, regions, and the private sector.

CHAPTER THREE

Decision Frameworks for Effective Responses to Climate Change

The earth's climate system is moving outside the range within which it has fluctu-
ated in recorded human history. As a result, decision makers will need to go
beyond conventional approaches and develop new ways to think about preparing for and adapting to change. This chapter reviews frameworks that decision makers
might use to address the unique combination of complexities that climate change
presents. We use the term frameworks to describe the underlying set of ideas and
principles that provide the overall basis for decision making.

There is an old joke about the person who was surprised to learn that he already
knew how to speak in prose. Similarly, most people already use a variety of decision
frameworks in their personal and professional lives. For instance, a pilot might use a
checklist to ensure the aircraft is ready for takeoff. A firm might use a hurdle rate (a
minimum required rate of return) to help determine what new products it will invest.
A court of law provides jurors with a strict framework for the information they can use
and the questions they must answer. Most people have an ethical code of conduct,
often derived from religious sources, that helps them determine what actions to take
and which to avoid. People make decisions based on an assessment of risk in their
everyday lives when deciding what level of insurance to purchase, whether to take
steps to improve their health, or where to invest their savings. In most cases, however,
people do not spend much effort considering what decision framework to use in a
particular situation, because habit, custom, law, and external framings (see Chapter 1,
"Barriers to Effective Decision Making," for further discussion in relation to the climate
problem) may dictate their choice and response.

At present, however, the appropriate framework for climate related decisions remains
an open question. As discussed in the preceding chapters, climate change presents a
host of novel challenges that may require many organizations to change their standard operating procedures in order to consider new types of information and to incorporate that information into their decisions in new ways. Such choices help define a
decision framework.

Our panel finds that an *iterative risk management framework*, suitably modified to address some of the novel characteristics of the climate challenge, represents the best available decision framework for climate related decisions. This finding mirrors that of the Intergovernmental Panel on Climate Change (IPCC) *Fourth Assessment Report* (IPCC, 2007b), which states: "Responding to climate change involves an *iterative risk management* process that includes both adaptation and mitigation, and takes into account climate change damages, co-benefits, sustainability, equity and attitudes to risk." An iterative risk management perspective uses a suite of tools to approach problems. It does not look at a single set of judgments at one point in time. Rather, the approach provides a basic conceptual structure to make choices that reduce risk, despite uncertain knowledge of the future. An iterative risk management approach also actively updates and refines strategies about complex issues as new information emerges. This kind of decision making is similar to moves in a backgammon or chess game, where pieces are repositioned and risk is reassessed in reaction to the roll of dice or a response from an opponent. In this same way, iterative risk management can be defined as an ongoing process in which the potential but uncertain consequences of climate change and climate policy are identified, assessed, prioritized, managed, and reevaluated in response to experience, monitoring, and new information. The advantage of an iterative risk management approach is that it includes a strategy for responding to climate related risks as conditions change and we learn more about them.

Some scholars use the term *adaptive risk management* to describe the process of learning from experience and adjusting management in response to new information, with policies sometimes designed as experiments. With either terminology, the panel recognizes that in practice decision makers will employ a variety of frameworks in decision making. They will merge results from multiple sources in refining their intuition and arrive at an informed decision. Overall, iterative risk management is an advisable strategy for climate change decision making because it uses a broad set of concepts and many other frameworks and tools, such as cost-minimization, cost-benefit, and integrated assessment, within its rubric.

Iterative (or adaptive) risk management has been adopted as an overarching approach to the climate change problem by groups that include the United Nations Development Programme (UNDP, 2002), the World Bank (2006), the Australian Greenhouse Office (AGO, 2006), and the U.K. Climate Impacts Programme. A risk management approach is also increasingly adopted, however imperfectly,[1] by the private sector, including insurance, agriculture, and in the management of greenhouse gas

---

[1] Many commentators attribute the 2008 meltdown of the U.S. and global financial systems to improper risk management.

emissions by major corporations. In the following sections, the main elements of itera-tive risk management identified by the panel are discussed in detail. Some of the more important qualities of iterative risk management appear in the real life case studies, which illustrate the ways this framework is used and can address different areas of cli-mate change decision making. The next chapter discusses some of the specific meth-ods and tools available to implement the decision frameworks discussed here.

WHY RESPONDING TO CLIMATE CHANGE NEEDS A DECISION FRAMEWORK

For many years, federal government efforts related to climate change have em-ployed a decision framework that uses scientific research to address questions such as whether Earth's climate is changing, in what ways it is changing, and whether the changes are attributable to human activity. Efforts to understand these questions are appropriately addressed within an analytical framework in which new surprising phenomena are not confirmed until observations have been demonstrated with high statistical confidence (typically 95 percent). The findings are then reviewed by scien-tists to check their accuracy, evaluate the validity of inferences, and rule out alternative explanations of the reported observations. The phenomenon of climate changes in the observational record have been investigated within this framework for decades and, as described extensively in other *America's Climate Choices* (ACC) reports, has lead to a preponderance of evidence that human actions are changing the Earth's climate.

However, responding to climate change now requires a decision-making framework that addresses an expanded set of questions. For instance, a decision framework might help policy makers consider how much greenhouse gas emissions might be reduced, and by whom, or help them consider how the design of a new bridge might take into account the potential for future climate change. To address such questions, a decision framework must help illuminate tradeoffs among often competing values and objec-tives. It must also provide a means for considering appropriate actions in the face of uncertainty. Inevitably, decision makers need to act in the face of uncertainties, which is a ubiquitous characteristic of our knowledge of the world around us. An appropriate framework can help avoid paralysis in the decision making process and help suggest where prudent action is appropriate and where it is not (see Boxes 3.1 and 3.2).

Decision frameworks can provide a variety of benefits. In some cases they can support a particular approach for a formal process by comparing alternative decision options. Decision frameworks can also provide decision makers with methods and rules of thumb that can help them determine alternatives. For instance, a large body of quan-titative analysis of greenhouse gas reduction policies by the integrated assessment

BOX 3.1
An Example of Rick Management Approach at the City Level: Case Studies from New York City and Chicago

New York City (NYC)

In December 2006, the NYC Mayor's Office announced PlaNYC, a comprehensive sustainability plan for the city. As part of this effort, the NYC Panel on Climate Change (NYCPCC) was created consisting of leading climate change and impact scientists, academics, and private sector practitioners, such as legal, insurance and risk management experts. Their role has been to respond to city level decision makers' demand for actionable climate science and climate change impacts assessment information. In February 2009, NYCPCC released a "Climate Risk Information (CRI)" report which called upon these various experts to identify the risks to and quantify the impacts from climate change on NYC infrastructure.

"The CRI is designed to help New York City decision makers better understand climate science and the potential consequences for city infrastructure. The CRI contains information on key climate hazards for New York City and the surrounding region, likelihoods of the occurrence of the hazards, and a list of initial implications for the city's critical infrastructure" (NYCPCC, 2009).

Working alongside the Climate Change Adaptation Task Force, NYCPCC has been successful in advancing useable science and implementing adaptation strategies across NYC. Several lessons have been learned since NYCPCC's inception that can inform other local-level governments in crafting an effective climate action plan (NYCPCC, 2009):

1. Climate change, impacts, and adaptation strategies should be regularly monitored and re-assessed as part of any climate change adaptation strategy. Iterative risk management is also presented in *Adapting to the Impacts of Climate Change* (NRC, 2010a).
2. Adaptation plans should be assessed regularly to determine whether they are meeting their intended objectives and to discern any unforeseen consequences.
For example, by monitoring trends in population, the economy, policy, operations, management and material costs, future adaptation strategies can be iteratively tailored to ensure they remain consistent with broader citywide objectives.

Chicago

The Chicago Climate Action Plan (CCAP) was formed in November 2006 by the Mayor's Office in response to the recognition that climate change is occurring and can have substantial and costly impacts on the city of Chicago. In contrast to the NYC model, Chicago first requested a Climate Impacts Report conducted by expert scientists. This report fed scientific information to an Economic Impact Analysis of Climate Change carried out by a risk assessment firm that applied costs to these potential impacts. This process ultimately allowed Chicago to prioritize its mitigation and adaptation efforts.

Mindful that creating a plan as complex as CCAP is challenging to large urban areas, much less smaller ones with fewer resources, a Lessons Learned report was developed (Parzen, 2009). A variety of useful tools, checklists, and guidelines for ensuring the creation of an effective climate action plan are provided in this report. The authors are cognizant, however, of the fact that their findings were specifically tailored for the city of Chicago and will likely require modifications to best inform decision making in other localities. Here, we provide a few examples from the Lessons Learned report that illustrate the demand for information from city-level decision makers, as well as how to provide it to them.

Key Lessons Learned from creating the Chicago Climate Action Plan (Parzen, 2009)
1. Mitigation and Adaptation Belong in the Same Plan (see figure below)
2. Strong Support from the Mayor and Mayor's Office Paves the Way
3. Support from Government, Civic, and Business Leaders Fuels Action
4. Dedicated City Staff is Essential for Program Success
5. A Strategic Nonprofit Partner Can Help Keep the Process Moving
6. Solid Research Helps Leaders Choose Credible 2020 and 2050 Goals and Actions
7. Dedicated Funds Are Needed to Support Research and Planning and, Later, Implementation
8. A Task Force of Local Leaders Adds Enormous Value and Legitimacy
9. City Commissioners and Sister Agencies Need Their Own Process to Provide Input
10. Frequent Climate Summits Keep Stakeholders Informed of Progress and Provide a Way to Get Input
11. A Research Advisory Committee Adds Knowledge and Credibility
12. Start on Implementation Early in the Process
13. Have an Aligned Communications Strategy
14. Build on Existing Initiatives
15. Successful Climate Action Depends Upon Long-Term Public-Private Partnerships
16. The Way to Ensure Success is to Track Progress and Continually Reassess

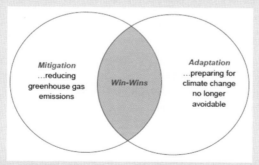

A schematic illustrating how carefully crafted mitigation and adaptation strategies can provide synergistic, or "win-win" outcomes. An example would be keeping rainwater on site to help reduce flooding (i.e., an adaptation action), and the need for pumping water which saves energy (i.e., a mitigation action). SOURCE: City of Chicago, 2008.

continued

BOX 3.1 Continued

To address the city's growing concerns over managing the risks from climate change on the city's infrastructure and livelihood, CCAP implemented a risk management framework for developing, implementing, and managing a climate action plan. The Chicago climate action plan checklist can prove valuable to other cities and states trying to craft their own climate action plan and risk management framework.

Chicago Checklist for Climate Action Planning, Part of a Risk Management Framework (Parzen, 2009)

- Create a staff and organizational structure to carry out work and manage funds.
- Find a nonprofit partner.
- Engage a group of funding partners.
- Create a climate planning task force.
- Create a research advisory committee and research plan.
- Perform or gather research on climate change impacts on the region and priorities for adaptation.
- Analyze baseline greenhouse gas emissions.
- Create a process for engaging municipal departments and sister agencies.
- Create a process for engaging local civic and nonprofit leaders.
- Assess and summarize existing city initiatives, resources, and capacities.
- Inventory best practices from other cities.
- Collect ideas for emissions reduction and adaptation from the task force, departments, and civic and nonprofit leaders.
- Analyze emissions reductions options, including size of potential reductions, cost-effectiveness, feasibility, and other benefits.
- Vet and prioritize climate mitigation and adaptation options with all stakeholders.
- Choose overall goals for emissions reduction and actions to achieve them.
- Develop implementation plans, structures, and partnerships for the highest priority actions (and a timeline for the rest).
- Establish performance monitoring tools.
- Develop and implement an ongoing communications strategy.
- Launch a climate action plan.
- Continue ongoing planning, monitoring, and reassessment.

modeling community, often conducted in a cost-effectiveness decision framework, emphasizes the importance of the following general rules:

- Reduce emissions where it is cheapest to do so (for example, by allowing for trade in emission rights, the Clean Development Mechanism, or emissions offsets);

- Allow banking and borrowing of emissions rights through intertemporal trade;
- Focus on the full suite of greenhouse gases;
- Develop a portfolio to allow poorer countries to develop using less carbon intensive technologies; and
- Plan for learning and midcourse corrections.

Finally, a decision framework, particularly one that has been well developed and widely used, can provide general insights and concepts that can help guide decision makers' intuition. For instance, decision makers have been using a cost-benefit framework at least since the time Benjamin Franklin suggested weighting choices by writing down lists of pros and cons. The vast body of formal cost-benefit studies generated in recent years makes the basic principles behind this framework even more accessible and useful to decision makers.

OTHER WAYS OF MAKING DECISIONS

A wide variety of decision frameworks have guided decisions about how to respond to climate change (Table 3.1). Some have proven more helpful than others. As a result, a set of criteria has been useful in choosing among such frameworks. Generally, an effective decision framework for climate related decisions should:

- Help to relate actions to consequences in a way that decision makers can compare the extent to which alternative actions achieve various objectives and goals;
- Provide a way to address uncertainty, in particular the deep uncertainties that often characterize many climate related decisions;
- Provide a way to handle multiple, often competing objectives; and
- Provide a process and results seen as legitimate by stakeholders to the decision.

These criteria suggest that iterative risk management, suitably modified as discussed in the section "Fundamental Elements of a Risk Management Framework," provides the best available framework.

Precautionary Frameworks

The precautionary principle takes many forms, but the key concept is that decision makers take steps to prevent future harms and identify key vulnerabilities (Schneider et al., 2007) even, and especially, when the causal chain between action and outcome

BOX 3.2
Responding to Hazards, Adapting to Climate Change Variation:
An Example from Tulsa, Oklahoma

Many disaster loss reduction programs (e.g., those concerned with land use management and building codes) are already well established, and they constitute an important line of defense against both near and longer-term adverse climate trends. For example, some communities have developed aggressive flood loss reduction initiatives in the face of climate-induced extreme events. The city of Tulsa, Oklahoma, is one such community. Located on the Arkansas River, Tulsa has a long history of flood disasters, including major floods in 1923, 1970, 1974, and 1976, as well as an especially deadly and damaging flood in 1984. The city's flood losses have increased over time, in part because the city relied on levees and dams for flood protection and allowed intensive development in the floodplain. In the mid-1970s, Tulsa began implementing a series of measures to reduce flood losses, including acquiring land in the floodplain, passing a moratorium on building in the floodplain, developing comprehensive floodplain and stormwater management programs, and establishing a flood early alert and warning system. Over the next two decades, these measures were strengthened and hundreds of buildings were relocated (see Haddow et al., 2008; Meo et al., 2004; Patton, 1994). Due to the city's flood hazard management efforts, flood insurance rates for Tulsa residents are significantly lower than those in other flood-prone communities around the country.

Tulsa's actions arose through a combination of drivers. Repeated flooding made the hazard difficult to ignore and led to the formation of citizen groups that pressured local government to act. The involvement of a member of Congress helped gain additional support. The passage of the Water Resources Development Act, which was championed by the same Congress member, provided a stimulus for further action. The 1984 flood (see figure below) occurred only 19 days after the election of a new mayor, who subsequently organized a flood hazard mitigation team for the city. The mayor was assisted in these efforts by other committed local officials, including a city attorney, and by engineering consultants. Later, federal funds provided support for coordinated local disaster loss reduction activities, and local businesses stepped in to continue those efforts when federal support ended (Meo et al., 2004).

The Tulsa case shows how institutional arrangements, programs, and collaborative networks designed to protect communities from specific hazards can also reduce their vulnerability to climate change.

Tulsa flood of 1984. SOURCE: Tulsa World (1984).

TABLE 3.1 Some Commonly Used Frameworks to Make Decisions about Climate Change

Framework	Example Principles
Muddling through	Ad hoc decisions
Scientific evidence	Scientific observation and analysis using statistical confidence
Economic	Least cost (or maximum value)
Precaution	Avoid harm
Political	Responds to voters, special interests, political beliefs
Risk	Iterative approaches learn from experience and respond to new information to reduce, control, or manage negative outcomes

is unclear and the likelihood of these outcomes is uncertain. Many decision makers use precaution in practice. For instance, in advising patients, doctors may ignore estimated probabilities and act as if a specific medical risk will in fact become reality (Van Asselt and Vos, 2006). The United Nations Framework Convention on Climate Change also contains precautionary language in stating its goal of preventing dangerous interference with the climate system. The precaution concept sets a threshold of acceptability in some observable parameter, and then prohibits any policy action that might cause that parameter to exceed the given threshold. The threshold may be zero, as in the doctor example above, or some other value, such as the target levels often proposed for dangerous climate interference (see Box 3.7).

Precautionary approaches have become especially important to the business community when it comes to environmental impacts because of the potential for litigation, and because of regulatory requirements for certain pollutants.

The precautionary approach, however, does not always address several of the criteria for an effective decision framework. Precaution provides no way to balance among competing goals (Sunstein, 2005). For instance, one group might use precaution to argue for rapid emission reductions to reduce the risk of adverse impacts from climate change while another group might use precaution to argue against any limits on emissions to reduce the risk of adverse impacts on the economy. Precaution also offers no systematic way to consider uncertainty, such as the confidence scientists have that rising above the 2°C temperature threshold is truly dangerous (Lempert and Collins, 2007).

Muddling Through and Political Frameworks

Decisions are often made in less systematic and structured ways. Extensive research on the role of heuristics, or commonsense problem solving techniques, in judgment under conditions of uncertainty reveals numerous "cognitive shortcuts" that influence decision making. For example, decision makers may focus only on the short-term consequences of their decisions; be influenced by recent dramatic but atypical occurrences; or fail to take into account high consequence, presumed low, but actually unknown, probability risks (Gilovich and Griffin, 2002; Gowda and Fox, 2002; Kahneman et al., 1982). The careful analysis and deliberation that are required for effective decision making may be ignored. Organizations and decision makers may also opt to follow trends, responding to pressure to jump onto the latest bandwagon, rather than thinking through the consequences of their decisions (Abrahamson, 2009; Collins, 2000; Jacobson et al., 2005; Kaissi and Begun, 2008). And while there may be advantages to "muddling through" a decision strategy (Fortun and Bernstein, 1998; Lindblom,1959), the common tendency toward sequential ad hoc decision making contains many pitfalls. Given the novel challenges and opportunities posed by many climate related decisions, and the often long time lags between actions and consequences, such ad hoc decision making could be a particularly poor means to address the criteria for an effective decision framework. In particular, it often fails to provide a systematic way to connect decision makers' actions to their potential consequences.

Economic Decision Frameworks

The use of economic analysis to support decision making about climate change has a long tradition (Cline, 1992; Nordhaus, 1977) but gained considerable attention with the publication of the Stern Review and subsequent debate about the findings that the benefits of early action on climate change considerably outweigh the costs (Stern, 2007). An economic decision framework focuses on the costs and benefits of alternative actions. Economic frameworks are often supported by common tools or methods such as cost-benefit analysis, risk analysis, and integrated assessment models.

Stern used economic analysis to compare the costs of reducing emissions with the damages associated with inaction. Economic decision frameworks have the advantage of providing a systematic structure for understanding the consequences of alternative decisions, in particular in complicated human systems when some actions may have surprising or counterintuitive consequences. In practice, however, economic analyses are often criticized for oversimplifying important aspects of climate related decisions, for example, by ignoring (or poorly representing) non-market values such as life or ecosystems (Fankhauser, 1995). Additionally, the framework is criticized for making assumptions about discounting the future and equity, and for inadequate attention to uncertainty (Helm and Hepburn, 2010).

For business, economic return is the traditional bottom line for decision making because profitability is often the most important criteria for executives, shareholders, and even workers who may lose employment if a business fails. The environmental impacts of business operations—externalities—have often been excluded from balance sheets and may only influence the economics of business through the cost of regulations, permits, or litigation. Many corporations now recognize that other factors are important to business decisions and that attention to environmental and social issues can benefit corporate performance and brand reputation.

Cost-Benefit Analysis

Cost-benefit analysis (CBA) is a commonly used economic framework to help decision makers at all levels evaluate whether or not to take a particular course of action (Boardman et al., 2001). In brief, CBA compares all of the costs of taking the action with all of the benefits. For example, a community considering building a new road might tally the funds needed to build and maintain the road as well as any adverse impacts the road might cause to the environment and to the quality of life of nearby residents. As benefits, the community might tally the economic and quality of life gains from reduced congestion and improved access that the road would provide. Cost benefit analysis recommends taking the action if benefits exceed the costs.

Cost-benefit analysis has been used in the United States at least since the 19th century by the Army Corps of Engineers in evaluating their public works projects. The 1936 Flood Control Act explicitly required CBA of proposed projects. In recent years, it has become increasingly used in the public and private sectors. While CBA provides a conceptually elegant and compelling framework, significant challenges often arise in practical applications of the approach. These include the need to quantify all the costs and benefits in a common metric and the need to estimate future costs and benefits with sufficient accuracy. It also requires specification of a rate of time preference—discounting—which is a normative exercise in easily monetized categories. These challenges prove fatal when applying CBA to many climate related decisions, because the uncertainties about the costs and benefits often prove too large and because the impacts are too diverse and extensive for all the parties to the decision to agree on a common metric for comparison or how future generations should be discounted relative to present ones. Nevertheless, such input is a legitimate part of the analysis of climate policy alternatives, though few would argue it should be the sole basis for decision making.

FUNDAMENTAL ELEMENTS OF A RISK MANAGEMENT FRAMEWORK

Risk management involves a broad, two-step approach to making decisions about events. The first step involves identifying, assessing, and prioritizing risks. Then, re-sources are coordinated and economically applied to minimize, monitor, and control the probability and/or impact of adverse events. It is important to pay attention to the first two stages of specifying the problem carefully, setting objectives and establish-ing criteria for making decisions, as these steps are often overlooked and can lead to later problems. For instance, a transportation agency might survey the potential risks from climate change by estimating the potential impacts of future sea level rise and increased coastal storm surges on its coastal highways, and the likelihood of occur-rence of damage to this infrastructure. Then, the agency might evaluate responses that could reduce the potential impacts on its roads, such as raising the roadways during their next major renovation, and evaluate whether the resulting decrease in risk would be worth the cost (NRC, 2008a). In the private sector, risk management frameworks dominate decision making in insurance and finance, and are commonly used in other areas of business as a basis for project management, engineering, financial, and mar-keting decisions.

An iterative risk management framework (Figure 3.1) defines risk as the impact of some adverse event multiplied by the probability of its occurrence (see *Adapting to the Impacts of Climate Change*, NRC, 2010a) for further discussion of a risk management framework).[2] High risk might result either from a significant impact virtually certain to occur (e.g., a serious auto accident disrupting traffic in a metropolitan area during commute hours) or a catastrophic event with a very low probability of occurrence (e.g., a tsunami washing away a freeway full of cars).

An *iterative* risk management perspective recognizes that the process does not consti-tute a single set of judgments at some point in time but rather ongoing assessment,

---

[2] Some literature follow the lead of Knight (1921) and distinguish risk from uncertainty, where the former indicates adverse events with well-determined probabilities of occurrence while the latter indicates events whose probability of occurrence is poorly defined, unknown, or unknowable. This and other ACC reports use the term uncertainty to denote both these concepts; they use the terms imprecise probabilities and deep uncertainty to denote situations where the probability of occurrence is poorly defined, unknown, or unknowable; and they use the term risk to denote impact multiplied by probability, whether or not the latter is precisely known.

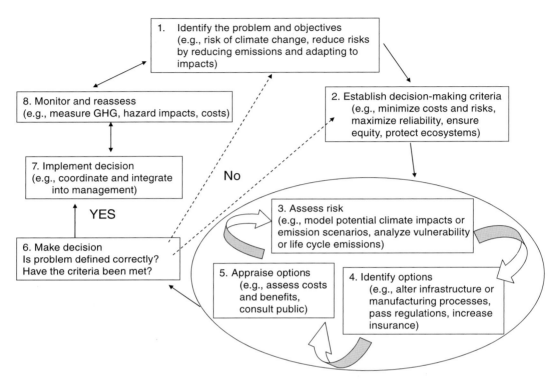

FIGURE 3.1 This illustrates that an iterative risk management and adaptive governance approach for climate change in which risks and benefits are identified and assessed, and responses implemented, evaluated, and revisited in sustained efforts by multiple levels of government, public, and private sectors to develop more effective policies or to respond to emerging problems and opportunities. SOURCE: Adapted from Willows and Connell (2003).

action, reassessment, and response that will continue—in the case of many climate related decisions—for decades if not longer, which will require documentation so that each iteration learns from previous iterations. For instance, the transportation agency in the example above might recognize that sea level rise will occur over many centuries so any decision about raising a road can wait until some future renovation. However, the agency might also conclude that future sea level rise should weigh more heavily in any near-term decision about siting new roads to reduce the risk of expensive remedial action in the future.

The most effective risk management strategies call for the use of a range of risk management strategies and tools. An effective societal response to climate change will require actions designed both to limit future climate change and to adapt to changes that do occur. Within each broad category, policy makers will also pursue portfolios of policies. For instance, limiting greenhouse gas emissions will require policies that support energy research and development, place a price on carbon and other greenhouse gases, set standards for fuel efficiency and enhanced efficiency in buildings, and implement clean, renewable sources of energy.

This iterative risk management approach has several advantages for climate related decisions. The approach emphasizes that:

- Action in the face of uncertainty is unavoidable. All assessment and management efforts involve uncertainty, and while it is important to assess and reduce uncertainties where possible, significant uncertainty can rarely be eliminated.
- Eliminating all potential risks is impossible. Even the best possible decision will entail some residual risk.
- Determining which risks are acceptable (and unacceptable) represents an integral part of the process of risk management. Different stakeholders will inevitably hold different views.
- Risk management actions can achieve an appropriate balance among the potential costs and benefits from the broadest range of potential outcomes, taking full consideration of available information on the likelihood of occurrence. These actions can be reassessed and rebalanced in an on-going process over time.

In recent years, iterative risk management has become widely used throughout the public and private sectors. This experience and familiarity provides an important foundation for applying this framework to climate change. However, many climate related decisions confront a number of especially difficult challenges that include the expectation of surprise, the need for urgent action, the need for long-term decision making, the potential demands of crisis response, and the overall characterization of climate change as a complex problem[3] (Box 3.3). Overcoming these challenges requires augmenting the basic iterative risk management framework in two important ways:

1. Recognize and manage the deep uncertainties facing many climate related decisions.
2. Embed iterative risk management in a broader process of institutional learning and adaptive governance.

As emphasized in recent NRC (2009a) and U.S. government (CCSP SAP 5.2, 2009) reports, the uncertainties associated with many climate related decisions are larger than and often have different characteristics than those involved with other risk manage-

---

[3] Also referred to in the literature as a "wicked problem."

BOX 3.3
Addressing the Special Challenges of Climate Related Decisions with an
Augmented Iterative Risk Management Framework

Complex Problem

Climate change is often characterized as a complex problem because it lacks both a definitive assessment and a clear point at which the problem is solved (Dietz and Stern, 1998; Rittel and Webber, 1973). Complex problems involve intense conflicts over definitions of the problem, objectives, and even what issues and topics are relevant to the decision. They also confront significant uncertainty, so that parties involved in problem solving must rely on highly imperfect, often conflicting information about what is known and not known. Even more difficult, values are intertwined with assessments of fact. Complex problems are commonly thought of as unique; although some aspects of the problem may have been seen before, each complex problem involves a distinctive constellation of constituent problems, meaning that prior experience with other problems may offer little guidance. An iterative risk management framework with a heavy emphasis on learning and embedded in a distributed institutional capacity to make sensible reforms can help address such complex problems (NRC, 2009a).

Managing Surprise

The notion of surprise is rooted in expectations. Governments and other organizations will often be surprised in part because their formal processes of informing and making decisions re-enforce the most commonly held and best-understood expectations (Lempert, 2007). Those faced with climate related decisions should expect to be surprised (NRC, 2009a; Schneider et al., 1998). The climate system is extremely complex, with innumerable parts and relationships among them. If current trends persist, the system will begin to diverge more and more significantly from historical experience, entering a realm where scientific understanding rooted in past observations will decreasingly hold. Moreover, any energy revolution that significantly reduces greenhouse gas emissions is virtually certain to spawn numerous social and economic changes beyond any current expectations. A number of iterative risk management methods and tools, including scenario, foresight, red-teaming, and horizon scanning exercises, can help decision makers widen their range of expectations. Decision analytic methods can place surprise in a formal quantitative framework by systematically describing those conditions where a decision is likely to fail. Such methods include tolerable windows, robust decision making (Lempert and Collins, 2007), and various forms of vulnerability analysis.

ment challenges. This is because the underlying probabilities are imprecise or the structure of the relationships that relate actions to consequences are often unknown. With complex, poorly understood systems like many of those involved in climate related decisions, research may enrich our understanding over time. However, the amount of uncertainty, as measured by our ability to make specific, accurate predic-

Long-Term Decisions

Many greenhouse gases have centuries long residence times in the atmosphere and the oceans take decades to warm. Thus, many decisions on limiting emissions will have their most significant impacts on the environment far in the future. Many current decisions, such as those regarding the location and design of roads, ports, urban development, and other infrastructure, will significantly affect future generations' ability to adapt to climate change.

Despite frequent claims to the contrary, policy makers often strive to factor events that may occur decades in the future into long-term decisions (Lempert et al., 2003; Meuleman and Veld, 2009; Princen, 2009), but there are significant barriers to doing so. In particular, making effective long-term decisions is hard for two deeply linked reasons: people's general preference for gratification in the present to that in the future and deep uncertainty about the long-term consequences of today's actions. Economists use the concept of discount rate to describe the former challenge, though there is considerable debate whether high rates that grant little value to the long-term future represent a reality to accept (Beckerman and Hepburn, 2007; Nordhaus, 2007; Roser, 2009; Weitzman, 2001) or a problem to address (Stern, 2007; Summers and Zeckhauser, 2009). The psychological literature emphasizes the connection between such discounting and uncertainty about the long-term future. People do not always have firmly established preferences between near and long-term rewards, but rather they construct their preferences in the context of each decision (Weber, 2006). For example, it is common for people to make long-term decisions if there is a clear connection with near-term actions (Princen, 2009).

Iterative risk management methods and tools that can help policy makers make better long-term decisions include visioning, foresight, and scenario exercises, which can help make images of the future more concrete (Georghiou et al., 2008). Summers and Zeckhauser (2009) emphasize the need for improving approaches to discounting, disaster management, distinction between the broad types of policy actions that people support and those they do not, and the treatment of uncertainty. Lempert et al. (2009) reviewed several classes of decision support approaches that could improve long-term policy analysis, including various statistical methods, adaptive control approaches, agent-based and multi-agent modeling, and robust decision making which seeks near-term actions that address long-term goals over a wide range of plausible futures.

tions, may grow larger (CCSP SAP 5.2, 2009). For instance, climate research may reveal previously unanticipated impacts if global mean temperature increases grow beyond 2°C, thus increasing the range of potential risks (see *Adapting to the Impacts of Climate Change,* NRC, 2010a). Technology research may reveal unanticipated possibilities that broaden the range of options to limit the magnitude of future climate change. These types of uncertainty are often termed deep uncertainty, occurring when decision makers "do not know or cannot agree upon the system model that relates actions to consequences or the prior probability distributions of the inputs to the model" (CCSP SAP 5.2, 2009).

In response to such deep uncertainties, many climate related decisions should seek to be robust, that is, to perform well compared to the alternatives across a wide range of plausible future scenarios, even if they do not perform optimally for any particular stakeholder's view of the most likely outcome.[4] The iterative risk management framework can implement this concept by characterizing probabilities by a range of plausible values or by a set of plausible probability distributions (CCSP SAP 5.2, 2009). Although many risk assessment tools provide optimal strategies, such strategies may prove brittle if the probabilistic expectations on which they are based are sufficiently imprecise. They may also prove overly contentious if different stakeholders have sufficiently different expectations about the future. As noted earlier in the report, people have different values and objectives that will guide different strategies. Robust uncertainty management strategies may address these difficulties by performing adequately and enabling multiple decision makers to agree on a portfolio of actions, even if they disagree about values and expectations.

The context for decisions about climate alters over time in response to changes in scientific knowledge, political, social and economic conditions, and perceptions of actual change in the climate change and effectiveness of policy. The objectives and values of the many different actors and decision makers may also change over time (NRC, 2009a). Given the likelihood that climate change and the response to it will affect people in new and unexpected ways, iterative risk management must involve a process of individual and institutional learning, which not only includes learning about changes in the climate but also about the array of possible response strategies (Boxes 3.4 and 3.5). The panel acknowledges that many companies and individuals are opposed to climate policy proposals for a variety of reasons. Overall, institutions and other systems that support and are affected by climate related decisions ought to be made more resilient. Furthermore, they should acquire the capacity to absorb disturbances, undergo change, and still retain the same basic function, structure, identity, and feedbacks.

---

[4] Or, more precisely, contingent on any particular stakeholder's view of the probabilistic distribution across outcomes.

BOX 3.4
Decision Making and Electrical Utilities

American Electric Power (AEP), one of the largest electric utilities in the United States, relies on coal for the majority of its power generation and is the largest coal burning electrical utility in the Western Hemisphere.[a] Climate change risks and policy pose a serious challenge to AEP, requiring decisions at the highest level about whether to respond to climate change, how much to invest in the response, whether to engage in policy debates, and how to choose between alternative responses. They have responded to stakeholder concerns, internal analysis, consulting reports, and recommendations from the World Business Council for Sustainable Development by taking on voluntary greenhouse gas reductions, joining the Chicago Climate Exchange, the Environmental Protection Agency Climate Leaders program, the International Emissions Trading Association, the Pew Business Environmental Leadership Council, reducing emissions of a potent greenhouse gas (SF_6), investing in forest carbon sequestration, buying carbon offsets from methane capture, greening corporate buildings, offering smart meters to customers, and starting new initiatives in renewables. Emerging cap-and-trade programs in several U.S. states, the establishment of the European cap-and-trade system, and the potential for Congressional climate legislation all suggested to AEP that the costs of operating traditional coal fired power plants may rise in the future. In responding to climate change, AEP has chosen a broad portfolio of responses; used a wide variety of information and decision support tools, including life cycle assessment, cost-benefit analysis, consumer and market surveys; and has piloted the Global Reporting Initiatives principles for electrical utilities which includes information on energy use, emissions, recycling, and water use. They collaborate and fund research at MIT, the Electric Power Research Institute (EPRI), and with the Department of Energy (DOE), including a major investment in a pilot project to evaluate carbon capture and sequestration (CCS) technology that, if successful, could eventually allow AEP and other utilities worldwide to continue to burn coal but without emitting carbon dioxide into the atmosphere.[b]

Consistent with the iterative risk management framework, AEP recognizes that it cannot eliminate all potential risks associated with climate change. Even though AEP's investment in the technology represents a substantial risk, the firm views its CCS pilot project as an opportunity to learn more about the technology and it will reassess its risks and adapt its plans as the project moves forward. The firm owns substantial capital stock associated with transporting and burning coal which would become more difficult to operate in a carbon constrained world. AEP thus judges its investment in CCS as a risk worth taking, both because commercially viable carbon capture technology could significantly enhance the value of the firm's existing capital stock and because gaining a leadership role in this new technology could open large new domestic and overseas markets for AEP in the years ahead.

---

[a] AEP delivers electricity to more than 5 million customers in 11 states. AEP's fuel is about 66 percent coal/lignite and it consumes about 77 million tons of coal each year (*http://www.aep.com*).

[b] On May 4, 2009, the West Virginia Department of Environmental Protection issued its first carbon dioxide sequestration permit which will allow AEP to capture and inject up to 165,000 metric tons of CO2 per year at its Mountaineer plant for a period of 4 to 5 years.

BOX 3.5
Decision Making in Conservation NGOs

Conservation non-governmental organizations (NGOs) are beginning to turn their attention to issues of emissions reduction and adaptation to the impacts of climate change. The ability of NGOs to implement this "new conservation paradigm" (Staudt et al., 2009) depends on addressing uncertainties in the extent of future climate impacts as well as in the efficacy of various proposed response strategies. Managing these uncertainties requires an iterative risk management approach. A recent survey of climate change adaptation literature identified the following five overarching principles for conservation and biodiversity management in the face of climate change (Glick et al., 2009):

1. Reduce other non-climate stressors.
2. Manage for ecological function and protection of biodiversity.
3. Establish habitat buffer zones and wildlife corridors.
4. Implement "proactive" management and restoration strategies.
5. Increase monitoring and facilitate management under uncertainty.

The last of these principles explicitly enables an iterative adaptive management approach (Heinz Center, 2008). Such an approach recognizes that there will always be uncertainty about future climate impacts and the effectiveness of proposed management strategies, so both ecosystem health and the success of any management strategies will have to be monitored, and decision makers will have to be prepared to modify their management plans in response new observations.

Adaptive management has long been practiced in environmentally related fields (Allan and Stankey, 2009; Holling, 1978; Holling and Meffe, 1996; Lee, 1993, 1999; Walters, 1986). The approach rests on the notion that policy interventions should be viewed as experiments and learning opportunities and requires well-conceived interventions combined with systematic monitoring procedures to track outcomes. It also assumes the ability to accept and learn from both successes and failures of risk management. In addition to these measures it may also be necessary to recognize that there are some things we cannot save. While the concept of adaptive management is ideal for the challenges of climate related decisions, it often proves difficult to implement because organizations find it difficult to design actual interventions as experiments; to document failures with the detail, transparency, and clarity needed to facilitate learning; and to spend sufficient resources on monitoring (NRC, 2009a).

Decision Making in the Insurance Industry

The insurance industry anticipates dramatically increased costs due to climate change, including changes in the frequency and severity of natural disasters and in disease vectors and mortality rates. The number of events and magnitude of losses has in-

Major natural catastrophes, 1972–2008

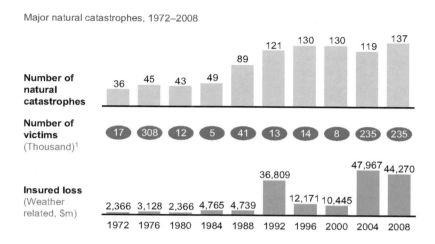

FIGURE 3.2 Major natural catastrophes from 1972 to 2008 and the associated insured losses. SOURCE: A report of the Economics of Climate Adaptation Working Group (2009).

creased in recent decades (UNEP FI, 2002; Figure 3.2). Whereas many companies are motivated by the risk or opportunities of future climate change regulation, those in the insurance sector are most concerned with the physical impacts from climate change (Hoffman, 2006). Allianz, the largest insurer in Europe, has estimated that climate change will increase insured losses from extreme events by 37 percent by 2017 (MacDonald-Smith, 2007). The insurance industry is thus faced both with the challenge of dealing with rapid changes, but also with a potential opportunity to innovate products and services to meet drastically changing global needs.

The industry regularly uses iterative risk management to assess the long-term implications of the activities they insure. It is uniquely positioned to manage climate risks, including potential losses from extreme events, health impacts, and other insured risks. If insurance operates as intended, it will influence decisions and actions by providing incentives for risk-wise behavior. In the climate change arena it can provide practical solutions to address currently intractable issues confronting the policy makers in developing frameworks to address climate change.

In general, the U.S. reinsurance and insurance industries have lagged in comparison with their European peers to effectively respond to climate change. The two largest

BOX 3.6
Climate Response in the Insurance Industry: The Case of Swiss Re

Swiss Re, a global reinsurer, derives 49 percent of premiums from its North American operations. Reinsurers create value by analyzing risks and providing coverage for those they judge to be insurable (Swiss Re, 2004). As a vital link in the risk chain, a reinsurer needs to be aware of how these risks may ultimately end up on its balance sheet. Climate change is a central concern because it undermines a fundamental assumption upon which (re)insurance is based: that the Earth's systems, though somewhat unpredictable in the short term, are stable in the long-term. Thus, if insurers fear their risk estimates are increasingly imprecise, it may undermine their ability to properly price their products.

Swiss Re has employed climatologists to work with its catastrophic business unit since the late 1980s and has interacted extensively with the climate science community. In 1994, it produced its first publication on climate change, *Global Warming, Elements of Risk* (Swiss Re, 1994), which was ground breaking because (1) it came from a financial services company and (2) it argued that the repercussions from climate change "could be enormous, with threats posed not only to citizens and enterprises, but also to whole cities and branches of the economy, even entire states and social systems."

Swiss Re's risk management strategy includes the following elements:

- Advance knowledge and understanding of climate risks and, where relevant, integrate them into risk management and underwriting frameworks. As insurance companies define the parameters of climate change risk, they will potentially be able to partner with government to provide incentives to change behaviors.

global reinsurers—Swiss Re and Munich Re—have been the most active on the issue by incorporating climate science into their models of natural catastrophes (see Box 3.6). Figure 3.3 illustrates computer-based catastrophe models being used by many private insurers. The increase in insured losses is a result of more people moving into harms way, higher property values, and to changes in the frequency of events.

Government Insurance Programs

The goals of major federal insurance programs differ from those of private insurers. Whereas private insurers seek to maintain their financial sustainability, the statutes governing the National Flood Insurance Program (NFIP) and the Federal Crop Insurance Program (FCIP) promote affordable coverage and broad participation by individuals at risk. The failure to apply a risk management model and to consider the implications of climate change may limit the effectiveness of these programs. Federal programs are not required to limit catastrophic risk strictly within the programs' ability

- Develop products and services to both mitigate and adapt to climate risk. An example is Swiss Re's collaboration with the World Bank for a demonstration aid project that pays Malawi farmers up to $5 million, based on an index, if they suffer from a drought related shortfall in maize production (A report of the Economics of Climate Adaptation Working Group (2009).

- Raise awareness about climate change with clients, employees and the public and advocate a worldwide policy framework for climate change. Swiss Re sponsored the study *Climate Change Futures: Health, Ecological and Economic Dimensions* (Center for Health and the Global Environment, 2006), which explains the links between climate change and human health in 2005,[a] and *A Report of the Economics of Shaping Climate Resilient Development: A Framework for Decision Making*[b] in 2009 (Economics of Climate Adaptation Working Group, 2009).

- Transparent annual emissions reporting to tackle the company's carbon footprint. In October 2003, Swiss Re was the major company in the financial services industry to announce that it would reduce or offset its greenhouse gas emissions with a goal of becoming carbon neutral in 2013.

---

[a] In partnership with the UNDP and the Center for Human Heath and the Global Environment at the Harvard Medical School.

[b] In partnership with ClimateWorks Foundation, Global Environment Facility, European Commission, McKinsey & Company, The Rockefeller Foundation, and Standard Chartered Bank.

to pay claims on an annual basis. One implication of this risk management approach is that there is little incentive to develop information on the potential risk of climate change. The government does not have the incentive to figure out what its losses will be. For example, if the insurance programs were to raise rates in areas that are at higher risk for the impacts of climate change, then this may in turn suppress development in those areas and be politically difficult to implement. However, escalating exposures to catastrophic weather events are already leaving the federal government at increased financial risk. According to a 2007 Government Accountability Office (GAO) study, taxpayer exposure has increased 26 fold to $44 billion since 1980 under the FCIP and quadrupled under the NFIP to nearly $1 trillion in 2005. The GAO (2007) report found that:

> Many major private insurers are incorporating some near-term elements of climate change into their risk management practices. One consequence is that, as these insurers seek to limit their own catastrophic risk exposure, they are transferring some of it to policyholders and to the public sector... Federal

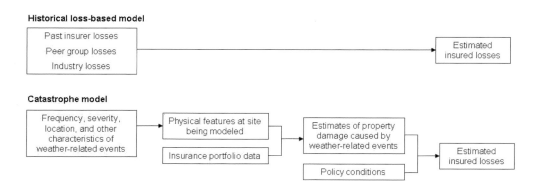

FIGURE 3.3 The historical loss-based model and the catastrophe model are two computer-based models being used by many private insurers. SOURCE: GAO (2007).

insurance programs, on the other hand, have done little to develop the kind of information needed to understand the programs' long-term exposure to climate change... Consequently, neither program has had reason to develop information on their long-term exposure to the fiscal risks associated with climate change.

This can come at high cost, and as costs increase there may be other, more effective policies, such as land use regulation, that can reduce vulnerability to risk. It is difficult, however, for policy makers to withdraw insurance support that helps particular interest groups or regions. In a changing climate, government insurers will need to analyze the implications for future insurance rates and identify prevention measures that may be taken to reduce climate exacerbated risks, such as floods, to prevent the average claimant from being a repeat claimant.

Finance Sector: Risk Awareness and Management

As noted in Chapter 2, the finance sector is using a risk management approach to include information about climate change in its investment strategies. To be included in financial statements, climate change risks must be quantified and given a transparent financial valuation. Financial reporting systems are the means by which investors, creditors, and others obtain the credible, transparent, and comparable financial information to make investment and credit decisions. A key element in the current financial crisis is inadequate and inconsistent regulation of financial markets, and particularly the insufficient availability of accurate information on risk exposure. Unfortunately, current U.S. accounting rules, disclosure requirements, and rating agencies do not adequately factor in climate change into usable financial information (Doran and Zimmerman, 2009).

Financial accounting offers a range of decision frameworks, methods, and tools for both emission reduction and adaptation strategies (KPMG, 2009). While major accounting firms have established climate change oriented advisory services, the accounting guidance for reporting contingent liabilities from potential climate exposure and for emissions reductions for U.S. companies is unclear. As a result, companies use different approaches from one another and sometimes even for different business units (IETA and PWC, 2007).

Accounting frameworks are required for all types of emission reduction efforts. In 2008, the U.S. Financial Accounting Standards Board (FASB), whose mission is to improve accounting standards to assist decisions by the public, insurers, and other stakeholders, announced it was considering proposed rules for handling undisclosed potential liabilities. Climate and carbon exposure could fall under these proposed rules, but, as of April 2009, FASB had not reached any conclusions on the accounting questions related to measurement of tradable offsets in cap-and-trade emissions trading schemes. The International Accounting Standards Board is also considering accounting for assets and liabilities in emissions trading.

Further integration of U.S. reporting systems into international accounting standards for carbon is required to enable multinational corporations to harmonize their accounting and to account for potential offsets from other jurisdictions, such as the Clean Development Mechanism.

The Securities and Exchange Commission (SEC) provides key information for climate related decision support. Investors, banks, customers, risk managers, and regulators are increasingly seeing climate change as a threat and thus requesting disclosure of climate related risks directly from companies (Mills, 2009). [5] Since 2004, a number of leading institutional investors coordinated by CERES[6] have called on the SEC to eliminate any doubt that publicly traded companies should be disclosing the financial risks of global warming in securities filings, and they recently petitioned the SEC to require that material climate risks be disclosed under existing law (Young et al., 2009). In 2008,

---

[5] For instance, investors have been filing shareholder resolutions requesting climate risk exposure for a number of years. Such resolutions hit an all-time record of 57 in 2008, as well as an all-time high of 25 percent of shareholders voting for the resolutions.

[6] CERES is a national coalition of investors, environmental groups, and other public interest organizations working with companies to address sustainability challenges such as global climate change. CERES directs the Investor Network on Climate Risk, a group of more than 80 institutional investors from the United States and Europe managing approximately $7 trillion in assets.

two of the largest emitters of greenhouse gases in the United States agreed in settlements with the New York Attorney General's Office to provide investors with detailed information on the financial risks posed by climate change. The settlements are the first binding agreements between government and private industry regarding climate change disclosure10-Ks (Kerschner, 2009).

A Form 10-K is required to describe all issues material to a company. According to a survey of SEC filings (Fishel, 2006), nearly 100 percent of the electric utility sector and 80 percent of companies in the oil industry discuss climate change in their 10-K forms. In contrast, only 15 percent of U.S. insurers even mention climate change, leading investors to file a number of shareholder resolutions requesting disclosure of potential climate change exposure. Insurers are not heavy emitters of carbon; their financial exposure is mainly on the impact climate change will have on the property, flood, weather, crop, forestry, and business interruption policies they issue.

In March 2009, the National Association of Insurance Commissioners (NAIC) adopted a mandatory requirement that insurance companies with annual premiums of $500 million disclose each year starting in May 2010 the financial risks they face from climate change and the actions the companies are taking to respond to those risks, including steps taken to engage and educate policy makers and policy holders.

Independent ratings agencies are well established instruments for enhancing the transparency and efficiency of financial markets. Mainstream Rating agencies, such as Standard & Poor's 500 (S&P 500), have a global low-carbon index to meet growing investor demands for environmentally focused indices. However, climate exposure ratings have not been factored into municipal or corporate bond ratings or in evaluation of Real Estate Investment Trusts in any substantial form. Determining the long-term viability of bonds or investment real estate is vital for the financial stability of the economy. In 2008, the world's first independent carbon credit ratings service was launched, which provides credit ratings for carbon offset assets in both the international Kyoto mechanisms (the Clean Development Mechanism and Joint Implementation) and voluntary offset markets. Each asset studied is given a rating based on an analysis of the underlying project, leading to an assessment of the likelihood of it delivering its stated emissions reductions in the stated time period.

In January 2010 the SEC decided to provide public companies with guidance on disclosure relating to climate change. The guidance suggests that climate change triggers disclosure in relation to the potential direct and indirect impact of climate change legislation, regulation and international accords, and to the potential physical impacts of climate change (SEC, 2010). If a company relies on fossil fuel based energy, legislation on greenhouse gases could affect future positions and should be disclosed.

If a company owns property or uses inputs which could be vulnerable to changes in climate, this is a relevant disclosure. The panel judges that this initial guidance on climate change risk disclosure requirement from the SEC will facilitate transparency and comparison of corporate exposure and provide information relevant to policy choices.

THE UTILITY OF ADAPTIVE GOVERNANCE IN DECISION MAKING ABOUT CLIMATE CHANGE

Adaptive management addresses uncertainty about the environment and human systems by consistently testing, monitoring, and revising policy assumptions and has strong links to iterative approaches to risk management. Adaptive governance extends these practical, problem-solving frameworks to policy institutions themselves (Box 3.7). Thus, adaptive governance can be a useful tool for Congress and state legislatures. The concept of adaptive governance is a foundation of American politics. The nation's federal system provides 50 state laboratories for policy and institutional experimentation. For example, recent state-level efforts to implement cap-and-trade systems, and long-standing programs for energy efficiency, provide valuable experience that Congress can draw upon in fashioning a federal program for limiting emissions of greenhouse gases. Frequent elections and a commitment to unfettered public debate provide ample opportunities for assessing what policies and institutions are working and then changing those that are not.

In practice, however, significant challenges face Congress and state legislatures in implementing adaptive governance for climate related decisions. By their nature, institutions are meant to cement particular policies and practices into place and to resist alteration. In addition, public officials address constituency needs and may not serve in office for long periods of time. Such stability is essential to the operation of markets, interactions among individuals, and other activities that form the basis of social life. In this vein, many U.S. firms are now advocating for federal regulation of greenhouse gas emissions. A key motivation is to obtain the "policy certainty" (USCAP, 2009) they need to more effectively plan their own long-term investments in technology, products, and infrastructure. But this need for certainty conflicts with the need for continuing learning and innovation. For example, most paths to a zero carbon economy will require significant regulatory innovation. The United States and other nations may create carbon markets and associated rules, regulations, verification processes, and related service industries. Nations may link these markets globally. The trade system may become involved, for example, through carbon tariffs, along with development agencies, public and private financial institutions, and other agencies. Implementing such policies will take decades and involve significant learning at every step to determine what

BOX 3.7
Targets Can Help Communicate Policy Goals and Motivate Appropriate Actions

Adaptive governance often involves the use of targets to frame near and long-term policy goals, and targets are being used extensively in efforts to mitigate and respond to climate change. For instance, in 2009, the G-8 endorsed a target for not allowing the global mean surface temperature to rise more than 2°C above pre-industrial levels (G8 Fact Sheet on Climate Change). The United Nations Framework Convention on Climate Change calls for stabilization atmospheric levels of greenhouse gas concentrations at a level such as 450 ppm. States such as California have pledged to cap greenhouse gas emissions at 1990 levels by 2020 and 80 percent below 1990 levels by 2050 (Assembly Bill 32: California Global Warming Solutions Act of 2006; Caponi et al., 2008). Bills before the U.S. Congress employ similar emission reduction targets.

Numerous scientific studies have explored the range of impacts that might be expected from various temperature, concentration, and emissions targets. However, there are also important questions regarding the ability of alternative types of targets to help communicate the goals of policy and to help motivate appropriate actions to achieve those goals. Such questions represent an important issue in any study of the best means to inform effective climate related decisions and actions.

A number of modeling studies have examined the ability of different types of targets to provide appropriate feedback that can guide the evolution of an adaptive decision strategy for limiting greenhouse gas emissions. Given the large uncertainties in climate sensitivity, it is generally understood that a target that focuses on temperature suggests a very wide range of potential emission reduction paths and thus provides weak feedback for an adaptive strategy. Dowlatabadi and colleagues have shown that adaptive decision strategies that rely on temperature, as opposed to concentration targets, can prove unstable over time in the sense that the evolving science can first suggest increasing then decreasing then increasing again the necessary emission reduction rates.

The advantage of temperature targets is that they may be more closely linked to actual impacts from climate change than targets based on emission reductions or atmospheric concentrations of greenhouse gases. Concentration and emission targets have the advantage of being more closely

policies work and which do not, what technologies and practices are cost effective, and what society tends to regard as dangerous levels of climate change. While many actions need to be made in the near-term, efforts will be require sustained commitments many decades into the future.

Lazarus (2009) notes the difficulty of setting the proper balance between policies sufficiently rigid to endure for the decades needed to stabilize atmospheric concentrations of greenhouse gases and sufficiently flexible to learn. He writes:

> The legislation [regulating U.S. emissions of greenhouse gases] must be sufficiently steadfast to resist, over the longer term, the constant barrage of pressures launched by economically and political powerful interests seeking

linked to human actions, that is, specific policies designed to reduce emissions. However, some have criticized some long-term emissions and concentration targets as appearing disconnected in most people's minds from any necessity to take near-term actions. For instance, given the long lifetime of energy producing and using infrastructure, it may take significant actions over the next few decades, starting now, to meet an 80 percent emissions reduction goal by 2050. However, this fact may not be readily apparent to most people. In response to this potential problem and to emphasize equity issues between developed and developing countries, some have proposed using targets based on cumulative emissions (Allen et al., 2009). To our knowledge, there is little research that explores any differences in the ability of such targets to help people understand the climate change challenge and to motivate appropriate actions.

Finally, there is some debate as to strengths and weaknesses of using any targets at all to motivate appropriate action. Lempert et al. (2009) describes the tension between stretch and legitimacy building goals. The former are intended to motivate people to achieve some difficult-to-obtain objective. Those setting the goals would be disappointed if the goals were often achieved, since that would be a sign that the bar was not set high enough. The latter are designed to demonstrate the competence of the goal setting organization. Thus, those setting the goals will endeavor to ensure that goals are set and responsibility assigned to reduce the chance of being blamed for missing the goals. Since limiting climate change will likely require stretch goals, but many political organizations will pursue legitimacy seeking goals, the use of targets as a primary means of communicating and implementing climate policy may suffer a serious tension between risking failure by promising more than can be delivered or failing to exploit potential opportunities by promising too little. Policy makers might reduce this tension by focusing more on creating strong incentives for emissions reductions, and building constituency, rather than particular targets to be reached.

A recent NRC study, *Climate Stabilization Targets: Emissions, Concentrations, and Impacts over Decades to Millennia* (2010c), evaluates the implications of different atmospheric concentration target levels and describes the types and scales of impacts likely associated with different ranges, including discussion of the associated uncertainties, time scale of impacts, and potential serious or irreversible impacts.

to delay and relax the law's proscriptions for their own short-term gain. But it would be no less of a mistake for the law to be wholly inflexible and not subject to revision. Precisely because the effectiveness of any climate change law depends on its success over the long-term, the law must admit the possibility of significant legislative or regulatory change in light of new information and changing circumstances.

Additional constraints result from the division of responsibility among different agencies, legislative committees, and branches of the Federal government and among different levels of federal, state, and local governments. Lazarus (2009) calls climate change law "no less than environmental law's worst nightmare," arguing that "[b]y fragmenting lawmaking authority and relying on short-term election cycles, we make it almost impossible to form the political coalitions necessary to address long-term issues." In particular, those most at risk from climate change, including future generations and those living in today's developing countries, are much less well-represented by today's elected officials than those who will pay any near-term costs. In addition, opponents of climate change policy and legislation, including those who will benefit from both non-action and from climate change itself, possess significant economic and political power; and the success of any U.S. emission reductions will ultimately depend on reductions in large developing nations like India and China.

Addressing these challenges requires leadership, use-oriented science, networks that facilitate information flow across diverse agencies and organizations and across levels of governance, and institutional arrangements that are able to operate at different scales (Olsson et al., 2006). The panel describes in Appendix B that there are lessons to be learned from past experiences on responding to complex problems that face our nation such as ozone depletion, clean air, small pox, and the transcontinental railroad. Adaptive governance structures have emerged to deal with sector-specific climate related problems such as ecosystem, coastal, and water management (Scholz and Stitfel, 2005). New York, Los Angeles, Chicago, and many other communities have developed climate change mitigation and adaptation strategies that involve a mix of governmental and private sector entities (NRC, 2009a, 2010d,a).

In other policy areas, Congress and state legislatures have successfully reformed institutions that remain both flexible and enduring over time. For example, the U.S. Social Security program and the 1965 Voting Rights Act (P.L. No. 89-100) have endured because each empowered groups with a strong interest in maintaining the program. Nurturing constituencies with an interest in addressing climate change may help enable effective adaptive governance in this area as well (Patashnik, 2003). Examples of constituencies that stand to benefit from a low-carbon economy include developers and vendors of low-carbon energy systems and investors in long-lived permits under a cap-and-trade system. Constituencies that could be mobilized to support climate change adaptation include the insurance and reinsurance industries, which stand to gain from programs that reduce losses from extreme events; health care professionals, particularly those concerned with the health of already at-risk populations; as well as social movements concerned with environmental quality, natural resource conservation, and ecosystem maintenance and restoration. Wiener (2009) argues that a key difference between a policy based on carbon taxes and one based on cap-and-trade is that the latter creates assets that can be used to provide incentives to nations that might otherwise be reluctant to join an international climate protection regime.

CONCLUSIONS AND RECOMMENDATIONS

A variety of frameworks can be used for making decisions about climate change ranging from ad hoc response to politics and events to those based in only economics or
science. Based on previous NRC reports, the research literature, and on the recent practices of both public and private actors, we have identified *iterative risk management*
as a key element in the decision making process for making sound policy related to
climate change because of the opportunities it offers for considering uncertainty and
adjusting decisions to experience and new information. For government, as well as
organizations such as environmental conservation groups, we find that adaptive governance also provides a useful approach to managing climate change risks, because it
allows a revision of policies in response to evolving conditions, new information, and
lessons learned.

Climate related decisions confront a number of especially difficult challenges that
include the expectation of surprise, the frequent need for long-term decision making,
and the potential demands of crisis response. Addressing such challenges requires
augmenting the basic iterative risk management framework, incorporating the objectives of multiple actors, using robustness criteria to help manage the deep uncertainties facing many climate related decisions, and embedding the framework in a broader
process of institutional learning and adaptive governance.

We find that the assumptions of a number of current decision frameworks may need
to be revised given the risks of climate change. For example, the GAO (2007) has
already noted that federal insurance programs such as Federal Crop Insurance and
National Flood Insurance should take account of the long-term fiscal implications of
climate change. The viability and costs of these programs will be seriously affected
by climate change as well as policies and programs to manage and adapt to climate
impacts and should be reviewed and revised in response to climate science and shifts
in policies. Conservation activities by both governments and NGOs must also include
climate change and climate policies in their decision frameworks, especially those
for adaptation, in order to protect investments and respond to evolving climate risks.
Businesses will need to address climate risks in their overall corporate decision strategies. Because most actors must respond to stakeholders, incorporating their views and
feedback into decision making processes about climate can also be a useful component of decision frameworks.

Our review of evolving frameworks for risk management in the financial sector suggests that many firms still do not take account of climate risks and that the lack of
uniform accounting, disclosure, ratings, and reporting requirements is creating confu-

sion. We find that the proposals that the SEC (Securities and Exchange Commission) develop a clear financial disclosure requirement for climate change risks are likely to facilitate transparency and comparison of corporate exposure.

Recommendation 3:

Decision makers in both public and private sectors should implement an iterative risk management strategy to manage climate decisions and to identify potential climate damages, co-benefits, considerations of equity, societal attitudes to climate risk, and the availability of potential response options. Decisions and policies should be revised in light of new information, experience, and stakeholder input, and use the best available information and assessment base to underpin the risk management framework.

There are important areas in which iterative risk management is already being used to manage climate risks. For example, the Federal government uses the Federal Crop Insurance Corporation (FCIC) and National Flood Insurance Program (NFIP) to share and reduce the risks of current weather variability for farmers and homeowners. However, the insurance programs do not take into account climate change, its impact on likely losses, and the fiscal implications. In the private sector, some firms already report on their management of environmental impacts to government and shareholders, but reporting can be inconsistent, and many firms still do not take into account climate risks (e.g., responsibility for emissions, policy uncertainty, and climate impacts) in their planning and disclosure.

Recommendation 4:

The federal government should review and revise federal risk insurance programs (such as FCIC and NFIP) to take into account the long-term fiscal and coverage implications of climate change. The panel endorses the steps that have already been taken to by federal financial and insurance regulators, such as the SEC, to facilitate the transparency and coordination of financial disclosure requirements for climate change risks.

CHAPTER FOUR

Resources for Effective Climate Decisions

What tools are useful in informing decisions about climate change? This chapter discusses how decisions regarding complex organizational, institutional, and individual choices are generally made and places climate-related decisions within that framework. While the *America's Climate Choices* (ACC) *Advancing the Science of Climate Change* (NRC, 2010b) discusses the state of the science of decision making and the latest research on decision support, this chapter focuses on specific resources and tools, ranging from simple maps and graphs to more complex models, that are used in decisions and actions about climate change (see Table 4.1). Of course, many actors make many climate-relevant decisions without the aid of complex tools. Some decisions are made through the use of sophisticated or data rich computer-based decision-structuring techniques, but others are made through informal methods that might include conversations with experts, personal opinions about costs and benefits, or fragmented and incomplete information that may or may not be relevant to the local situation. In formulating courses of action, people and organizations respond to many different kinds of signals, including evidence of institutional norms of conduct, social influences, and relatively simple but persuasive information products derived from scientific research. Decision tools generate results based on the assumptions and data, which will vary depending on the user. For example, models that estimate the costs of climate change that heavily discount future values tend to produce results with lower costs and less urgency for immediate action, and graphs that only show short-term trends and variability may suggest lower risks than those with longer time scales. Those who hold doubts about the necessity of taking action to reduce emissions or invest in adaptation may rely on tools that include assumptions that minimize the risks and costs of climate change and on scientific literature that supports these assumptions. In contrast, those who are more concerned to act may select tools that allow for the exploration of possible extreme changes or place a high value on future damages.

These choices are easily illustrated by how different decision makers interpreted the model results published in the Stern Review on the Economics of Climate Change (Stern, 2007). Figure 4.1 shows the model-based estimates of average global losses in income per capita using several sets of assumptions, including (a) whether climate

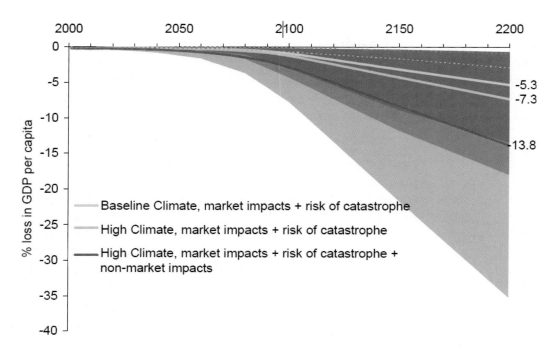

FIGURE 4.1 The impact of climate change on global GDP per capita. SOURCE: Stern (2007).

has a medium (baseline) or high sensitivity to greenhouse gas emissions, (b) whether impacts are only those which can be monetized (market impacts) or whether non-market impacts such as loss of species are included, and (c) whether there is a risk of rapid climate change (risk of catastrophe) or if climate will change slowly. The graph also includes a shaded area that represents the probabilities (or chance) of impacts from a 5 to 95 percent level.

A conservative interpretation of this graph, a decision support tool in itself, might select the baseline climate, where only market impacts and the lower end of the probability of impact such that the loss of gross domestic product (GDP) in 2200 would be less than 5 percent. However, a decision maker who is worried about high climate sensitivity and the chance, however small, of serious impacts, would conclude that the costs could be as high as 35 percent of GDP per capita. The varying interpretations of such graphs and model outputs are one of the sources of disagreement about how to respond to climate change. In addition, when the Stern Review summarized the damages, future damages were not discounted, estimating them at up to 14.5 percent of future consumption. Conversely, those who consider it more rational to discount future costs would conclude that damages would only be about 4.2 percent (at a 1.5 percent discount rate).[1] The debates over the Stern Report are more than academic because the analysis became the basis for the U.K. government's decisions about emission reduction targets and adaptation policy. What the case illustrates are the enormous challenges in providing clear and useful support tools for decision makers, and the importance of transparency about the assumptions that underpin the results.

WHOSE DECISIONS? WHICH RESOURCES?

As chapters 1 and 2 make clear, informing climate-related decisions involves many kinds of activities, products, and services, including identifying decision makers' information needs, producing decision-relevant knowledge and information, creating information products based on this information, disseminating these products, and encouraging and facilitating their use. Because responding to climate change necessitates so many different decisions, many groups in society can benefit from decision support tools, including officials in the executive branch of government, members of Congress, agency personnel at federal, state, and local levels, and persons in leadership positions in large corporations, small businesses, and non-profit organizations. They also include residents of communities and neighborhoods, households, and individuals. Decision support tools and resources must thus be adapted for a broad range of decision makers and decision-related challenges. Additionally, strategies to aid decision making must recognize what is distinctive and challenging about climate-related decision making while at the same time drawing upon knowledge developed in comparable decision arenas. The sections that follow first discuss how climate-related decisions can be conceptualized and then move on to discuss special challenges associated with climate-related decision making and resources that can help inform and improve decision making among public, private, and non-profit sectors.

Institutions such as the U.S. Congress and organizations ranging from large federal bureaucracies to corporations and small businesses are faced with numerous decisions on an ongoing basis, including various climate-related decisions. General research on decision making in organizations provides insight into what drives decision making for organizational and institutional actors. There is also a solid empirical basis for understanding household and individual decision making on environmental issues that can inform climate-related decisions at those levels of analysis. Box 4.1 provides examples of social science research needs to support decision making, including

---

[1]See *http://webarchive.nationalarchives.gov.uk/+/http://www.hm-treasury.gov.uk/media/8A3/83/Chapter_2_A_-_Technical_Annex.pdf.*

BOX 4.1
Social Science Research Needs

Research for and on decision support would improve the design and function of public and private decision support systems (NRC, 2009a). Science for decision support provides information that decision makers need and includes both "fundamental research on human processes and institutions that interact with the climate system (e.g., risk-related judgments and decision making, environmentally significant consumption, institutions governing resource management)" (NRC, 2009a), including the following:

- *Climate change vulnerabilities.* Improve understanding of the vulnerability of people, places, and economic activities as a function of climate-driven events, and improve analysis of likely future vulnerability due to the intersection of climate change with demographic, economic, and technological change (see also NRC, 1999, 2007b).
- *The potential for limiting climate change.* Improve understanding of the human drivers of climate forcing; the potential to alter these drivers with particular kinds of policy interventions; and the costs, benefits, and non-climate consequences of such policy interventions. Policy interventions to limit emissions can benefit from finer-grained knowledge (see also NRC, 2002b, 2005).
- *Adaptation contexts and capacities.* Develop indicators of adaptive capacity by type of disruptive event, improve understanding on why adaptive capacity is or is not fully utilized, and assess the ability of specific adaptation options to reduce impacts of climate change while taking advantage of opportunities (Brooks and Adger, 2005).
- *Interactions of limiting and adapting.* Improve understanding of climate response options in terms of their interrelationships and their joint effects on the human consequences of climate change (see also Klein et al., 2007).

---

research into human courses of action as well as how to most effectively communicate the information needed by decision makers.

The Basis for Decision Making in Organizations and Institutions

Many tools that exist to support organizational and institutional decision making rest either implicitly or explicitly on rational choice and assumptions. The *rational choice* perspective sees actors making decisions in order to actualize their preferences in an efficient and calculated manner—based mostly on an estimate of the economic costs and benefits of actions. Bargaining and negotiation are seen as involving various forms of exchange, which are again driven by preferences for particular outcomes.

- *Emerging opportunities.* Improve information to support climate-related decisions that can be beneficial and profitable.

The science of decision support builds knowledge about how to inform decisions effectively, including the following:

- Identify the kinds of information decision makers want and the kinds that would add greatest value for their climate-related decisions (see also NRC, 1999, 2005).
- Develop useful and decision-relevant indicators (e.g., of human pressures on climate, vulnerability, adaptive capacity, actions to limit or adapt to climate change, and decision quality) (see NRC, 2005).
- Understand how people interpret climate-related information and develop novel ways of framing and presenting information about climate risk and scientific uncertainty for climate-sensitive decisions. Most decision makers want to consider not only the probability and magnitude of risks but also qualitative aspects, tradeoffs among values, and the context of choices (NRC, 1999).
- Improve processes for informing decisions (e.g., channels and organizational structures for delivering information; fitting information into decision routines; the use of networks in distributing information; determinants of whether useful information is actually used; ways to overcome barriers to information use; improved approaches to integrating analysis with deliberative decision processes) (NRC, 2005, 2008b,c).
- Improve the decision tools, messages, and other products, and their use, to enable decision-relevant information to be conveyed and understood in ways that enhance decision quality (e.g., models, simulations, mapping and visualization products, and websites) (NRC, 2005).

Assumptions about organizational rationality, instrumentalism, and concern with costs and benefits form the underpinning for many approaches to decision support, including those discussed in this chapter. Such approaches are useful, particularly when limits on rationality are acknowledged; when the values at stake and the consequences of decisions are conceptualized broadly; and when considerations that are not easy to quantify, such as the cultural meanings associated with iconic species and places, are taken into account.

There are alternative ways of thinking about decision making that can supplement and sometimes even supplant models based on rational choice. Some alternative approaches are rooted in scientific knowledge concerning naturalistic and actual decision making, based on studies of how organizations and institutions decide on courses of action in real world situations (March, 1994). This emphasizes non-instrumental and non-economic drivers of decision making, such as beliefs, norms, and "logics of appropriateness" (March and Olsen, 2004) that are embedded in and reinforced by cultural practices within entities that are faced with making decisions. Countering the classical rationalistic approach to decision making, scholarship on naturalistic decision making emphasizes that under certain conditions action can precede reflection; that decisions may be only loosely linked to the quantity and quality of available information; and that historically developed rules and routines constitute a stock of knowledge upon which actors draw when they are faced with making decisions. Indeed, even the use of formal decision support tools to inform decisions about climate change and other issues is embedded in cultural practices that are characteristic of some organizations, but not others.

The social science perspective known as *institutionalism* (DiMaggio and Powell, 1983, 1991; Drori et al., 2006; Meyer and Rowan, 1977; Scott, 2001; Suddaby and Greenwood, 2005) also offers insights on decision making. Institutional theories tend to downplay the rationalistic and instrumental sources of organizational practices, including decision making. One insight is that organizational decision makers may choose a particular course of action not because they have systematically weighed its costs and benefits, and not because the decision increases efficiency and profits, but rather because of other factors, such as the imposition of new regulations, or pressures created by formal and informal standards developed within groups of similar entities, or even the diffusion of similar decisions and practices within specific organizations and professions. Institutionalists would argue that the desire to adhere to "green" building standards, obtain Leadership in Energy and Environmental Design (LEED) certification, reduce carbon footprints, or build structures that exceed hazard loss reduction codes and standards may be partly instrumentalist in nature, but it may also stem from the desire to achieve status or reputational capital within a particular organizational field, or even from simple bandwagon effects. A key institutionalist insight is that organizations quite often do not decide and act alone but instead are influenced by broader "decision making ecologies" in which they are embedded. Put another way, by virtue of their network ties, individual organizations are susceptible to influence by network partners, and such ties also influence decisions (Cyert and March, 1992).

For example, small organizations that are part of a supply chain that is dominated by a large retailer and that are financially dependent on that retailer are likely to comply with the large retailer's rules and requirements, including those associated with climate change mitigation and adaptation, without having to go through complex cost-benefit calculations or other formal decision support exercises. For such organizations, even if they are not inclined to comply, requirements articulated by a dominant supply-chain partner are sufficient to induce changes in behavior. Recognizing the importance of symbolism, shared norms governing conduct, other elements of organizational and institutional culture, and network-based sources of influence is a requirement for providing support for decisions and actions in the climate change arena.

The Basis for Public Decision Making

Decisions by members of the general public are critical for climate change mitigation and adaptation. Too often, members of the public are viewed merely in terms of their role as consumers. From this point of view, decision support is equated with providing information so that the public can make informed choices about which automobiles or appliances to purchase, or whether to drive to work or take public transportation. Such decisions are of course important in shaping responses to climate change. Equally important, however, is the power that the public has to influence decisions that are made by governmental, corporate, and non-profit actors. Like organizations public decisions can be seen as based on rational or cultural principles and influenced by factors, such as networks and status aspirations, which stem from an institutionalist perspective on decision making. Decision support activities must recognize the dual role of members of the public as both consumers and citizens who can take an active role in influencing the decisions made by other entities (Nerlich et al., 2010). Public influence can take a variety of forms, including voting, lobbying, and social movement activity that seeks to influence policy agendas. Historically, both better-off and less-privileged segments of the U.S. population have mobilized on a variety of environmental issues and controversies. Concern with environmental issues is sometimes greater among higher-status groups in the United States, but lower-income and minority groups also mobilize to take action on environmental issues, particularly when such issues are framed as reflecting environmental inequities and questions of fairness. The fact that climate change is increasingly being viewed as having disparate and inequitable effects is influencing political positions on climate change issues, including positions taken by publics in the United States and around the world (Roberts and Parks, 2006).

Risk and Decision Support

In Chapter 3 we recommend an iterative risk management approach to responding to climate change and this has implications for the resources and tools needed to support effective decisions. A risk management approach assumes that decision support tools, whether simple graphs or complex models, provide information about the level of uncertainty and error, the chances of occurrence, and the amount of confidence associated with analysis of climate change, its impacts, and the effectiveness of responses. The Intergovernmental Panel on Climate Change (IPCC) Working Group I, for example, provided estimates of probability (e.g., very likely is equivalent to 90 percent likelihood of occurrence) and of confidence (e.g., high confidence is an 8 out of 10 chance of being correct) for each of their main conclusions (IPCC, 2005). Because these terms can be confusing it is important that decision tools be as clear as possible about how error, uncertainty, probability, or confidence is defined or expressed.

Research suggests that, even when risks are communicated clearly, other factors such as emotions are important in shaping decisions with respect to various risks (Finucane, 2008). This is not to say that decision makers behave irrationally in the face of risk-related information. Rather, research stresses that positive and negative emotions of various kinds are bound up with cognitive calculations concerning risk. Emotions that enter into risk calculations include fear and dread, outrage, feelings of distrust or protectiveness, love, and empathy. Views on decisions related to climate change may thus be colored by emotional responses to a wide variety of objects of concern, including nature in general, particular species at risk from climate change, ideologies and what they imply for social and political action, government, free markets, and regulation. This is not meant to imply that emotions somehow diminish decision making capabilities. Rather, the point is that many if not most decisions cannot be separated from the emotions that accompany them, and that many points of view on climate change are not just about climate. Public receptiveness to risk-related information is influenced by a range of factors, including psychological attributes such as fatalism and religiosity; social characteristics such as race, class, and gender; and a host of other influential factors. Providing support for decision making is, in other words, a complex task that must include both attention to the information that is provided and attention to relevant social and cultural characteristics of those who are the intended recipients of the information. Other factors, such as the time and energy required to acquire, process, and understand new information, must also be taken into account in decision-support efforts. Technical reports like this one contain executive summaries for just that reason: members of some audiences to which this report is directed lack the time to read the entire report, but will instead read the executive summary and will potentially make decisions based on that condensed information.

DECISION SUPPORT TOOLS: THEIR CHARACTERISTICS AND USES

Decision tools are structured methods for evaluating the results of different decisions and provide a way of assessing the impacts, costs, and benefits or different decisions and strategies (including the option of not making a decision and allowing "business as usual"). Table 4.1 demonstrates an array of tools commonly used to aid effective decisions and actions related to climate change. Decision tools are as old as the human race itself, ever since the days when the peoples of the earth prognosticated about the future by studying the motions of the stars and planets, interpreted messages hidden in the entrails of animals, and consulted oracles. In modern times, decision methods based on expert judgments, deliberative consultations, historical records, and actuarial analyses slowly replaced those earlier methods in many regions of the world. Currently, computer-based information systems are extremely significant in helping decision makers use data and models to improve their decision making capabilities. In line with contemporary society's reliance on information technology and with advances in the art and science of visualization, there are now a wide variety of computer-based tools to help inform effective decisions and actions related to climate change. These include earth system models, impact models, various economic modeling techniques (including cost-effectiveness and cost-benefit analyses), integrated assessment models, and a range of other computer-based tools and products for engaging users and the public in deliberative decision processes or for helping them access and evaluate information related to alternative strategies. Many tools now include explicit consideration of uncertainties and are able to incorporate spatial detail through the use of Geographical Information Systems (GIS).

But many decision makers use a basic set of accessible decision support tools that include graphs, maps, images, GIS, and spreadsheets. One example of the demand for decision tools is that of local water managers. At a 2008 workshop, hosted by the Arizona Water Institute, participants identified a need for tools that provide information on how the accuracy of hydrological variability, patterns of seasonality, and groundwater might change with climate warming, improved snowmelt/runoff models, strategic monitoring of summer precipitation, groundwater recharge, and water quality. Participants also requested tools with better visualization and explanation of data limitations and more personal engagement with scientists providing decision support (Jacobs et al., 2010).

Although a wide spectrum of tools currently exists, few have the capacity to work across international, national, regional, and local scales. The fact that so many tools exist can also create confusion on which tools are the most appropriate for particular decisions. Additionally, the same tool used with different assumptions or design specifications may result in different results. Decision makers often turn to federal or state agencies, local universities, and national or international assessment reports to provide information on the merit of such tools to support climate-related decisions.

TABLE 4.1 Tools Commonly Used to Aid Effective Decisions and Actions Related to Climate Change

Tool	Main Uses in Decision Making
Basic toolbox	Graphs, maps, spreadsheets, images, GIS—used in local analysis of climate change and to communicate trends, patterns, impacts and alternatives
Earth systems models (e.g., general circulation models, carbon cycle models, climate forecast models)	Predict climate (e.g., seasonal forecasts, past climate) Estimate how emissions (and alternative emission paths) will affect global and regional climate Understand how changes in climate or other factors (e.g., land use) might affect global carbon and biogeochemical cycles Explore and communicate key uncertainties Assess the global climate implications of some geoengineering options
Impact models (e.g., ecosystem models, crop models, water resource models, disease models, coastal models)	Analyze the impacts of changes in climate on the environment and human activity Explore the interactions of climate with other changes (e.g., in water demand, land use, agricultural technology, vulnerability) to understand range of impacts Examine the potential for adaptation to reduce impacts
Economic models (e.g., cost-effectiveness and cost-benefit analysis, individual choice modeling/agent-based models, input-output models)	Estimate and analyze the costs and benefits of various policies and assumptions to limit emissions, develop cost-effective energy policies Understand the results of individual economic decisions about use of energy, land, and other resources

Some decision tools are also highly technical, which requires training and also stakeholder engagement in the development of the tools to ensure the output is useful for decision makers. For example, the International Research Institute (IRI) runs training programs and online tutorials for users to understand climate forecast maps. A number of private sector companies and consultancies offer workshops in how to calculate GHG emissions or involve stakeholders in decisions.

Not only do decision makers have difficulty in interpreting and applying climate prediction in practice, there is often a mismatch between needs of decision makers at

TABLE 4.1 Continued

Tool	Main Uses in Decision Making
Integrated Assessment Models	Provide an integrated assessment of how alternative policies influence an interconnected system that links human and natural system activities, emissions, climate, impacts, technology options, and/or economics
Assessments	Bring together a broad range of qualitative and quantitative information to provide an overall state of the science (such as IPCC), policies, or climate change in a region
Tools to evaluate and incorporate opinions, judgments (e.g., surveys, expert elicitation, and structured deliberation)	Understand and integrate the views of experts and citizens about climate change and policies
Policy simulations	Explore the implications of alternative policies using games and heuristic methods
Decision matrices and use of criteria to search databases	Structure and weigh alternative options, identify options from database of available strategies (e.g., adaptation options, greenhouse gas reduction strategies)
Participatory decision techniques (e.g., participatory GIS, structured stakeholder involvement)	Collective decision making
Emission calculators (e.g., Life Cycle Analysis, GHG accounting)	Calculate emissions embodied in products, estimate emissions from firms, sectors, and regions

multiple levels and in different sectors and the available information resources. This also requires stakeholder engagement for the development of such tools to ensure that the output is useful (Nicholls, 1999). "Boundary organizations" that provide assistance in collaborations among scientists, decision makers, and practitioners, can help ensure that tools are structured in ways that meet decision makers' and end-users' needs, while at the same time ensuring that scientific results are accurately conveyed.

The effectiveness of any decision tool depends on whether it provides information that is relevant to decision makers. Tools need to be useful at space and time scales that are meaningful and relevant for specific decisions and decision makers, and they also need to be based on up-to-date and reliable information (see also NRC, 2009a). For example, water resource mangers require methods and tools that are able to provide early warnings for potential droughts, assess their potential impacts, and help evaluate potential responses. Drought information varies over time and space, and different users may require information at daily, weekly, or longer timescales. Droughts can span counties, states, and river basins, but those boundaries do not always coincide with management regions. Any decision tool must address the diverse requirements of regional decision makers.

This report does not attempt to summarize or evaluate all the tools that exist for climate-related decision support. It is difficult to identify sources of comprehensive information on the full range of decision support tools, including their appropriate uses and limitations. Rather, this report highlights examples of the use of resources for assessing options and making decisions in the climate change arena. This chapter discusses decision resources that are used for addressing the following illustrative problems faced by different types of decision makers. These problems include: local level decision making reducing emissions at national and international levels; informing state level emissions reductions; informing efficiency decisions at the firm or household level; understanding impacts and informing adaptation; assessing the value of information for resource allocation; and using assessments as tools for effective climate-related decisions. Representative tools for supporting such decisions are discussed in the sections that follow.

The Basic Toolbox

At a minimum, most decision makers need a basic and accessible set of information to understand and make choices about climate change. As an example, a local government official seeking to manage both the causes and the consequences of climate change in their jurisdiction might require the following:

- Background briefing materials on the science of climate change (especially diagrams and graphs that illustrate the links between emissions and temperature, temperature and sea level rise, trends in emissions and climate variables at global level, basic question and answer (Q&A) about areas of uncertainty and skepticism, documentary film, and web sites);
- Information about observed climate impacts and vulnerabilities (e.g., water supply, ecosystems, crop yields, fires)—graphs, maps, verbal reports from those

affected, case studies, and both remote and ground level photo images (see Figure 4.2);

- Information and analysis about climate conditions (e.g., temperature, precipitation, drought, storms, and sea level rise) and how they may be changing in the local area—most easily conveyed by graphs and maps with some statistics on trends, variability, and data reliability (see Figure 4.3);
- Projections of what climate change may mean for the local area (and for other regions of relevance such as trading partners and economic competitors)— graphs and maps showing temperature, precipitation, and sea level based on easily understandable best- and worst-case scenarios with confidence and probability estimates and examples of potential climate impacts (e.g., river flows, ecosystem shifts);

FIGURE 4.2 Lake Powell before and after 2002 drought. This type of image enables decision makers to see the effects of specific changes in climate without the need to interpret data or graphs. SOURCE: John C. Dohrenwend, U.S. Geological Survey.

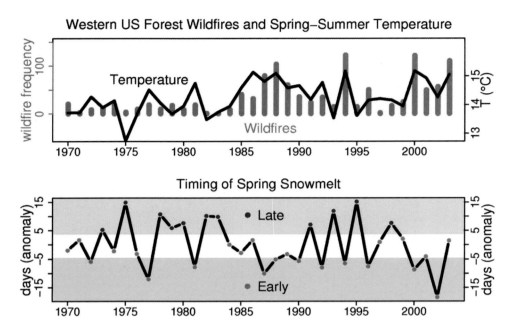

FIGURE 4.3 The top graph shows the positive relationship between annual frequency of large (>400 hectare) wildfires (bars) and average spring and summer temperatures (line) in western U.S. forests. Using the same x-axis, the bottom graph shows the first principal component of the center timing of streamflow in snowmelt dominated streams (pink = early, white = average, blue = late). This is an example of a graph that can provide useful information on observed climate impacts to decision makers. SOURCE: Westerling et al. (2006).

- Information on trends and patterns in GHG emissions and their drivers—graphs or spreadsheets for major facilities, land uses, population groups, zip codes, etc., expert opinion on future trajectories and the potential impact of policies; and
- Information on the economic and health impacts of climate variability and potential changes—current costs and benefits, morbidity, and mortality in spreadsheets or tables.

While many of these tools seem quite simple, they are by no means easy to provide or interpret. For example, there are many gaps in the understanding of regional climate trends, impacts, and vulnerabilities, especially at a level of detail that can generate accurate maps. Climate change projections, and associated impacts, are still uncertain, especially at the local level where many decisions are being made. Many locally available decision support products may not provide clear information on error, probability, or confidence in particular data sets or projections. Local case studies and reports from local experts and residents may lack quality control or careful documentation. Many local decision makers are now familiar with Geographic Information Systems (GIS) and use them for analysis and visualization in planning and communicating with the public. Geovisualization tools also include commercial web-based geoinformation such as Google Earth that has several tools relating to climate and environmental change. These tools can be used as support for decisions about climate change, especially in showing how coastlines, ecosystems, and settlements may be affected by climate change, and are already widely used in responding to climate related disasters (e.g., Federal Emergency Management Agency (FEMA); Greene, 2002; Shaw et al., 2009).

Figure 4.3 shows some examples of graphics provided to local decision makers by the Southwestern Climate Change Network with information on temperature trends, fire, and drought risks.

Decision Tools to Inform International and National Emission Reduction Strategies

Computer simulation models provide a crucial resource for supporting many climate-related decisions. The success of such decisions will often depend on the extent to which models can take into account the complicated interaction of physical systems such as the ocean and atmosphere, biological systems such as forests and estuaries, and societal systems such as migration, settlement patterns, and various forms of economic activity. Simulation models are especially useful in decision making in part because the choice of which model to use forces users to be specific about the options and potential consequences they are considering. Often, these models can be used by decision makers in various levels of government and at different political jurisdictions (international, national, and state or regional levels).

Earth System Models

Earth system models can aid decision making at both international and national scales. About twenty different large climate models (mostly general circulation models or GCMs) exist worldwide to help inform decisions about reducing GHG emissions. In general, these models take a trajectory of atmospheric GHG concentration and calculate the response of the global atmosphere and oceans over many decades to answer questions such as how global temperature and precipitation patterns might change if atmospheric GHG concentrations double over the 21st century. Several of these models are based in the United States, and most of the models have been tested, compared, and evaluated through the IPCC Working Group I (WG I), U.S. Climate Change Science Program (CCSP), and various model intercomparison exercises (e.g., Coupled Model Intercomparison Project—CMIP).

The results of simulations that examine the impacts of different GHG concentrations (e.g., 450 ppm) on future climate have been influential internationally by informing discussions of the UN Framework Convention on Climate Change, where the mandate to "avoid dangerous anthropogenic climate change" demands understanding the links between emissions and global and regional climate change risks. The model results informed decisions by the European Union and others to try and limit climate change below 2°C above preindustrial levels, as well as proposals by various countries, scientists, and NGOs to reduce emissions by up to 80 percent by 2050 or set targets such as 350 ppm. The global and regional climate projections produced by different model scenarios have been used to develop global and regional projections of impacts on water, ecosystems, and other sectors and as the basis for estimating economic losses associated with those impacts.

Taking into account both the scale of U.S. emissions and the size of the country, global climate models are somewhat useful in understanding the global climate impacts of alternative U.S. (and other major emitters) emission choices and for understanding the impacts of various emission futures on climate at the regional level. For example, the U.S. National Assessment (USGCRP, 2001), an assessment mandated by the U.S. Congress in the Global Change Research Act of 1990 (P.L. 101-606), used climate scenarios generated by climate models using a set of alternative emissions scenarios generated by IPCC. At the same time, efforts to support decision making based on climate models must recognize a number of limitations of the models. First, it is important to understand that models are abstractions of reality and that their scale, initial conditions, and assumptions about processes are simplifications that necessarily incorporate considerable uncertainty. Second, the scale of the models is such that the impacts of regional and local emissions choices are not usefully captured in the modeling process. The accuracy of regional and local climate projections is also severely limited by vagueness with respect to topographic details and by the fact that some key processes (such as precipitation) are not included or are oversimplified in the models.

The significance and complexity of the earth system requires a subset of models that focus on specific aspects of climate change and its impacts, many of which are relevant for informing different kinds of decisions such as policies on sequestration of carbon in forests and soils, for estimating overall carbon budgets at global and regional levels, and for understanding the feedbacks that may increase or decrease overall risks of climate change and the impacts of climate policies. For example, carbon cycle models (as well as models for methane and other gases and aerosols) have been developed to understand how carbon dioxide is released and absorbed by oceans and land, how these natural processes are affected by anthropogenic emissions and land use changes, and how climate change itself may alter the release and absorption of carbon dioxide and other GHGs.

The most commonly used global climate models are not actually used for long-term decisions about climate change, but rather for understanding how climate changes on a seasonal to decadal scale—time scales that are especially relevant for some decision makers. Global climate models are also used in experiments that try to downscale climate change scenarios using mesoscale models.

Integrated Assessment Models

Earth system models are often linked to economic models or to simulations of the evolution of the global economy over time, with a particular focus on how the economy in different regions of the world generates and consumes energy (CCSP, 2007). These integrative models can be used to answer questions concerning, for example, what mix of energy technologies (e.g., coal, oil, natural gas, wind, nuclear, and solar) might emerge in different regions of the world if policies were put in place to limit atmospheric concentrations of GHGs to different target levels. Such models can also provide information on how much would it cost to produce energy using such a mix of technologies. Integrated assessment models lie at the core of attempts by IPCC to link the work of Working Groups I, II, and III by using emission scenarios based on mitigation options to climate change projections and impacts. They also underpin global assessments of the costs and benefits of alternative mitigation and adaptation strategies, such as those conducted by the Stern Review (Stern, 2007). A network of integrated models has been created as the Energy Modeling Forum (EMF), a set of tools that uses different models to assess alternative futures and policy options. Simulations such as the Energy Information Administration's National Energy Modeling System (NEMs) address similar questions, with a focus on the United States and time horizons of a few decades.

Such climate, energy, and economic simulations can help support risk management decision making by providing information to aid learning over time. For instance, they can help describe options for and the implications of achieving various goals proposed for emissions reductions policies, such as the 2°C temperature limit beyond preindustrial levels recently endorsed by the G-8. Consistent with a risk management framework, these models cannot identify exactly what emissions path would be needed to hold temperatures below any given value but rather can suggest the scale of reductions needed to ensure a reasonable chance of achieving that target (Figure 4.4).

Integrated energy-economic simulations can also suggest the types of energy technologies that need to be deployed and the potential economic costs of meeting these emission reduction targets. Given the large uncertainties involving any projections of factors such as the cost and performance of future technologies and the growth of the global economy, these models give no definitive answer to such questions. However, the long residence time of some greenhouse gases in the atmosphere means that a certain amount of warming is already underway despite the future of technology and the economy (for ideas about difficulties in making such projections, see CCSP, 2007). Regardless of the inevitable uncertainties, these models reliably emphasize important themes that could prove very useful to informing a risk management framework. For instance, they suggest the importance of allowing flexibility over time, and geographic location in allocating emission reductions, and they also emphasize the need to take the needs of different economic sectors into account. The costs of meeting any given climate target are always lower when such flexibility is allowed. The models also suggest key contingencies that must be considered in meeting various reduction goals. For instance, despite the uncertainties, the models all suggest it is very unlikely that global temperatures can be held below 3°C without significant advances in energy efficiency, the widespread use of carbon capture and storage with coal-fired power plants, and the participation of the United States, India, and China (see *Limiting the Magnitude of Climate Change,* NRC, 2010d).

Climate, economic, and energy simulations can also prove helpful in informing choices about near-term emission reductions in the context of emission reduction decisions that might be taken in future decades. For example, the research community has identified attractive near-term limiting strategies that account for future learning about important factors such as the sensitivity of the climate to human emissions, the cost of new energy technologies, and the willingness of various countries to implement emission reduction policies. Experiments have also shown how outcomes can differ depending on the assumptions made. However, common themes emerge across all such experiments, such as the importance of beginning emission reductions in the near-term in order to reduce the possibility that sudden emission reductions will be needed in the future and to keep open the ability to decide later to hold climate changes below levels that appear especially risky. Models also differ in their treatment of uncertainty (see Box 4.2). The models that treat uncertainty as if it is resolved prior to decision making are referred to as deterministic; those that allow for adaptive management or iterative decision making over time are referred to as stochastic. Stochas-

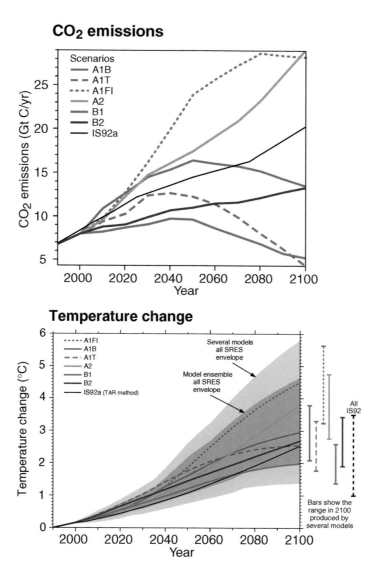

FIGURE 4.4 This figure shows the greenhouse gas concentrations and mean surface temperature from a set of climate models across a range of greenhouse gas emission scenarios. Such model results help show the range of emission paths that might prove consistent with a 2°C temperature limit. SOURCE: IPCC (2001).

BOX 4.2
Sequential Decisions in Iterative Risk Management

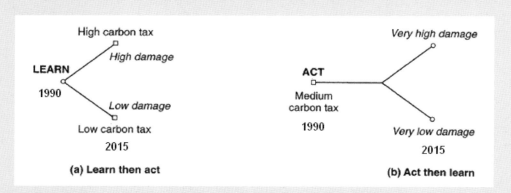

Examples of simple decision theory models. The circles indicate points where uncertainty is resolved and the squares indicate where a decision is made. SOURCE: Adapted from Nordhaus (1994).

Decision theory provides a simple model, called *sequential decisions*, that provides a stylized but often very useful description of how an iterative risk management approach might evolve. Many modeling resources, even ones that make no assumptions about uncertainty being resolved, often find it useful to employ these straightforward tools. In the figure above, the circle denotes a point where uncertainty is resolved and the square shows where a decision is made. We consider in this example only one uncertainty, the level of damages due to GHG emissions, and two states of the world, low and high damages. If we are fortunate enough to learn the true state of the world prior to acting with (a) "learn then act", we can adopt the appropriate carbon tax at the outset. Alternatively, with (b) "act then learn", we must hedge our bets and adopt a carbon tax somewhere in between that corresponding to the two states of world. With (b) "act then learn", we suppose that we learn the nature of future damage in 2015 and hence adjust the carbon tax accordingly.

tic models are particularly useful for exploring near-term hedging strategies in the face of such key uncertainties as the existence and nature of tipping points. Stochastic models can be used to address such questions as how much GHG emissions should be reduced in the near-term given the longer term uncertainties. Both deterministic and stochastic models can be used to support decision analyses that address uncertainty in a variety of different ways. Integrated assessment models can describe a set of scenarios or support a probabilistic analysis where subjective probabilities are elicited from experts regarding the likelihood of key uncertainties and the models then used to identify optimum policy responses.

Integrated Assessment models help decision makers anticipate and understand the consequences of decisions involving complicated, interacting systems and help them structure complex decisions. But like all models, integrated assessments present important challenges. First, those who use models need to understand the purposes for which a model was designed and how those purposes relate to the questions the user would like to answer. For instance, climate, economic, and energy models can provide much information about the potential benefits and costs of choosing a particular emissions reduction target, but determining the best course of action (to the extent that a single best course can be said to exist) under multiple uncertainties is much more complex. As discussed earlier, decisions about climate change are not based only on climate-related information. Practical policy choices require consideration of value judgments and political impacts well beyond anything included in current models. Second, analyses of problems involving multiple variables and sources of uncertainty present complex computational challenges. Optimization problems are straightforward to formulate but become increasingly difficult to solve as problem-related variables and dimensions increase. Researchers continue to seek ways of addressing these challenges.

Additionally, evaluating many climate-related decisions makes it necessary to make projections about the future behaviors of systems ranging from the climate to new technologies—behaviors that are inherently very difficult to anticipate. Unlike models used by engineers designing cars and airplanes or by weather forecasters predicting the weather, many models that might be very useful for informing climate-related decisions cannot be validated until decades after those decisions needed to be made. Unlike models that can be validated by historical records and real time observations, models of climate-related decisions that involve projection decades into the future are hard to validate in the near term (Collins, 2007). Even when a model has successfully reproduced past observations, that alone cannot guarantee it will successfully predict future changes. Even with these limitations, models still can inform decisions within an iterative risk management framework by demonstrating the implications of alternative assumptions and the conclusions that will likely hold despite uncertainties, such as the risks of allowing GHG emissions to continue to increase at current rates.

Further complicating matters, the economic components of models typically include key (and sometimes controversial) assumptions about the valuation of non-market impacts, such as the impacts of climate change on ecosystems or health, and also about discounting future costs and benefits. Overall results of modeling efforts can be quite sensitive to these assumptions. This is illustrated in the debates over the conclusions of the Stern Review (that the cost of mitigation now is much less than the costs of damages later), where varying the assumptions about non-market impacts and discount rates produced significantly different cost estimates (Beckerman and Hepburn, 2007; Dasgupta, 2007; Nordhaus, 2007; Stern et al., 2006).

Despite their limitations, integrated models offer some useful insights for decision making. They show that, under uncertainty, hedging our bets is a good strategy for current decisions, even though it could result in emission reduction choices that might not be optimal if perfect information about the future were available now. Models that incorporate subjective probability judgments of experts about points of uncertainty in climate science can be helpful by showing the sensitivity of outcomes of near-term decisions to long-term uncertainties.

Policy Simulations

In some situations policy simulation can result in powerful learning experiences that provide decision makers and the public with information they can use to better evaluate the decisions and implications associated with current policy issues (Geurts et al., 2007). Simulation exercises can also engage the emotions of decision makers by making them viscerally aware of the complex nature of the decision making process and of challenges they might have previously overlooked. The information in policy simulations is presented as a projection of what the future might look like with explicit assumptions about specific scenarios. Policy makers are thus able to consider how they might respond to a given situation once they understand the assumptions behind it. These simulations have long proved valuable to policy makers in the national security policy area (Mayer, 2009). A group of officials might spend a day role playing members of the National Security Council in the midst of a crisis. The experience of, for instance, working with colleagues in an exercise that calls for writing recommendations for the President in presumed aftermath of the detonation of a nuclear weapon on U.S. soil can help participants recognize weaknesses in current contingency plans and perhaps lead to a lasting appreciation for the importance of safeguarding nuclear materials.

In recent years, policy simulations have begun to make use of computer models to better predict the outcomes of policy choices. For instance, a number of computer simulations, available over the web and in other venues, provide a menu of alternative spending cuts or tax increase options and, by giving participants an opportunity to try to balance the federal or a state budget, may provide evidence about the difficulty of some of the tradeoffs involved. More sophisticated policy simulations may play out the evolution of complex systems and, by allowing participants to intervene and try to influence the evolution, provide insight into what type of interventions might prove effective and the ways in which some seemingly beneficial actions can go seriously awry. Such simulations range from widely available products like SimCity to a research field that produces "serious games." These games are educational and entertaining ways for interested citizens to become informed on important policy decisions and their associated implications.

Policy simulations are beginning to be used to support climate-related decisions. For instance, the Climate Rapid Overview and Decision-Support Simulator (C-ROADS), developed by a team at the Massachusetts Institute of Technology, Ventana Systems, and the Sustainability Institute, is a simple model that traces the implications of alternative GHG abatement policies and their ability to meet various types of emission reduction targets (see Figure 4.5). The model allows users to examine the impacts of a wide

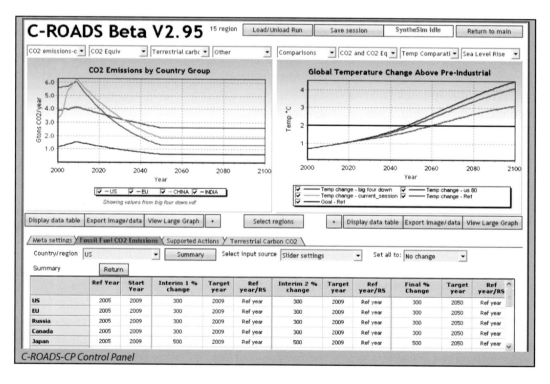

FIGURE 4.5 The control panel of the C-ROADS Climate Simulator is intended to help decision makers understand and interpret various emissions scenarios and climate responses. SOURCE: ClimateInteractive (2010).

range of policy proposals under consideration by governments or proposed by advocacy groups. The simulation is designed for easy, real-time use with groups of individuals. It has been used to support mock negotiation exercises with decision makers from government, business, and NGOs. These exercises help participants develop a better understanding of inertia, long time lags, and other often misunderstood fundamental dynamics of the climate system that make near-term action important to reach the many desired climate goals.

In general, models such as C-ROADS can be useful in providing insights into certain parts of the complex climate problem, such as the difficulty of maintaining various temperature limits. For example, what would it take in terms of emissions reductions by Annex B and non-Annex B countries in the Kyoto Protocol to limit temperature to 2°C above preindustrial levels? The model calculates this with the help of simplified data on atmospheric concentrations and climate models fitted to more complex physical systems models. The target is reached by posing a series of "what if" questions in terms of the reductions made by each country in a hypothetical coalition.

There are a variety of other simulations that can be used by policy makers and the public. These include the Framework to Assess International Regimes for differentiation of commitments (FAIR) (Den Elzen and Lucas, 2005) or the "good enough" tools provided by Socolow and Lam (2007). FAIR is designed to provide information on environmental impacts and the costs associated with projected mitigation efforts. Socolow and Lam (2007) provide tools for assessing the types of actions needed on various time scales and the consequences of not taking any action at all. These kinds of simulations have been successful in engaging many members of the public in assessing alternative climate futures. For example, more than 2 million players undertook policy simulations with Red Redemption's Climate Challenge video game on the BBC Science and Nature web site.[2]

Informing State and Regional Level Emissions Reductions

A number of states have taken action to reduce GHG emissions, both on their own and in regional partnerships with other states (see Chapter 2). In formulating their plans, states, much like national governments, have relied on a variety of tools to determine what reduction strategies to use and to estimate the ultimate environmental and economic impacts of these strategies. Some of these tools, as discussed in the previous section, are used by decision makers in many different levels of government and have

---

[2] See *http://red-redemption.com/portfolio/climate-challenge/*.

enabled states to plan for and begin to implement their GHG reduction programs. There are, however, some practical limitations to their effectiveness in some cases.

States looking to develop climate action plans need to assemble scientifically based knowledge in three areas: (1) baseline GHG emissions in the state, (2) projection of what those emissions would be by a designated target date without implementing any reduction strategies (a business-as-usual reference case), and (3) a compilation of the GHG reduction strategies the states could use, with the potential emissions reductions. This is normally done in the context of a target for GHG emissions reductions that has been established within political arenas. In developing their reduction plans, states and regions have commonly worked with a facilitating group, such as the World Resources Institute (WRI), the Pew Center, the Center for Climate Strategies (CCS), or the Great Plains Institute, to help guide them through the process.

Most states and regions have selected 1990 (which was used in the Kyoto discussions) or 2005 as a baseline year. They typically determined the GHG emissions in the baseline year by a combination of actual emissions data (principally from large emitters such as power plants), and with tools (usually formulas) provided by the facilitators for sectors such as agriculture, transportation, and smaller industry, where no precise measurements have been kept. They have estimated the business-as-usual emissions—which assumes no policy changes through the target date—by modeling done at the same time as modeling for the reduction strategies. The information that goes into the reference case may differ by jurisdiction, but it typically includes forecasts for population, employment and GDP, fuel cost, and technology performance and cost.

The universe of possible reduction strategies is developed in consultation with the facilitators and is not meant to be the list of what will be recommended. Generally, these lists include power plant reductions, transportation, agriculture, landfills, energy efficiency, renewable energy standards, and many other potential strategies. In analyses for cap-and-trade policies, much of the determination of who would be covered by the cap and what strategies to employ is driven by the jurisdiction and those they recruit to provide assistance. This has been done very differently in different jurisdictions. In some cases, groups were created encompassing a wide range of stakeholders (NGOs, potentially affected business sectors, academic experts, civic and religious leaders, labor, and others) to sort through the potential strategies and select those which they wanted to include, as well as to determine who they believe should be covered under the cap. The latter discussion typically considers the potential amount of reductions from inclusion of each class of emitters, as well as the ability to accurately administer the cap in each sector. In other jurisdictions, the decisions about what to include under the cap, and which strategies to employ, were made by a group of public officials and then taken through a stakeholder process, much as would be done in an environmental rulemaking procedure. In either case, stakeholders are involved in the process, and time is needed to allow for their involvement.

Within the United States, many environmental laws and regulations also require processes of public participation. Acquiring public participation is often challenging due to the public's understanding of the climate change issue (see Chapter 8). Careful use of tools can improve public understanding and provide a structured analysis of public concern and willingness to respond. When the determinations have been made as to the targets, which emitters would be included under a cap, and which potential strategies would be examined, additional modeling is employed. Modeling is done by a fairly limited number of firms, so many of the jurisdictions employed the same modelers.

Here again, the outcomes of modeling efforts depend on the information fed into the models. As a result, a great deal of time is spent developing the assumptions that underlie the model (for example, about whether new coal plants will be built or retired, what natural gas prices are likely to be, what role energy efficiency and renewable resources will play in the jurisdiction, what new laws will affect energy usage, etc.). The modelers develop the reference case, with which a comparison can be made to the results modeled when factoring in the GHG reduction strategies the group has settled upon. The models are used as guides to find the most efficient or cost-effective strategies and are not presumed to provide a definitive solution.

States also use modeling to help determine the economic impacts of particular strategies. Models can forecast costs to consumers of implementing policy strategies, as well as impacts on jobs (even by particular sectors), and on the GDP of the jurisdictions. As with emissions modeling, the results depend on good assumptions and data, and this modeling is also expensive and time consuming. However, for the stakeholder processes, as well as for communicating the benefits of a particular public policy, having this information is essential.

Modeling is valuable because it enables decision makers to consider both individual strategies and interactions among them. For example, if a policy to increase the fuel economy of cars and another to increase the amount of mass transit are examined separately, double counting of the reductions may result. The modeling takes these overlapping strategies into account.

The kinds of processes described here give decision makers modeling results, assistance from a facilitating group to provide them with perspective of other jurisdictions, and input from stakeholders. Done correctly, such analyses can guide decision makers with difficult policy choices.

Informing Efficiency Decisions at the Firm and Household Levels

As awareness about climate change and responses to climate change increase through the private sector, many businesses are assessing how their activities affect the environment and the environmental performance of their products. One of the key tools used to assess the role of GHGs in a firm's operations and products, and to identify steps to reduce them, is *life cycle analysis* (LCA). LCA is a technique to assess the environmental aspects and potential impacts associated with a product, process, or service by compiling an inventory of relevant energy and material inputs and environmental releases; evaluating the potential environmental impacts associated with identified inputs and releases; and interpreting the results to improve the quality of decision making. The process includes an assessment of raw material production, manufacture, distribution, use, and disposal, including each transportation step necessary by the product's existence. LCA can be done for the full life cycle, from manufacture to use to disposal (cradle-to-grave) or from manufacture to the factory gate (cradle-to-gate), with variants that include cradle to cradle (where materials are recycled), and well-to-wheel (used for transport fuels).

Several LCA studies have aided manufactures in the decision making process regarding products or processes that have the least impact to the environment or society. For example, an early study by the Tellus Institute demonstrated that over 95 percent of environmental costs are created by the production of packaging, including energy used and toxins created, rather than in the disposal of products (Tellus Institute, 1992). LCA is also at the core of eco-labels that are designed to inform consumers about the environmental impacts of products (Cowell et al., 2002).

Most LCA is undertaken using dedicated, often private sector, software packages such as GaBi, SimaPro, and TEAM. The U.S. National Renewable Energy Laboratory hosts a U.S. Life Cycle Inventory Database that provides information on cradle-to-grave accounting. The Environmental Protection Agency (EPA) provides several LCA tools on its web pages, including a Waste Reduction Model that estimates GHG impacts of solid waste (WARM); the Recycled Content Tool, which calculates the life cycle GHG and energy impacts of certain materials (ReCon); the National Recycling Coalition (NRC) Environmental Benefits Calculator (which determines the GHG content of waste materials); the Northeast Recycling Council (NERC) Environmental Benefits Calculator (for states and organizations to measure benefits from recycling); and the Durable

Goods Calculator (DGC) which calculates the GHG content of disposing and recycling of household goods. The U.S. Department of Energy provides a fuel cycle model for evaluating transport engine and fuel alternatives (GREET), and internationally the eco-invent portal (www.ecoinvent.com) provides tested and science based LCA data for more than 4,000 activities in industry, agriculture, and waste management.

LCA has been used to optimize the environmental performance of both individual products and individual companies. For example, the University of Michigan Center for Sustainable Systems conducted a study in 1999 on the life cycle assessment of the Stonyfield Farms product delivery system. The study found that choice of container size for products places the greatest burden on the environment, with smaller containers having more impact than larger containers. The study also made several recommendations to the Stonyfield Farms Company to reduce impact. LCA has been a valuable tool for a company's decision making.

LCA analysis is useful when parameters can be quantified and then reduced, and is an important step in some of the GHG management strategies discussed in Chapter 6. Once again, the quality of an LCA is only as good as the data that go into it, and it is especially important to use up-to-date information, especially where production processes change rapidly (Ayres, 1995; Reap et al., 2008a,b). Uncertainty pervades LCA where data are not available and when subjective decisions are made about what costs to include and exclude. For example, insufficient data on the GHG emissions of a production method or product can lead to large uncertainties in LCA analysis (see Chapter 6 for further discussion on GHG management). Comprehensive LCA can be expensive, and it can also be difficult when relevant information is proprietary or sensitive. Some classic LCA controversies have involved everyday goods such as paper versus plastic grocery bags and disposable versus washable diapers. Standards are important but often are of limited usefulness for LCA. For example, ISO 14048 sets international standards for nomenclature and frameworks for life cycle inventory but provides little guidance on specific data models.

UNDERSTANDING IMPACTS AND INFORMING ADAPTATION DECISIONS

Although impact models can be included as elements in the integrated assessment models described in the previous section, they are more commonly used to estimate the impacts of climate change on particular sectors and to assess how adaptation might reduce these impacts. Also, many are already used in existing management frameworks for analyzing the impacts of current climate variability and other decisions on water resource management, agriculture, ecosystems, coastal regions, energy management, disasters, and health. Many of these methods are described in the reports of IPCC Working Group II and by the companion ACC reports *Adapting to the Impacts of Climate Change* (NRC, 2010a) and *Advancing the Science of Climate Change* (NRC, 2010b). Because of their operational use in current decision making, many impact focused models have been customized for particular places and groups of decision makers, account for uncertainty, and allow for interactive exploration of policy options and future scenarios. In line with discussions at the beginning of this chapter, research on actual decision making shows that, although computer tools may be used to evaluate alternatives, many decisions are made without models and are based, for example, on individual judgment, collective discussion, institutional trends, and political pressure. Decisions may differ from what model results would recommend, and results may be cited only when supporting decisions that have been made on other grounds (Brewer, 1975).

Tools supporting adaptation decisions are numerous and varied. For example, many different simulation tools and optimization models are used in the management of water resources. Such tools can estimate impacts on hourly, daily, and seasonal time scales and may include forecasting and modeling of runoff, groundwater, irrigation, reservoirs, water quality, demand, and water management.

Similarly, both simulation and statistical models are used to understand impacts and inform agricultural decision making, taking into account biophysical conditions (e.g., climate and weather, soils), input availability (water, fertilizer, improved seeds, labor, machinery), and economy (e.g., costs, prices, subsidies). Such models can allow farmers and other decision makers to anticipate the effects of climate extremes and seasonal forecasts, adjust the use of inputs and production to local conditions and international trade, and explore policy options, including emergency relief, incentives, and environmental regulation.

Coastal management also uses a range of modeling and decision tools to prepare for severe storms, assess alternative land uses, and analyze management options that might include coastal protection, warning systems, and building regulations. These models must be carefully calibrated to take into account local conditions because of the ways in which particular coastal formations and human activities influence the impacts of climate and management options.

Managing ecosystems in the context of climate variability has also led to the development of both general and customized tools for predicting and managing the effects of drought, fire, and other climate related influences on forests, grasslands, wetlands, and other landscapes. These are sometimes used for real time prediction and management

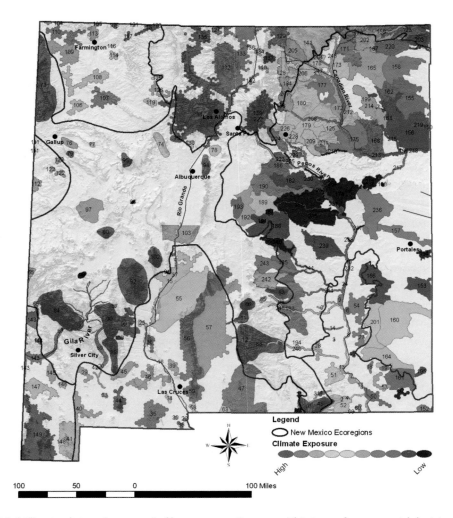

FIGURE 4.6 Climate change "exposure" of key conservation areas. This type of map can aid decision makers as they explore options for future conservation or adaptation efforts. SOURCE: TNC New Mexico Conservation Science Program (2008).

(e.g., for fire and insect infestations) but are also used to explore longer-term alternative management options (see Figure 4.6).

Urban impact assessments can involve modeling and integration of multiple elements of the urban system, including the sensitivity and vulnerability of water resources, energy use, industrial production, and transport to climate conditions. With respect to public health, there is a growing emphasis on models that can indicate how various climate conditions, such as extreme heat and excess moisture, can influence the severity and extent of health problems such as asthma, infectious diseases such as influenza, and various vector-borne diseases. Models can be used to anticipate the evolution of epidemics and also to allocate resources such as vaccines and health care professionals.

Using these kinds of tools for informing decisions on adaptive responses to climate change has considerable potential, but users must also consider model limitations. Models based on current conditions or those of the recent past may be limited in their ability to predict impacts resulting from future climate changes, such as those that are outside the range of past experience. Other model limitations stem from deficiencies in the ability to predict future non-climate related changes such as social and economic trends. And even if projections turn out to be correct, options for adapting to future climate conditions—for example, strategies for protecting endangered species—may be limited. Table 4.2 provides a useful overview of key advantages and disadvantages relating to the use of various methods of constructing regional climate change scenarios in climate impact assessments.

Examples of decision support tools for climate change adaptation used by U.S. based projects in coastal and water management are described in Boxes 4.3 and 4.4.

UNDERSTANDING THE VALUE OF INFORMATION FOR RESOURCE ALLOCATION

Globally, governments are spending billions of dollars annually on research programs to improve the knowledge base for future climate related decisions. In an era of highly constrained resources, it is not surprising that policy makers are interested in the value of reducing uncertainty and in what kinds of research will yield the highest payoff. Decision tools can be used to estimate the value of new information, which can help decision makers plan research programs and determine which trends to monitor to best implement an iterative risk management strategy. The types of decision tools used to inform a particular climate-related investment will depend, in large part, on the nature of the decision under study and the quality of available information (see Box 4.5). In the climate area, investments are often parsed into five areas, related to: limiting the magnitude of climate change; adaptation, or reducing vulnerability to climate change; research and development to expand the range of mitigation and adaptation options; improved scientific information to provide the foundation for better informed decisions in the future; and geoengineering, or technologies and activities aimed at changing the heat balance of the earth.

In decision theory, the value of information in these and other areas is calculated in

TABLE 4.2 Options for Constructing Regional Climate Change Scenarios Listed in the Order of Increasing Complexity and Resource Demand

Method (application)	Advantages	Disadvantages
Sensitivity analysis *Resource management, Sectoral*	1. Easy to apply; 2. Requires no future climate change information; 3. Shows most important variables/ system thresholds; 4 Allows comparison between studies.	1. Provides no insight into the likelihood of associated impacts unless benchmarked to other scenarios; 2. Impact model uncertainty seldom reported or unknown.
Change factors *Most adaptation activities*	1. Easy to apply; 2. Can handle probabilistic climate model output	1. Perturbs only baseline mean and variance; 2. Limited availability of scenarios for 2020s.
Climate analogues *Communication, Institutional, Sectoral*	1. Easy to apply; 2. Requires no future climate change information; 3. Reveals multi-sector impacts/vulnerability to past. climate conditions or extreme events, such as a flood or drought episode.	1. Assumes that the same socio-economic or environmental responses recur under similar climate conditions; 2. Requires data on confounding factors such as population growth, technological advance, conflict.
Trend extrapolation *New infrastructure (coastal)*	1. Easy to apply; 2. Reflects local conditions; 3. Uses recent patterns of climate variability and change; 4. Instrumented series can be extended through environmental reconstruction; 5. Tools freely available.	1. Typically assumes linear change; 2. Trends (sign and magnitude) are sensitive to the choice/length of record; 3. Assumes underlying climatology of a region is unchanged; 4. Needs high quality observational data for calibration; 5. Confounding factors can cause false trends.
Pattern-scaling *Institutional, Sectoral*	1. Modest computational demand; 2. Allows analysis of GCM and emissions uncertainty; 3. Shows regional and transient patterns of climate change; 4. Tools freely available.	1. Assumes climate change pattern for 2080s maps to earlier periods; 2. Assumes linear relationship with global mean temperatures; 3. Coarse spatial resolution.

TABLE 4.2 Continued

Method (application)	Advantages	Disadvantages
Weather generators *Resource management, Retrofitting, Behavioural*	1. Modest computational demand; 2. Provides daily or sub-daily meteorological variables; 3. Preserves relationships between weather variables; 4. Already in widespread use for simulating present climate; 5. Tools freely available.	1. Needs high quality observational data for calibration and verification; 2. Assumes a constant relationship between large-scale circulation patterns and local weather; 3. Scenarios are sensitive to choice of predictors and quality of GCM output; 4. Scenarios are typically time-slice rather than transient.
Empirical downscaling *New infrastructure, Resource management, Behavioural*	1. Modest computational demand; 2. Provides transient daily variables; 3. Reflects local conditions; 4. Can provide scenarios for exotic variables (e.g., urban heat island, air quality); 5. Tools freely available.	1. Requires high quality observational data for calibration and verification; 2. Assumes a constant relationship between large-scale circulation patterns and local weather; 3. Scenarios are sensitive to choice of forcing factors and host GCM; 4. Choice of host GCM constrained by archived outputs.
Dynamical downscaling *New infrastructure, Resource management, Behavioural, Communication*	1. Maps regional climate scenarios at 2050km resolution; 2. Reflects underlying land-surface controls and feedbacks; 3. Preserves relationships between weather variables; 4. Ensemble experiments are becoming available for uncertainty analysis.	1. Computational and technical demand high; 2. Scenarios are sensitive to choice of host GCM; 3. Requires high quality observational data for model verification; 4. Scenarios are typically time-slice rather than transient; 5. Limited availability of scenarios for 2020s.
Coupled AO/GCMs *Communication, Financial*	1. Forecasts of global mean and regional temperature changes for the 2020s; 2. Reflects dominant earth system processes and feedbacks affecting global climate; 3 Ensemble experiments are becoming available for uncertainty analysis.	1. Computational and technical demand high (supercomputing); 2. Scenarios are sensitive to initial conditions (sea surface temperatures) and external factors (such as volcanic eruptions); 3. Scenarios are sensitive to choice of host GCM; 4. Coarse spatial resolution.

SOURCE: Wilby et al. (2009).

BOX 4.3
Decision Support for Coastal Responses to Climate Change

Decision makers in coastal areas face a daunting set of challenges associated with climate change such as sea level rise; habitat destruction; invasive species; damage to natural protective systems such as wetlands, dunes, and barrier islands; land loss; increased vulnerability of critical infrastructure facilities such as ports and transportation systems; and property and population vulnerability. Coastal regions also face a variety of population and development pressures as growing numbers of Americans migrate to those areas in search of the amenities they value. Many tools and strategies are being used to assist decision makers in coastal regions. Three examples of initiatives and the decision support resources offered include the following:

1. The Environmental Protection Agency's Climate Ready Estuaries Program (CREP) provides a range of tools for communities seeking to adapt to climate change impacts. Estuaries are vulnerable to climate change and variability and are jurisdictionally complex, often encompassing more than one state and numerous cities, towns, and counties. The programs enable stakeholders in estuary regions to analyze their climate change vulnerabilities, develop and implement strategies for adapting to climate change and variation, communicate with various audiences about climate-related risks, and promote information sharing and the dissemination of lessons learned.

   The program provides grants and technical assistance to support adaptation efforts in estuarine settings, actively seeks to develop networks that can serve as conduits for information on best practices and convenes workshops for grant recipients, publishes newsletters, and provides space on its web site for inter-project communication.

   CREP maintains an extensive web portal that includes access to a "Climate Ready Estuaries Toolkit" that contains a suite of GIS-based risk and vulnerability assessment tools and databases for monitoring climate change. The site also enables users to access CCSP Synthesis and Assessment Products, materials that can be used in education and outreach programs, and information on how to obtain funds for local programs. CREP also assists decision makers through publications that structure problems and lay out options for climate change adaptation, including maintaining and restoring wetlands; maintaining sediment transport; preserving coastal lands development and infrastructure; maintaining shorelines through both "soft" measures such as marsh creation to slow shore erosion and "hard" measures such as the construction of sea walls and breakwaters; and maintaining water quality and availability.

2. The National Oceanic and Atmospheric Administration Coastal Services Center (CSC) assists coastal management organizations in locating decision-relevant information and developing

climate change adaptation programs. For example, its "Digital Coast" data resource contains links to a wide variety of datasets containing orthoimagery, coastal elevation and land cover data, bathymetry and topography data, and data on demographic trends affecting coastal regions. The CSC provides training in the use of "Digital Coast," conducts workshops on vulnerability assessment techniques and applications ("VATA"), and operates a listserv for information sharing. It also maintains a climate change adaptation web site that includes guidelines for adaptation planning, reports on policy and legislation, case studies, and other informational resources.

3. PlaNYC (described in *Adapting to the Impacts of Climate Change*, NRC, 2010a), and also *Informing Decisions in a Changing Climate* , NRC, 2009a) represents a more locally based coastal decision support program in the New York City region which targets three priority activities for adaptation: formation of an intergovernmental task force for the protection of the city's critical infrastructure, development of strategies for protecting especially vulnerable neighborhoods, and development and implementation of a citywide strategic planning process for climate adaptation. PlaNYC uses a variety of strategies to aid decision making, providing decision makers with information on a range of climate-related indicators, including climate change scenarios, downscaled regional scenarios, projections regarding future extreme events, and physical and social vulnerability indicators. The New York metropolitan region faces significant hazards related to sea level rise—in particular storm surges from extreme weather events, which will become more severe as sea level rise progresses. In studies carried out for the New York City Department of Environmental Protection, researchers at the Goddard Institute for Space Studies (GISS), using the GISS Atmosphere-Ocean model, were able to predict sea level rise over time for the New York metropolitan area under different emissions scenarios. As indicated on the web site of the Columbia University Center for Climate System Research, this set of studies found that in a major hurricane "[a]reas potentially under water include the Rockaways, Coney Island, much of southern Brooklyn and Queens, portions of Long Island City, Astoria, Flushing Meadows-Corona Park, Queens, lower Manhattan, and eastern Staten Island from Great Kills Harbor north to the Verrazano Bridge" (for more details, see Rosenzweig and Solecki, 2001).

All three programs discussed here seek to address decision makers' needs in a variety of ways. Such approaches include providing information on decision-relevant topics (e.g., climate impacts, model adaptation plans, model legislation, and policy initiatives); making analytic tools and databases more widely available; establishing and sustaining networks for information sharing; engaging in public outreach and education activities; and employing a variety of other stakeholder engagement strategies.

BOX 4.4
Resources for Implementing Iterative Risk Management in the Water Resources Sector

A variety of data sources, simulation models, and decision support methods exist to help water managers incorporate climate change into their operations and plans. As one example, this case study describes how Southern California's Inland Empire Utilities Agency (IEUA) has used a water management simulation model, down-scaled climate projections, and decision support software in a participatory stakeholder process to implement an iterative risk management approach to improve its ability to respond to climate change.

The IEUA, a wholesale water and wastewater provider in Riverside County, California, is legally required every few years to prepare or update a plan demonstrating how they will ensure their community's access to water. At present, IEUA serves slightly fewer than one million people relying primarily on local groundwater and imports from Northern California. To serve its growing population, IEUA in 2005 completed a 25-year water plan that called for the agency to increase the agency's groundwater use by 75 percent and its recycled water use by 600 percent by 2025.

But as recently as 2005, IEUA had not considered the potential impacts of climate change on its long-range plan. However, in 2007 the agency conducted—with the assistance of a RAND-led team funded under the National Science Foundation's (NSF's) climate change decision making under uncertainty (DMUU) program[a]—a vulnerability and response option analysis to determine whether and how the potential for future climate change should cause them to alter their 2005 plan.

To conduct this analysis, the RAND team combined a water management model (WMM) with downscaled regional climate projections from an ensemble of atmosphere-ocean generation circulation models (AOGCMs). The water management model, built using the Water Evaluation and Planning (WEAP) modeling environment (see *http://www.weap21.org* for more information), simulated the IEUA region's hydrology, water supply, and water demand. To address the challenge of planning under uncertainty the simulation was designed to evaluate the performance of IEUA plans under a wide range of future scenarios, each of which reflects plausible trends in climate change and other planning assumptions. The model reported two measures of plan performance: the reliability of the IEUA system in meeting all projected demand and the cost of implementing the agency's base plan and any additional actions needed to improve reliability in some scenarios.

Using an iterative risk management framework, the WEAP simulation was explicitly designed to consider adaptive strategies, those designed to monitor changing conditions and respond over time. In particular, the model began with a specified set of near-term actions IEUA might take, such as investments to increase the use of wastewater recycling or improved water use efficiency. Beginning in 2015 and every 5 years thereafter in the simulation, the model evaluates whether supply has been sufficient to meet demand over the previous 5 years. If the gap between demand and supply exceeds some specified threshold, the simulation implements additional actions as specified by the strategy under consideration.

Climate change is not the only important uncertainty facing IEUA, so the study also considered a wide range of cases representing different assumptions about the agency's ability to implement its aggressive new groundwater and recycling programs, as well as different assumptions about events outside the agency's service area such as those affecting supplies of imported water.

The RAND team then used a decision analytic approach called robust decision making to implement the iterative risk management approach using this simulation model and ensemble of future climate projections. With decision support software designed for this purpose, the study used the simulation model to follow its current plan into several hundred different futures, each characterized by one of the future weather sequences and one set of assumptions about the agency's future level of success in implementing its plans, and future supplies of imported water. Each of these cases explores how the candidate strategy will perform given some particular set of "what if" assumptions about the future state of the world. The study then used statistical analysis to identify the key factors that would cause the agency's plans to fail to meet its performance goals. This analysis suggested IEUA's 2005 plan would fail in the future if all the following factors occurred simultaneously: a significant decrease in precipitation, any decrease in the share of precipitation that infiltrated into the groundwater basin, and significant impact of climate change on the availability of future imports. Other failure modes were identified which included other important factors such as the need for IEUA's recycling program to meet its ambitious goals.

The agency then used this information to create visualizations describing the strengths and weaknesses of alternative plans and the tradeoffs among them. The project collaborators and IEUA used these results to help identify and evaluate potential ways to augment its plan to improve its ability to address these challenging conditions.

The simulation model, climate projections, and decision analysis were developed through a series of workshops with IEUA managers and technical staff, local elected officials, and representatives of local business, environmental, and other groups in the IEUA region (Groves and Lempert, 2007; Groves et al., 2008). These workshops were interspersed with in-depth technical reviews with IEUA technical staff and the RAND team developing the model and climate data. Based on this analysis and interactions, IEUA decided to augment its 2005 Urban Water Management Plan by increasing its current water use efficiency programs in the near-term and by monitoring and updating if necessary its plans in the future.

---

[a] See *http://www.rand.org/ise/projects/improvingdecisions/*.

BOX 4.5
The Value of Information to Help Guide Resource Allocation Decisions in the United States

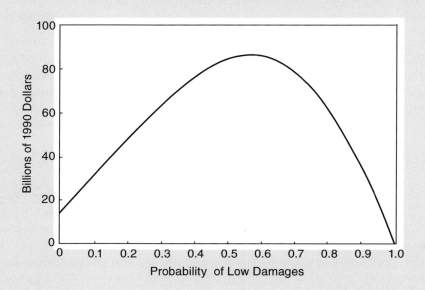

The results of model runs that are used to estimate the dollar value of information that could determine the probability of a low-cost, unlimited emissions scenario. SOURCE: Adapted from Manne and Richels (1992).

The figure above represents the results of model runs that estimate the dollar value of information that could precisely and accurately determine the probability that the costs of expected climate change if emissions were not limited would be low, as a function of what the determined probability is. That value—the value of information that could define the actual probability—is useful for resource allocation decisions for research. The model results indicate that if the research determined that low damage was a certainty, the value of perfect information would be zero. As uncertainty about the future increases, so does the expected value of perfect information. The model showed a maximum value of information when the probability of low damages from unlimited emissions is 0.6. In terms of macroeconomic consumption, the discounted present value is $81 billion (in terms of constant 1990 dollars). The curve, however, is not symmetrical, since even if the probability of low damages with unlimited emissions were zero, there is still uncertainty about whether damage would be moderate or high.

the context of the decision it is intended to inform. In the realm of climate change decision making, there are not only multiple uncertainties, but also multiple decisions to be made by diverse entities, as well as multiple outcomes of each decision. Decision theoretic techniques can also be applied to estimate how much outcomes could be improved if additional information could reduce uncertainty about the future, even if the resulting information is imperfect. It is also important to note that information that may have little or no value for decisions about limiting climate change may neverthe-less have high value for adaptation. Additionally, information that does not necessarily reduce uncertainty, in the sense of narrowing the width of the probability distribution of outcomes, such as information generated from deterministic models and scenarios, can still have a high value for improving decisions.

ASSESSMENTS AS TOOLS FOR CLIMATE-RELATED DECISION MAKING

Integrated assessment models are just one method used in the process of developing broader assessments of environmental issues which bring together a wide range of scientific information and analysis to provide state of the art summaries for decision makers. Assessments are collective, deliberative processes by which experts review, analyze, and synthesize scientific knowledge in response to users' information needs relevant to key questions, uncertainties, or decisions (NRC, 2007a) and as such con-stitute an important interface between science and policy. The U.S. Global Change Research Act (1990) mandates regular (4 year) assessments of global change impacts on key sectors. However, only two major U.S. assessments of climate change have been conducted—a national assessment in 2001 and the recent CCSP synthesis and as-sessment exercise (USGCRP, 2009). Hundreds of U.S. scientists have participated in the high profile assessments of the IPCC, and climate change has also been an important component of international assessments of ecosystems (MEA, 2005), Arctic climate im-pacts (ACIA, 2004), and stratospheric ozone (WMO, 2007). Reports by the Congressio-nal Research Service also serve as focused assessments for policy makers. Assessments can establish the basic significance of an issue and communicate it to decision makers. They can also respond to particular scientific questions of high policy relevance and can evaluate whether policies are delivering expected benefits.

The NRC (2007a) has identified 11 elements of effective assessments (Box 4.6), where effectiveness is defined in terms of salience (ability to communicate relevant informa-tion to users), credibility (high-quality technical basis), and legitimacy (fairness and impartiality). The NRC also observed that the most common weaknesses in assess-ments are a mismatch between the scope of the assessment, inadequate funding, and the inability to match assessment goals with the needs of decision makers.

BOX 4.6
Elements of Effective Assessments

1. A clear strategic framing of the assessment process, including a well articulated mandate, realistic goals consistent with the needs of decision makers, and a detailed implementation plan.
2. Adequate funding that is both commensurate with the mandate and effectively managed to ensure an efficient assessment process.
3. A balance between the benefits of a particular assessment and the opportunity costs (e.g., commitments of time and effort) to the scientific community.
4. A timeline consistent with assessment objectives, the state of the underlying knowledge base, the resources available, and the needs of decision makers.
5. Engagement and commitment of interested and affected parties, with a transparent science-policy interface and effective communication throughout the process.
6. Strong leadership and an organizational structure in which responsibilities are well articulated.
7. Careful design of interdisciplinary efforts to ensure integration, with specific reference to the assessment's purpose, users needs, and available resources.
8. Realistic and credible treatment of uncertainties.
9. An independent review process monitored by a balanced panel of review editors.
10. Maximizing the benefits of the assessment by developing tools to support use of assessment results in decision making at differing geographic scales and decision levels.
11. Use of a nested assessment approach, when appropriate, using analysis of large-scale trends and identification of priority issues as the context for focused, smaller-scale impacts and response assessments at the regional or local level.

SOURCE: NRC (2007a).

The *Informing* panel endorses these elements and key recommendations of the report, and includes the following:

- Those requesting assessments should develop a guidance document that provides a clear strategic framework, including a well-articulated mandate and a detailed implementation plan realistically linked to budgetary requirements. The guidance document should specify decisions the assessment intends to inform; the assessment's scope, timing, priorities, target audiences, leadership, communication strategy, funding, and the degree of interdisciplinary integration; and measures of success.

- The burden of assessments on the scientific community should be proportional to the aggregate public benefits provided by the assessment. Alternative modes of participation or changes to the assessment process—such as limiting material in regularly scheduled assessments or running "nested" or phased multiscale assessments—should be considered. As appropriate, U.S. assessments should acknowledge the work of the international community and avoid redundant efforts.
- The intended audiences for an assessment should be identified in advance, along with their information needs and the level of specificity required. In most cases, the target audience should be engaged in formulating questions to be addressed throughout the process in order to ensure that assessments are responsive to changing information needs. Both human and financial resources should be adequate for communicating assessment products to relevant audiences. Clear guidelines and boundaries should ensure both salience to those requesting the assessment and legitimacy, especially with respect to the perceived influence of those requesting the assessment might have over the scientific conclusions drawn. A strategy for identifying and engaging appropriate stakeholders should be included in the assessment design to balance the advantages of broad participation with efficiency and credibility of the process. Capacity building efforts for participants from various disciplines should be undertaken in order to develop a common language and a mutual understanding of the science and the decision making context. This capacity building may be required to ensure the most salient questions are being addressed and to meaningfully engage diverse stakeholders in assessment activities.

Building on the NRC study (2007a), our panel identified other considerations that should be taken into account when assessments are used as decision support tools, such as the following:

- Assessments, such as the IPCC and CCSP, have become overwhelming in their scope, size, and demands on the scientific community. It is often hard for decision makers to identify the key messages and information that are relevant to the choices they face. More focused assessments to support specific questions and decisions may be more effective, especially if they are concise and clearly responsive to decisions and stakeholders.
- Assessments tend to be focused on information of relevance to governments at national and regional scales, and they often fail to address concerns and decisions of local governments, the private sector, and civil society. As discussed in Chapter 2 of this report, given the importance of non-federal actors as both

users and sources of information, greater attention should be paid to their decision needs and to their inclusion in the production of assessments.

- There is value to viewing assessment as an ongoing process of engagement with stakeholders which provides regular updates on climate, impacts and responses and responds to the information needs of both federal and non-federal decision makers. However this requires a commitment to supporting the process, to listening and responding to stakeholders, and to the information systems that are needed for the assessments.

- As the United States and international communities make decisions that have significant economic and development implications for countries, business interests, and other communities, the assessments (such as IPCC) on which these decisions are based become matters of "high politics" with much greater scrutiny of their legitimacy and of review processes. This demands even greater care in the preparation, transparency, and communication of assessment products, especially in the communication of uncertainty, social, economic, and ecological impacts, and results of relevance to particular interest groups and regions.

CONCLUSIONS AND RECOMMENDATIONS

A variety of tools and resources exist for informing decision making about climate change. Each of them has advantages and disadvantages, but many are overlooked or misunderstood in the portfolio of decision tools used by decision makers. It is frequently argued that a major purpose of analysis is insights, rather than numbers. Decision tools work best when they provide decision makers with an analytical framework for thinking about a particular problem. With a problem as multifaceted as the climate problem, issues can quickly become intractable. Without systematic procedures for "working the problem," decision makers often become confused and reluctant to act even in cases where action is needed.

Among all the tools that are available, decision makers need to select tools that are capable of providing the information they need. This points to the necessity of providing information within time frames and geographic scales that are relevant to decision makers as well as information on the uncertainties associated with those time scales. Communicating tool results is also important and this requires partnering with stakeholders when making decisions.

The *Science* Panel report (NRC, 2010b) has identified key research needs in developing decision support tools (see also Box 4.1). There is clearly a need to develop tools for responding to climate change, and this need will continue to evolve as tools are designed to be decision-specific. Our review suggests several important challenges in the use and development of decision tools and methods to inform decisions about climate change. These include a mismatch between the global, aggregate, or national scale of climate and energy models and the needs for decision making at more local or sectoral scales; controversies over how to handle economics, uncertainties, and subjective judgments; user misunderstandings about the assumptions and limits of methods; major information gaps; and the need to ensure that assessment activities are effective, are focused, and respond to user needs.

Observational systems and databases are critical to developing tools and the evaluation of methods for modeling, mapping, networking, and decision making. The Federal government has an important role in supporting such information systems as we discuss in subsequent chapters. We find that "value of information" techniques may be helpful in order to inform decision makers on the relative value of investments to improve understanding across key unknowns in the climate system. Where such expertise does not reside in particular agencies, experts should be engaged from outside these agencies (e.g., academia) to provide the requisite skills.

The discussion of assessments as a decision support tool is based on the NRC (2007a) report on lessons learned from assessments and we endorse the recommendations of this report and its suggestions for effective assessments. We judge that future assessments may need to be more focused on specific questions and decisions developed in consultation and collaboration with decision makers.

The panel, in preparing this chapter, also found it difficult to identify good reviews and clear unbiased discussions of the full range of decision support tools, their appropriate uses and limitation. We therefore conclude that there could be a stronger role for the Federal government to provide better guidance on decision support tools for climate decisions, perhaps through a climate tools database, network, and best practice examples. This could be considered part of a broader attempt to provide climate and carbon management services.

At the same time, the panel also recognizes that formal decision-analytic procedures may not constitute the tools of choice for many decision makers. Support for decisions comes from a wide range of sources that include mandates, standards, and regulations; informal norms that govern procedures and practices adopted by decision-making entities; priorities and practices that are diffused within interpersonal and interorganizational networks; and institutional pressures that produce alignments among entities pursuing similar goals. While solutions to climate-related problems should never rely on these kinds of sources alone, it is important to note their significance as drivers of decision making, both in the climate arena and more generally. Formal decision tools may be used to illuminate choices, but they may also be used to validate strategies that have already been decided upon on other grounds. Resources that support decision making are myriad and varied, ranging from sophisticated computer simulations, to scenarios of climate futures presented in the form of GIS visualizations, to films and documentaries, and to less elaborate materials that merely inform decision makers about what measures their counterparts have decided to undertake. Decision makers themselves determine which decision support resources are most relevant in the context of the dilemmas they face, and for that reason all efforts to provide such resources must begin with an understanding of decision maker needs.

Recommendation 5:

a) **The federal government should support research and the development and diffusion of decision support tools and include clear guidance as to their uses and limitations for different types and scales of decision making about climate change.**

b) **The federal government should support training for researchers on how to communicate climate change information and uncertainties to a variety of audiences using a broad range of methods and media.**

CHAPTER FIVE

Climate Services: Informing America About Climate Variability and Change, Impacts, and Response Options

A sked to consider the roles of federal, state, and local governments and other groups in providing effective "climate services,"[1] the panel approached this task by considering the information needs of different stakeholders that might be met by climate services and the functions of a national information system that best integrate our knowledge to better inform decisions. Our task was complicated by a rapidly changing institutional landscape for climate services including steps by several agencies to improve provision of climate information. For example, the Department of the Interior announced the creation of Climate Change Response Councils and regional response centers to facilitate information sharing and response strategies. The World Meteorological Organization (WMO) announced an international agreement to establish a global framework for climate services (WMO, 2009). The National Oceanic and Atmopsheric Administration (NOAA) announced intentions to create a "NOAA Climate Service" within its agency with a redesigned prototype web interface.[2] These initiatives involve substantial reorganization and investments before the services are fully functional and at the time of writing were not coordinated with each other or other federal climate services (e.g., climate information in NASA or the U.S. Department of Agriculture [USDA]) into a "National Climate Service." Regardless of timeline, implementing a national climate service will require careful deliberation including all major federal and non-federal partners. The task of redesigning detailed climate services, especially given ongoing initiatives, is outside the scope of this study but we have provided explicit suggestions on the functions of climate services and the criteria for evaluating effectiveness. The panel draws from previous studies that have focused on the various models for climate services (Miles et al., 2006; NOAA SAB, 2009;

---

[1] A previously defined vision of a National Climate Service is to "provide information to the nation and the world to assist in understanding, anticipating, and responding to climate, climate change, and climate variability and their impacts and implications" with a mission to "inform the public through the sustained production and delivery of authoritative, timely, and useful information to enable management or climate related risks, opportunities and local, state, regional, tribal, national, and global impacts" (NOAA SAB, 2009).

[2] See *http://www.noaa.gov/climate.html.*

NRC, 2001; Overpeck et al., 2009). The panel has relied on invited presentations and expert judgment of the panel to identify a series of functional components, institutional considerations, and principles of operation for successful climate services.

To date, the ongoing national conversation about the establishment of a new entity called a "National Climate Service" has focused on the provision of information about the impacts of climate change and variability and has not addressed how best to provide broader information such as services related to greenhouse gas emissions and reduction strategies. The nation needs climate services that include both kinds of information. Although a case can be made for an overarching climate change information service, especially from the perspective of local decision makers who manage both greenhouse gas emissions reduction and adaptation decisions, the panel chose to discuss the two major components of climate information (information related to climate change, impacts, and adaptation; and information related to greenhouse gas emissions and reduction strategies) separately because of the complex sets of agencies, actors, and scales involved, and in order to clearly identify the functions associated with each.

This chapter focuses on the role of the federal government and others in providing information about current and future climate change and variability, impacts and vulnerability, and response options for reducing risk. The following chapter focuses specifically on the information needed to support emission reductions. Each chapter discusses the potential functions of these institutions to provide these information services. As noted, the panel recognizes that many decision makers either manage or are seeking options that can both reduce greenhouse gas emissions and adapt to the impacts of climate change. Therefore, it is necessary to coordinate, and over time integrate, these information systems across the federal government, other scales of government, and with other public and private actors for an informed national response to climate change. The *America's Climate Choices* (ACC) panel report *Adapting to the Impacts of Climate Change* (NRC, 2010a) makes important points about decision makers and their information needs for adaptation to climate change. We have worked closely with that panel in considering the functions of climate services. The primary goal in this chapter is to identify the functions that must be part of effective climate services, building on previous reports, and institutional considerations based on currently available services from different agencies (see "Potential Functions of Climate Services" in this chapter). Among the key functions of climate services highlighted are the following:

- A user-centered focus which responds to the decision making needs of government and other actors at national, regional, and local scales;

- Research on user needs and skills, effective information delivery mechanisms, and response options;
- Development and timely delivery of credible, authoritative information and products to decision makers at multiple scales (e.g., local, state, regional, national, and global) about how climate is changing (e.g., observations), how it may change in the future under different socioeconomic scenarios and policy decisions (e.g., climate model projections at multiple time scales), and information on current and projected impacts of climate change;
- Collection and integration of information to support national, sectoral, and regional impact and vulnerability assessments and adaptation planning, including socioeconomic and environmental trends and projections;
- A system for sharing strategies and options for adaptation and providing useful decision support tools across a range of regional and time scales;
- A comprehensive web interface to facilitate access to information and products; and
- An international information component.

Successful climate services require an institutional design involving multiple agencies that includes strong research components (e.g., in climate science, vulnerability analysis, decision support, and communication), operational activities (e.g., communication and delivery of decision relevant information and assessments), and ongoing evaluation to ensure response to user needs and new science at national and regional scales. As discussed in this chapter, successful and effective climate services need

- Leadership and coordination at a high level to ensure focused engagement of relevant federal agencies;
- Responsiveness to user needs, including the ability to make scientific information understandable and useful;
- Reliable observations and modeling that provide decision-relevant information at the space and time scales of decision making;
- The ability to support and incorporate research that delivers improved information, assessments, and decision tools;
- Provision of information on an equitable basis to all decision makers, including citizens, communities, states, sectors, and tribes;
- Adequate capacity for the development and delivery of climate information; and
- The provision and support of relevant international information in support of decision making by U.S. stakeholders and the international community.

THE NEED FOR CLIMATE SERVICES

The need for climate information as well as the utility of climate information through-out societal decisions is expanding worldwide. The basis for U.S. national climate services is well established and dates back to the National Climate Service Act of 1978, when Congress recognized that the nation's "ability to anticipate natural and man-induced changes in climate would contribute to the soundness of policy decisions in the public and private sectors" and that "information regarding climate is not being fully disseminated or used, and Federal efforts have given insufficient attention to as-sessing and applying this information."[3]

Table 5.1 provides an overview of the types of decisions made by different stakehold-ers that might be informed by climate services with a focus on the provision of infor-mation about climate, impacts and adaptation in the Unites States. Further examples of climate information needs can be found in the boxes scattered throughout this report and in the companion ACC reports (NRC, 2010a; b).

Recently, the World Climate Conference 3 (WCC-3) in September 2009 agreed to a Global Framework of Climate Services (GFCS) as a concept to be undertaken by the world's nations. This concept called for major strengthening of the essential elements of a global framework for climate services, including

- the Global Climate Observing System and all its components and associated activities with provision of free and unrestricted exchange and access to cli-mate data;
- the World Climate Research Programme, underpinned by adequate comput-ing resources and increased interaction with other global climate-relevant research initiatives;
- climate services information systems taking advantage of enhanced existing national and international climate service arrangements in the delivery of products, including sector-oriented information to support adaptation activi-ties; and
- mechanisms for climate users and producers to interact, building linkages and integrating information, at all levels, between the providers and users of cli-mate services; and efficient and enduring capacity building programs, includ-ing education, training, and strengthened outreach and communication.

The sharing of data and expert knowledge on the global and regional climate through the GFCS would be a benefit for U.S. climate service activities for both adaptation and

---

[3] P.L. 95-367.

TABLE 5.1 Information Needs Provided by Climate Services

Decisions to Respond to Climate Change	Example Information and Analysis
Federal Government	
Setting targets for emission reductions to avoid dangerous climate change	Baseline emission trends and carbon cycle analysis; modeling the climate impacts of alternative targets and timetables
Prioritizing federal investments in adaptation (wide range of agencies, especially those managing or supporting water, agriculture, ecosystems, health, transport, and emergency management)	Regional vulnerabilities and scenarios for climate change; observations of how climate is changing, sea level rise, storm surges, coastal inundation
Targeting international development and disaster relief (e.g., State, Defense, U.S. Agency for International Development) and responding to human rights and migration concerns	International vulnerabilities, climate trends and scenarios, existing adaptations, seasonal forecasts, satellite remote sensing, and field reports of population movements and humanitarian crises
Forest management: What resources will be needed for fire response?	Seasonal outlooks, longer-term climate change scenarios, fire-climate-drought-pest modeling
Public health: Are patterns of disease likely to change as a result of climate?	Seasonal and longer-term climate projections and impacts on major disease vectors and vulnerabilities
State and Local Governments	
Planning: Are changes needed in environmental and land use regulation to reduce the risks of climate change and facilitate adaptation? Should infrastructure or people be relocated?	Regional analysis of vulnerability and possible climate changes including temperature, precipitation and sea level, water and energy utilization
Private Sector	
Agricultural producers: What to produce and how much to invest in insurance, water and other inputs	Seasonal forecasts for drought and other climate conditions (in both the United States and for international competitors); information on likely pest and disease outbreaks, commodity futures; advice and assistance with longer-term adaptation strategies

continued

TABLE 5.1 Continued

Decisions to Respond to Climate Change	Example Information and Analysis
Tourist industry: How will climate variability and change affect revenues and longer-term investments in facilities?	Seasonal and longer-term forecasts of temperature, precipitation, snow, storms, and sea level rise
Energy and utilities: How will climate variability affect supply and demand for energy? What weather or climate derivatives will help manage risks? Will climate change influence longer-term investments and siting decisions?	Seasonal forecasts of heating and cooling degree days, severe weather; longer-term scenarios of climate impacts on water availability, wind, solar energy, and hydropower
Urban planning: Should building codes and land use controls be changed or implemented to reduce the risks of climate change?	Changing risks of storms and sea level rise in relation to both climate and vulnerability resulting from socioeconomic changes; changing public perceptions of risks
Finance sector: How does climate change alter insurance exposure and the long-term viability of firms?	Changing climate risks to firms and sectors
Retail sector: How will climate affect supply of products and demand from consumers? Should we change sourcing and marketing in response?	Seasonal forecasts, changes in regional climate
Conservation organizations: Does climate change require rethinking conservation plans and the location of protected areas?	Observations and scenarios of marine and terrestrial ecosystem change in response to climate

emission reduction strategies. Equitable access to these data and expert analysis are critical for the success of such an entity.

An informed and effective response to climate change requires comprehensive, authoritative, and useful climate information on current and future climate change, climate impacts, the nature of extremes and vulnerabilities, and response options, including how adaptation and emissions reductions interact to reduce risks. This information needs to be made available to the widest possible range of people and organizations. Climate services have the potential to sustain the application of current and future climate information for government, industry, and individuals. To address decision makers' needs, scientific data must be presented at the appropriate geographic

---

BOX 5.1
Information Needs for the U.S. Navy's Arctic Roadmap

The U.S. Navy recognizes the importance of a positive and active presence in the Arctic maritime environment, for Arctic security and stability, especially in a climate change regime where these regions are experiencing significant and rapid changes (see TFCC/Oceanographer of the Navy, 2009). The capacity to anticipate and manage future changes in this environment and the associated impacts is crucial for shaping future naval missions, maintaining appropriate infrastructure, and advancing strategic opportunities (Commander Gallaudet, personal communication, 2009).

In November 2009, the Navy developed a suite of objectives and action items in the Arctic region known as the Navy Arctic Roadmap. It emphasizes the need for accurate, timely, and useful information on the changing Arctic environment. The Roadmap calls for a number of desired effects, including increased partnerships with interagency and international stakeholders, an active contribution to Arctic safety and stability, the capability to meet combatant commander requirements, and an understanding of and ability to anticipate access for Arctic shipping and other maritime activity.

The ability to ensure that these effects are achieved is dependent upon reliable data and information. Among the many scientific and technological needs of the Navy, model resolution, uncertainty management, and model physics have been identified as three priority information needs required to adequately address climate change in this region. Model resolution must be increased to include a higher regional scale spatial resolution and decadal scale temporal resolution. The Intergovernmental Panel on Climate Change (IPCC) model scenario resolution is insufficient for the decisions that need to be made. Improving information on processes that are not yet well understood, such ice melt and sea level rise, is also essential. It is equally important to address variability across a range of spatial and temporal scales (Commander Gallaudet, personal communication, 2009).

In order to facilitate the success of the identified objectives, the Roadmap lists a number of action items related to the advancement of environmental assessment and prediction. It includes methods for increasing the amount and quality of data collected and calls for additional scientific operations, such as the deployment of unmanned systems for monitoring and research. The Roadmap also supports additional observations, mapping, and modeling to improve capabilities in the Arctic. Through increased partnerships and frequent assessments and evaluations, the Roadmap provides a comprehensive strategic plan to address Arctic-specific needs.

---

scale and time scale to aid effective decisions and actions (see, for example, Boxes 5.1, 5.2, and 5.3).

Decision makers are now expecting and demanding up-to-date reliable climate information for them to integrate into management decisions. Today, we have the weather forecasts provided by the National Weather Service (NWS), and climate projections

---

BOX 5.2
Information Needs of a Transportation Official

Rising sea levels, storms surges, and land subsidence will likely lead to the greatest impacts on transportation systems (NRC, 2008a). The coastal transportation official is faced with a variety of decisions driven by the threat of these impacts on roadways and infrastructure and must seek information to carry out effective responses. Transportation officials in coastal states must be able to identify vulnerable areas and understand the linkages between sea level rise and other problems, including erosion, flooding, and damage to infrastructure. Planning is typically on 10 to 30 year time scales, so model projections of how much sea level will rise will need to be modified accordingly to be useful to the transportation official. In addition, transportation planners and managers in some regions will have to prepare for more intense precipitation events associated with warmer climates and increased water vapor in the atmosphere. Today, many of these information needs are not currently being met. There is a lack of access to the types of information needed, in a format that can be readily understood and interpreted by coastal and transportation managers, and this hinders effective decision making. Transportation officials can coordinate with agency officials, climate scientists, and other transportation officials to facilitate the exchange of information and best practices, as well as the development of methods to maintain adequate infrastructure. This information can be used to implement coastal protection measures and adaptation strategies that are physically and economically feasible.

---

on 100-year time scales. Neither addresses climate information at seasonal to decadal timescales in an authoritative and fully operational manner necessary for many societal decisions. Decision makers need information tailored to their particular needs, communicated clearly, and accompanied by decision support tools that allow the exploration of alternative risks and pathways, local priorities, and flexible responses to new information. Dissemination of climate information, however, can often be inadequate to serve user needs because it is either delayed or not at the right spatial scale. This is due in part to the many participants involved, as well as unclarified institutional roles inhibiting the timely dissemination of climate information. Through stakeholder engagement, climate services can foster the integration of climate information into planning efforts at the local, state, and federal agency levels and help develop management strategies to deal with socioeconomic consequences of climate change and variability. Furthermore, it should be noted that climate information goes beyond pure climate science. It includes social and economic sciences, as well. For example, in the realm of vulnerability assessment, there is a great need for better coordination of datasets that go beyond census data to better reflect the structure of local economies, measures of resilience at the community level, and the type of ecosystem services that may be impacted.

Climate services are already provided in various forms by the NOAA Regional Integrated Sciences and Assessments (RISA) program (see Box 5.7), NOAA's regional climate centers, NOAA's Climate Prediction Center and National Climatic Data Center, private consultants, state climatologists, and the NWS. Other USGCRP agencies are also providing climate-related services such as USDA's Soil Conservation Service, USGS's river and soil moisture monitoring, the Environmental Protection Agency's Climate Ready Estuaries Program (CREP), NASA's satellites and application programs, and others. Many states have state climatologists who also provide services. Climate services (provided though regional groups) have been meeting user needs by providing climate information to improve planning, risk management, resource allocation, impacts assessment, and adaptation and emission reduction strategies.

However, in the United States there is no coordinated authoritative, credible, and useful source of information and products on the complete range of climate change, impacts, vulnerabilities, and response options. For effective decision making, it is necessary not only to continue to make the best and most comprehensive scientific observations, but also to improve significantly the integration of the information into the decision making process. A wide range of public and private entities in sectors such as transportation, insurance, energy, water, fisheries, and agriculture are increasingly demanding and incorporating climate information into their planning. The range of decision makers responding to climate change are motivated to promote sustainability, protect property, and make long-term investments to promote the economy (see Chapter 2 for further discussion). These demands and activities demonstrate the need for a permanent and clearly identifiable national climate service that can coordinate and integrate climate information to develop products and tools; provide access to comprehensive, up-to-date reliable information on current and future climate change, variability, and risks; and provide response options to inform decisions ranging from adaption and emission polices to education and communication initiatives.

National and regional assessments are a key component in guiding national decisions about responding to climate change, including assessments of the climate implications of alternative emission reduction policies and adaptation needs. The report, *Restructuring Federal Climate Change Research to Meet the Challenges of Climate Change* (NRC, 2009d), recommends the USGCRP to "initiate a national assessment process with broad stakeholder participation to determine the risks and costs of climate change impacts on the United States and to evaluate options for responding." Climate services, involving regional partnerships, have the potential to provide the valuable informa-

BOX 5.3
Information Needs of a Fisheries Manager

A fisheries manager must consider a range of impacts due to a changing climate. These include rising seawater temperatures, sea level rise, increased storm activity, ocean acidification, and increased saltwater intrusion on traditionally freshwater areas (GAO, 2007). An example of useable scientific information for fisheries managers is given in the figure below, which illustrates the effects of sea surface temperatures on fish egg distributions in El Niño versus La Niña years. Many aquatic species are adapted to specific conditions and even modest changes or shifts in those conditions could have a negative effect on fisheries resources and productivity. Fisheries managers must have access to information on the effects of climate change in conjunction with other environmental stressors to ensure the sustainability of their fisheries. Thus, integrated information is critical for decision making. Information on annual to interannual timescales is crucial for this type of decision making, especially when needing to make a decision on annual fishing quotas in certain regions. However, fisheries managers are unable to use climate models or scientific information on the scales that are necessary. For example, active monitoring of water levels, salinity, fauna, and vegetation is needed to reduce model uncertainties (GAO, 2007). These resource managers must work with federal agencies to address the various long-term planning challenges associated with the protection and maintenance of fisheries and ecosystems. This includes an evaluation of new technologies and regulations that may affect fisheries management. Furthermore, it is important to analyze how climate change impacts and subsequent adaptations might reverberate across borders and catchment areas. This might be facilitated by bolstering international and regional management regimes and agreements that help regulate fishery rights and synthesize information (World Bank, 2010).

tion needed in the national assessment process. The contribution of a national climate service in this case would be providing the leadership to work collaboratively with state and local governments to collect and store that kind of information.

Climate services can also identify gaps in observation systems and contribute to providing adequate coverage of the United States and other regions of strategic interest, including early warnings of abrupt changes as well as an effective on-demand climate modeling system that can provide timely answers about the impacts of alternative emission paths on global and regional climates. Advances in observations, data integration, and thoughtfully tailored dissemination of climate information provide a foundation for development of an effective national climate service. The development of systems and standards to deliver near real time products to meet national, regional,

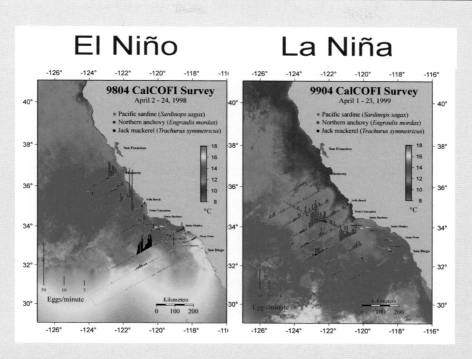

An example of the type of climate information that could be valuable to a fisheries manager. This figure illustrates fish egg distributions from state-federal California Cooperative Oceanic Fisheries Investigations cruises in 1998 (an El Niño year) and 1999 (a La Niña year) layered over sea surface temperature satellite imagery. SOURCE: Rich Charter, NMFS/SWFS.

and state needs are essential. For example, the National Integrated Drought Information System has provided a wealth of information to aid decision makers and this information is derived through a coordinated effort of federal agencies, led by NOAA, state and local governments, and non-governmental interests. The effort has even been extended across the borders to produce a North American Drought Monitoring capability.

Climate services nationwide could fulfill the rising demand for information to inform adaptation to the impacts of climate change. The timeliness of delivery of credible information is a key issue and the development of new technologies will provide opportunities for rapid and cost-effective dissemination of climate information.

POTENTIAL FUNCTIONS OF CLIMATE SERVICES

The overarching goal of climate services is to provide the essential information on climate conditions, variability, and change needed for effective decision making. Climate services need to ensure and sustain a core infrastructure to support products, tools, and services for informing responses to climate change. To meet the nation's climate information needs, climate services could have four essential functions:

1. Engaging users,
2. Central and accessible information,
3. Use and dissemination of international climate information and response options, and
4. Research, observations, and modeling.

Engaging Users

Working with and Listening to Communities, States, Sectors, Regions, Tribes, and Other Stakeholders

The ultimate product of climate services is the service component. Through stakeholder engagement, climate services can foster the integration of climate information into planning efforts at the local, state, and federal agency levels, as well as in the private sector (Figure 5.1). It can help different users develop management strategies to deal with socioeconomic consequences of climate change and variability. The engagement strategy needs to include ways to entrain, leverage, and expand existing operational capacity (including the NOAA RISA programs; science translation capacity within universities, including the Cooperative Extension Programs; natural resources management non-governmental organizations and a variety of private sector interests, and local and regional jurisdictions and interest groups). This interface needs to be managed on an ongoing basis to ensure that there is a focus on answering the right questions, two-way communication, and ongoing assessment of progress (in terms of both outcomes and process). There is a need for resources and incentives for independent actors to support the climate services on the ground. A climate service can also establish impact and vulnerability assessment methods to advise planning efforts of local, state, and federal agencies.

FIGURE 5.1 Illustrative examples of climate service experts engaging various stakeholders in education, outreach, and two-way learning. (top left) Agricultural engineers Daren Harmel and Clarence Richardson inspect soil cracks caused by severe drought to determine the effects on crop production. SOURCE: USDA, photo by Scott Bauer, National Wildlife Refuge. (top right) Manager Jill Terp explains how creating defensible space between the San Diego refuge lands and bordering residential areas can prevent potential spread of wild fire to private homes. SOURCE: U.S. Fish and Wildlife Service, photo by Scott Flaherty. (bottom right) National Weather Service recognizes San Lorenzo, Puerto Rico, as a StormReady® community. SOURCE: NWS, photo by NWS San Juan. (bottom left)). Lt. Sarah (Jones) Duncan gives a climate lecture at the community college in the Federated States of Micronesia. SOURCE: NWS Pacific ENSO Applications Climate Center.

Capacity Building and Training for Linking Knowledge to Action

Because there are a limited number of people qualified to communicate science in ways that are useful for specific policy applications, a deliberate effort is needed to expand the community of people who can tailor science information for specific applications. This can involve, for example, government, university, industry partnerships, as well as training programs for scientists, resource managers, and elected officials. Current experience suggests that, while there are good models for the necessary scientific basis (e.g., IPCC WGI) and for some timely delivery of services (e.g., NIDIS, RISAs, USDA Extension), there is a shortage of intermediaries who can help connect science with decision making in specific sectors (water, transportation, energy, fisheries, and natural resource management). Elements of such expertise can be found in areas and organizations not traditionally involved in climate services such as non-governmental environmental organizations, communications, management, education, extension, and the social sciences (including organizational behavior and governance). Building an effective climate service will require a new class of expertise in abundant quantities and one with sufficient knowledge of both scientific uncertainties and the risks that need to be addressed. For example, NOAA's Sea Grant program has very few climate specialists on staff to meet the needs of stakeholders.

Central and Accessible Information

Information at the Time and Space Scales that Decisions Are Made

An important component of providing services will be building a system that provides answers at the scale of decisions (e.g., reservoir operations at the watershed scale). Resource managers across the board are frustrated that climate model projections are at such a large scale that they have little utility for actual decision making. Although "downscaling" efforts are being initiated, they are far from answering policy-relevant questions fully.

Indicators of Relevance to Decision Making

An effective climate service provides information that is relevant to everyday decision making and provides early warnings of changes at local and regional scales. This means providing information that goes beyond historical and projected average temperature and precipitation changes but translates these into meaningful indicators such as moisture availability and heat stress and highlights critical information about changes in extreme events, thresholds, or rapid changes in climate and its impacts. Decision makers can also benefit from information about what others like them are doing in terms of providing and collecting information on climate and adaptation and such a clearinghouse function will be helpful, especially if it includes comparative information and indicators on, for example, the costs of information and adaptation and the effectiveness of programs.

A Central Portal for Information

New information systems, especially web-based platforms, should be fully utilized to ensure that information is accessible and understandable to all users. Our judgment is that most current agency web sites that provide climate information are poorly designed from a user perspective. An integrated web interface for the climate service should be an absolute priority. It should be interactive and focused on themes relevant to users (rather than institutional organization) and should include access to a variety of decision tools. A central portal can also provide information that is timely, relevant, and credible at a range of time and space scales. Better access to information for the wide array of climate-related decisions will be expensive. For instance, in order to provide the tools that local, regional, state, tribal, and sectoral decision makers need, major investments in information technology or "cyberinfrastructure" are required by government agencies and the private sector. In many cases, providing better information and decision tools over the internet and more useful ways to manipulate and visualize data will provide a path forward. Significant progress is being made along these lines in the context of the National Integrated Drought Information System (NIDIS) and at NOAA's National Climate Data Center, but even these systems lack cutting-edge, user-friendly, interactive web portals. A diverse group of users needs to find climate information easily without negotiating agency-specific sites or dealing with complex technical language. During this study, NOAA established a prototype climate services portal (*http://www.climate.gov*) with a commitment to gather user feedback and serve a broad range of users. The panel welcomes this development but observes that the portal is thus far focused on climate observations and shorter-term predictions with little information on decision tools, vulnerability, adaptation, longer-term future climate change, or information and services available from other federal or local agencies.

Timely Delivery of Climate Information

A climate service could increase agency capacity to provide near real time climate data, provide access to climate data from observations to archives, and integrate those data into planning and management at multiple government levels. A climate service could also provide valuable up-to-date information for assessments at regular intervals, with a clear articulation of the confidence levels and uncertainties; disseminate products and services in a timely manner; and provide policy-relevant inputs into decision-making processes.

Use and Dissemination of International Climate
Information and Response Options

A climate service can also serve as a national knowledge-sharing network to exchange information and share best practices between different levels of government on effective actions.

International Information

A national climate service can also have the capability to tap into international information to aid decisions at home. This can include international information on climate change, impacts, vulnerability, and adaptation options of use to U.S. stakeholders. Agricultural producers provide a clear example of stakeholders whose planning and livelihoods are affected by climate impacts and vulnerabilities elsewhere as price signals move through international markets. The United States also plays a very important role in the support of international observing systems and climate assessments and this support should be continued for an effective national and international response to climate change (also see Chapter 7).

Figure 5.2 illustrates the complex set of information that needs to be provided for an effective climate service, including the wide range of non-climatic information that is needed for effective decision support (e.g., vulnerability information, adaptation options, and decision tools). The whole system is underpinned by ongoing research and driven by stakeholder needs.

Research, Observations, Modeling

Ongoing Research Programs to Support the Services

Climate services should be underpinned by the best available science. This requires a well-supported national program of observations, monitoring, analysis, and modeling to provide information on how climate is varying and changing at global, national, and regional scales, with clearly articulated knowledge of uncertainties. To develop and improve decision support tools, to support impact and vulnerability assessments, and to identify best practices for communicating information, research is also needed in the social and economic sciences (including management information systems, communication, and planning). This includes research on user needs, effective information delivery mechanisms, and processes for sustained interaction with multiple stakeholders.

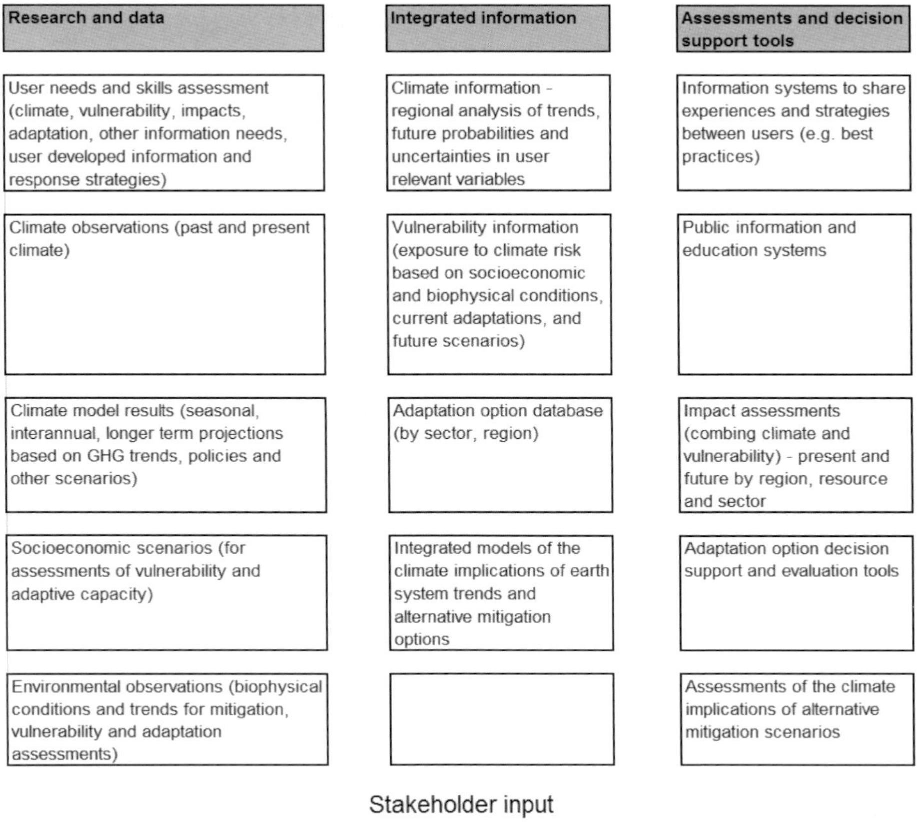

Research and data	Integrated information	Assessments and decision support tools
User needs and skills assessment (climate, vulnerability, impacts, adaptation, other information needs, user developed information and response strategies)	Climate information - regional analysis of trends, future probabilities and uncertainties in user relevant variables	Information systems to share experiences and strategies between users (e.g. best practices)
Climate observations (past and present climate)	Vulnerability information (exposure to climate risk based on socioeconomic and biophysical conditions, current adaptations, and future scenarios)	Public information and education systems
Climate model results (seasonal, interannual, longer term projections based on GHG trends, policies and other scenarios)	Adaptation option database (by sector, region)	Impact assessments (combing climate and vulnerability) - present and future by region, resource and sector
Socioeconomic scenarios (for assessments of vulnerability and adaptive capacity)	Integrated models of the climate implications of earth system trends and alternative mitigation options	Adaptation option decision support and evaluation tools
Environmental observations (biophysical conditions and trends for mitigation, vulnerability and adaptation assessments)		Assessments of the climate implications of alternative mitigation scenarios

Stakeholder input

FIGURE 5.2 Examples of the types of information to be provided by climate services.

Components of a climate service research endeavor should be linked to ongoing efforts of the USGCRP. The climate service could provide information to the USGCRP on new and emerging needs of stakeholders to help guide research or modeling priorities (Box 5.4).

Observations

Despite many calls for more focus on public engagement, there is a major disconnect between adaptation actions in regions and sectors and the types of monitoring that are currently under way. A more strategic design of a monitoring program could focus on answering important management questions and detecting trends in real time. Although the United States has made great progress in remote sensing of climate data,

---

BOX 5.4
Climate Services and U.S. Global Change Research Program

The USGCRP plays an invaluable role in providing basic climate science research and observations to support climate services. Concerns have also been expressed about whether the USGCRP itself is well configured for coordinating and providing information that can inform decisions and actions. The program tilts heavily toward basic science, as opposed to the development of scientific findings that would address the needs of decision makers, or even in some cases influence science policy itself (Dilling, 2007; Pielke, 2000a,b; Sarewitz and Pielke, 2007). Current strategic plans continue to place heavy emphasis on understanding basic climate processes and, to a lesser degree, the impacts of such processes on societies and ecosystems (see, e.g., CCSP and the Subcommittee on Global Change Research, 2008). Several National Research Council (NRC) reports have demonstrated the need for the USGCRP to focus some research on societal decisions through, among other things, human dimensions observational systems, or downscaling models for more reliable regional climate data (NRC, 2009d; see also *Advancing the Science of Climate Change*, NRC, 2010b).

---

the capacity to measure parameters on the ground or calibrate satellite sensors is inadequate in many fields (NRC, 2009e). For example, the breakdown of the USGS stream gauge program has hampered effective decisions at a time when gauge information is critical. In addition, more snow monitoring sites (especially at high elevations) and soil moisture measurements in certain regions would better inform the decisions related to climate change. A climate service can help identify critical gaps in the national observation system. Chapter 7 addresses the critical role of international observing systems. Of particular concern are adequate observations and data that would aid in assessing the effectiveness of adaptation strategies. There are many state and federal data collection programs that could be augmented by direct linkage to weather and climate events during the data collection process. These are not necessarily onerous new observing systems; they could be simple changes in protocol. For example, by adding to a database on whether a road was closed due to an environmental factor (e.g., high water, snow, etc.) one could more easily assess whether an urban or regional adaptation transportation plan was having a positive impact. Other examples include power outage statistics along with weather attributes, or the number of days with water restrictions during droughts and heat waves.

On-Demand Modeling in Support of Decision Making

A climate service needs the capacity to assess the climate implications of different policy options in order to progressively inform adaptation and greenhouse gas emissions
questions at the national level. The panel envisions that a climate service can serve to
inform federal policy makers through modeling to test and monitor the implications
and effectiveness of national climate policies that reduce greenhouse gas emissions
or foster adaptation. A national climate service cannot do on-demand modeling with
accompanying services for every type of user but given federal investment in climate
modeling the panel judges that the federally funded modeling enterprise could be
more responsive in informing federal choices about alternative mitigation paths
and their climate implications. However, the topic on the stabilization of the climate
system with the projections of future climate impacts has not been systematically addressed in the United States, and investigation of national policy choices will need to
use state-of-the-art climate models. This could involve global climate models, regional
scale models for specific aspects of the climate change problem (e.g., regional change,
hurricane predictions), and integrated assessment models that hone in on the socioeconomic impacts. As the nation begins to look increasingly at the envelope of information to be deployed for national emissions reduction policy options, a consortium
of the modeling centers needs to be coordinated to run the on-demand scenarios for
understanding and anticipating future climate change and impacts. The coordination of modeling centers will ensure that federal policy makers have steady access to
the advancing science in a consistent and reliable manner. On-demand modeling to
aid federal policy makers requires a smooth flow of information from the scientific
research and impacts assessments. Routine assessments of the models and the value
of the resulting impacts analyses will be required by the climate services framework so
that the information for decisions is sustained at the state-of-the-art level and therefore represents the best available knowledge.

Recently, the Department of Energy (DOE), in conjunction with the USDA and the
National Science Foundation (NSF), announced the launch of a joint research program
to produce high-resolution models for predicting climate change and its resulting
impacts, thereby helping decision makers develop better adaptation strategies to address climate change.

RECOMMENDATIONS FROM PRIOR REPORTS

Several reports have discussed the need for climate services and principles for their
design. Some of the functions and criteria that have been proposed are summarized

---

BOX 5.5
National Climate Services: Summary of Needed Functions and Criteria
Identified in Previous Studies

Stakeholder Engagement

- *A user-centric approach* should be taken for all activities and information, where users range from the individual in the public and private sectors to the international community.
- *Regional and long-term partnerships* are needed within communities as climate products will have a regional focus when considering impacts, vulnerabilities, and climate conditions.
- *Continuous evaluations and assessments* by both users and providers are needed on the relevance and quality of the data and climate products, as well as the risks and vulnerabilities in a changing climate regime.
- *Education and outreach* are important for information exchange to the public in order to improve climate literacy and to users as their needs evolve.
- *Participation is needed from government* (interagency partnerships must exist across federal, state, and local levels), *business, and academia* (interdisciplinary expertise from universities includes physical, natural and social sciences, as well as engineering and law) with clear central (federal agency) leadership that includes a source of sufficient funding.
- *Empowerment of existing successful adaptation efforts* is a clear way to move forward relatively quickly with establishment of services that are embedded in communities, regions, and sectors.

Observation Systems

- *Wide ranges of spatial* (local, state, regional, tribal, national, and international) *and time scales* and at diverse locales must be represented by observations.

---

in Box 5.5. For instance, *A Climate Services Vision: First Steps toward the Future* (NRC, 2001) identified the growing demand and considerable value in climate information ranging from extended outlooks and seasonal to interannual forecasts used in water and energy management to decadal and century scale climate scenarios for different concentrations of greenhouse gases. The NRC 2001 report identified the following five guiding principles for climate services:

- user-centric,
- supported by active research,
- include predictive and historical information on a variety of time and space scales,

- *Existing observational networks* can be expanded or combined in an effort to provide an overarching structure for a global observing system.
- *Natural variability* on seasonal and interannual to decadal timescales needs to be understood and monitored.

Research

- *Basic and applied research* (mission-oriented and scientifically credible) should represent what stakeholders need to manage their resources and regional vulnerabilities as well as what scientists view as necessary to understand coupled climate resource systems.
- *Operational delivery systems* are needed to transition from research to useful products and predictive capabilities to serve stakeholder needs and for effective decision making.
- *Instrumentation and technology* including new engineering and communications techniques are needed to support increasing stakeholder research interests in a changing climate system.
- *Comprehensive databases and archives* to manage data relevant to stakeholder needs should be maintained.

Modeling and Analysis

- *Models for decision support* are needed to inform various social, economic, and environmental decisions and to promote environmental stewardship and sustainability.
- *Forecasts on various time and space scales* to serve national needs should include analysis on probabilities, limitations, and uncertainties.
- *Analysis and interpretation of model results* are needed at appropriate spatial scales and may include regional or "downscaled" information.

- have active stewardship of the knowledge base, and
- have active and well-defined participation by government, business, and academia.

The report recommended an inventory and integration of existing observation systems and data, incentives for new systems at local levels, creation of user-centric functions in agencies and experimental partnerships, better delivery of research including interdisciplinary studies that include societal impacts and model results for long-term projections including ensembles and uncertainties, expansion of services to new sectors and data products, development of regional enterprises to address societal needs, and improved formal and public climate education. It noted the potential contributions of NOAA and the network of state climatologists, other federal agencies, and the private sector to overall climate services.

More recent NRC reports (2009a,d) also address the question of climate services and recommend that federal efforts be coordinated to provide climate services to decision makers and maintain strong links to the U.S climate change research program (NRC, 2009d), and that any form of national climate service should conform to principles of effective decision support.

A workshop organized in 2008 by the nine existing RISA centers (Overpeck et al., 2009) discussed their common and different experiences and drew conclusions from these for the design of an effective National Climate Service (NCS) in the United States. The lessons they drew included the following:

- A NCS must be stakeholder (user) driven, and accountable to stakeholders.
- A NCS must be based on sustained regional interactions with stakeholders.
- A NCS must include efforts to improve climate literacy, particularly at the regional scale.
- Multi-faceted assessment as an ongoing, iterative process is essential to a NCS.
- A NCS must recognize that stakeholder decisions need climate information in an interdisciplinary context that is much broader than just climate.
- A NCS must be based on effective interagency partnership—no agency is equipped to do it all.
- Implementation of a NCS must be national, but the primary focus must be regional, at the level where decisions are made.
- NCS capability must span a range of space and time scales, including both climate variability and climate change.
- A NCS design should be flexible and evolutionary and be built around effective federal-university partnerships.
- NCS success requires that an effective regional, national, and international climate science enterprise, including ongoing observations, model simulations, and diagnostics, exists to support it.

In addition to the above efforts, the panel also considered the recommendations of a report of the Climate Working Group (CWG) of the NOAA Science Advisory Board (NOAA SAB, 2009) which identifies four options (which could be combined in different ways) for developing a NCS:

- The Climate Service Federation of federal agencies and regional groups of climate information providers which would drive a national organization responsible for climate observing systems, modeling, and research. Options could include something like the USGCRP or a non-profit or chartered corporation.

- Non-Profit National Climate Service (similar to UCAR) sponsored by the federal government.
- NOAA as the lead agency with specified partners.
- Expand the NWS to include a Climate Service through merges with other components of NOAA.

The NOAA Science Advisory Board CWG report echoes others in stating that a climate service should promote interactions between users, researchers, and information providers, be user-centric, and provide useable information and decision support tools based on a sustained network of observations, modeling, research, and user outreach. The report recommends:

- the internal reorganization of NOAA with the objective to better connect weather and climate functions, research, operations, and users, but identifying NOAA as the logical lead agency;
- clearly defined roles for federal agencies as well as for state and local governments and the private and public sectors;
- leadership at the highest level, preferably within the White House; and
- a large dedicated budget.

The CWG also suggests metrics for success that include measurable impacts in terms of increased public understanding of climate and climate impacts, benefits to society, and improvements to decision making as well as outputs that can be shown to be accessible, credible, and useful to a broad range of regions and sectors, engaging a diverse community of users, and assessed by stakeholders as useful in making decisions. Process and input metrics would include high level leadership and authority, clear strategic planning and priorities, peer review, a strong research basis and infrastructure/financing, and robust observation and modeling systems.

INSTITUTIONAL CONSIDERATIONS

The reports noted above vary in scope for activities addressed by climate services. For example, some of the reports discuss information about climate change but not about vulnerabilities and response options. The climate service that this panel has in mind is broader in scope. Federal agencies such as NOAA, NASA, NSF, and United States Geological Survey (USGS) have taken responsibility for data and information on climate observations and climate projections. However, a much wider range of agencies and groups collect the environmental and socioeconomic data needed for vulnerability analysis and provide climate information to the public and decision makers. Other components of successful climate services, such as user needs and skills assessment, socioeconomic scenarios, adaptation information and option evaluation, and tools for sharing information among stakeholders are often poorly developed as operational activities with the exception of some regional pilot activities such as the RISAs.

The panel's vision for climate services is broad and thus one agency alone cannot perform all of the needed functions. Rather, the nation needs climate information and services based on partnerships involving federal to local levels, all appropriate agencies, the academic community, and the private sector. It is important for the credibility and functionality of a climate service that research questions are driven in part by the users. **A successful climate service should provide two-way communication and embrace learning from decision makers.**

There are many organizational, political, and technical challenges in designing climate services. For a variety of historical, political, and functional reasons, the United States is unlikely to establish a free-standing climate service agency in the near future. As previously noted, some of the information that is needed is already being offered by federal agencies, extension services, regional and state activities, and research and pilot activities based in universities and NGOs. There are also regional, state, and sectoral partnerships offering pieces of the climate services puzzle, although they are not well coordinated, have inadequate funding, and often lack high level vision and leadership.

To be successful, climate services need to engage existing institutions that have a track record of providing climate information (on both physical climate and impacts) to a wide range of stakeholders. In the past, these institutions have mainly provided information about current climate variability and extremes, and seasonal forecasts, rather than information on longer-term trends and impacts, and information on abrupt changes. In addition, there is little capacity at the non-federal level for providing intraseasonal to interannual climate predictions that can aid decision makers in planning. Long-term, steadily increasing global warming and accompanying climate extremes (e.g., heat waves) "forced" by greenhouse gas emissions have introduced a new element into information needs for many stakeholders. **A climate service could provide climate information on multiple timescales, from intraseasonal to century scale, to inform decision making and actions.**

Various federal agencies have expertise in environmental information and have established long-term relationships of trust with their stakeholders. The design of climate services could build on this capacity while recognizing the need that some new functions and expertise are needed. For example, Seattle relies on several federal agency monitoring and forecasting services to help inform their decision-making, some of which include

- USGS stream gauges;
- Natural Resources Conservation Service (NRCS) SnoTel sites; and
- NOAA NWS weather observations and daily and midrange weather forecasts, the Climate Prediction Center's 30 to 90 day and multi-seasonal climate outlooks, and remote sensing of snow cover.

To date, NOAA has taken a lead in providing climate information and in research on impacts; agencies such as NASA, the Environmental Protection Agency (EPA), USGS, USDA, and NSF have also funded important efforts in data collection, impact assessment, programs to deliver information to users, and climate modeling. Agencies with missions relating to specific sectors (e.g., USGS and the U.S. Bureau for Reclamation for water and ecosystems, the U.S. Fish and Wildlife Service for land and marine ecosystems, the U.S. Forest Service for forests, the U.S. National Park Service for parks, USDA for agriculture, and the U.S. Department of Health Service for health) have already initiated programs to collect information on climate impacts and assess impacts of climate change on the resources they manage. The Department of the Interior (DOI) has initiated an interagency program to establish regional climate centers, and several other departments and agencies have built outreach systems that focus on interactions with stakeholders at the regional level. The Agricultural Extension Services and the NRCS of USDA provide useful models for what might be needed for climate services given their engagement with stakeholders (Box 5.6).

No single agency currently has the capacity to collect and analyze the full range of information needed to assess vulnerability to climate change or plan for adaptation or has a consistent and coordinated approach to communicating climate change to stakeholders. There are also agencies that do not have sufficient expertise or resources for responding to climate change within their areas of responsibility (e.g., the Department of Housing and Urban Development (HUD), the Department of Transportation (DOT), and the U.S. Agency for International Development (USAID).

Coordination among federal agencies, state and local governments, and the private sector is fundamental to a successful climate service. The panel acknowledges the challenge of coordinating information among multiple federal agencies and regions, but such coordination is essential if climate services are to be perceived as reliable and to avoid confusion and duplication. Effective coordination requires strong leadership and a budget to support coordination efforts. A climate service design needs to be able to provide information for ongoing assessments, and adaptation and emission reduction efforts. This is especially important in the assessment of vulnerabil-

BOX 5.6
USDA'S Cooperative State Research, Education, and Extensions Service

The Smith-Lever Act of 1914 established a system of cooperative extension services connected to the land-grant universities. The purpose was to move information quickly and effectively from universities to farmers. Each U.S. state and territory established a state office at its land-grant university and a network of local or regional offices. The U.S. Cooperative Extension system employs nearly 15,000 people and already offers some climate change information to farmers. To be able to address adaptation, while providing an efficient interface between policy makers and local communities, extension services will need to be strengthened substantially. Options and strategies are already enhanced through interaction with local insights. Communication among extension experts and local communities is important. Individuals and communities want to learn (1) about linkages between individual actions and environmental impact and (2) how behavioral changes can mitigate those impacts.

Extension programs also focus on training agricultural extension agents to equip them with knowledge and tools to accurately translate climate information to advise farmers. One example program in the Northeast focuses on financial opportunities, illustrated by the program's tagline, "Promoting Practical and Profitable Responses." Because farmers are concerned with their bottom line, this framing engages them in a way that a strictly "environmental" approach might not. Some presentations focus on climate change's specific agricultural impacts: weeds, insects, pathogens, and heat stress. Because farmers deal with these issues on a daily basis, connecting climate change to these concerns makes the information relevant and useful. However, information alone is not enough. The interaction between extension agents and farmers is a critical component in the program to ensure that farmers are using up-to-date reliable information specific to their needs. The extension model could be especially helpful in conveying climate information because it uses trusted agents that farmers already consult for assistance. Hearing information that is specifically relevant to them from people they already trust may make farmers more likely to adopt mitigation and adaptation strategies than they would otherwise.

Steps have been taken toward a land-grant and sea-grant climate extension service, including

1. Joint development of Extension Professional of all types,
2. Collaboration of the Association of Public and Land-Grant Universities (APLU) Extension Committee on Organization and Policy (ECOP) and Sea Grant's Assembly of Sea-Grant Extension Program Leaders,
3. Joint Advisory Board for Interagency Stakeholder Input, and
4. Inventory of climate extension services across the country.

ity and options for adaptation, where information on biophysical and socioeconomic vulnerability and the decisions about responding to climate change are needed. In the absence of information provided by federal agencies about regional and local impacts, states and cities, conservation groups, and corporations have been collecting information on vulnerability, commissioning regional downscaling of climate scenarios, and using this information to development adaptation plans (see *Adapting to the Impacts of Climate Change*, NRC, 2010a). There is an emerging national interest in downscaling climate models to the regional level and there are numerous "bottom-up" activities, supported by groups such as RISAs.

What is powerful about this convergence of national and local interests is the potential to develop and validate regional scale information derived from the climate models. However, there needs to be careful attention to ensuring that there are realistic expectations about the degree of certainty of such downscaling activities. Leadership will be needed to carefully bridge the gaps between science and decision making in this area to make climate projections relevant to decisions at multiple levels.

Any effort to establish a climate service should build on, enhance, and avoid unnecessary damage to state and local efforts. The design of climate services needs to carefully articulate the division of labor between agencies and departments on research, operations, and evaluation. To respond to new science and emerging user needs, climate services should include or have access to a dynamic research component that can translate information into forms that are useful for decision makers. A service needs to establish metrics for robust evaluations and progress for operational activities.

Climate services need a process for incorporating the needs and views of stakeholders and for training personnel—especially field personnel who will help assess climate change and plan for adaptation in specific regions and sectors. The Land Grant agricultural extension systems (Box 5.6, or the related Sea Grant and Space Grant) provide some models for informing and training across broad regions and they are starting to incorporate consideration of climate change, including prototype climate extension components. These existing networks can be a tremendous resource for mobilizing the climate service in regions and sectors where there might otherwise be inadequate workforce capacity.

Climate services will also need to provide information about the climate effects of various emission reduction efforts and polices. To date, the U.S. contributions to the IPCC, and the USGCRP Synthesis and Assessment Reports, have focused on analyzing alternative emission scenarios and stabilization scenarios (e.g., at 450 ppm), rather than on the consequences of actual decisions and options (e.g., international commitments to reduce greenhouse gases, proposals to limit emissions within the United States). Climate services can link research to national and regional decisions about emissions reductions by providing comprehensive information on how the climate system would change as a result of emission reduction decisions at the national and international scale (see *Limiting the Magnitude of Climate Change,* NRC, 2010d). Understanding the interaction of overall trends with national and international policies to limit climate change is essential to informing decisions about responding to climate change.

An important consideration is whether climate services should provide services and information only about climate change, its impacts, and implications for adaptation, or if they should also provide information on emission reduction strategies. The panel believes that both kinds of information and services are needed. Even though many decisions already seek to manage emissions reductions and adaptation together, there should be a division of labor that provides focused services and information. We believe that climate services should provide services and information about current and future climate change and its impacts, vulnerability, and response options. Response options would be focused on adaptation responses but recognizing that some adaptation options also reduce emissions. The research provided through climate services is relevant to decisions about emission reduction strategies because it can clarify the effects of emission reduction policies and thus help decision makers set goals. However, decisions about how to limit greenhouse gas emissions will need other kinds of information, such as how to achieve those goals and about the effectiveness and costs of various technological options. These kinds of information should be developed and provided from other sources (see Chapter 6). **The implementation of a climate service should be explicit about how to link information services that support both adaptation and emission reduction strategies.**

Finally, climate services will need to be designed to adapt to regional needs. Experience with regional programs such as the RISAs and interagency conversations at the regional level suggest that user needs can vary considerably between regions. For example, in the western states, the convergence of climate change with other stresses, including land use change and increasing water demands is driving new demands as climate change impacts become more evident and interest grows in regionally downscaled information to undertake the planning. The RISAs in the West have been approached both jointly and separately by coalitions of regionally based agencies of USDA and the DOI for training in planning for climate change and are also working with tribal groups concerned about climate issues. In the southeastern states, user interest is driven especially by agriculture and concerns about sea level rise and storms. The latest observations of climate change in different regions, including the rapid changes in the Arctic, suggest hot spots for climate change (IPCC, 2007b) that have immediate demands for region-specific information. **A climate service needs to be responsive to the various regional demands and be adaptive as those needs change over time.**

Several reports and legislative proposals have offered various institutional designs for a formal NCS. For example, H.R. 2407, the National Climate Service Act of 2009, sets out a process whereby the executive branch, led by the Office of Science and Technology Policy (OSTP), creates a NCS, spells out how coordination is to be achieved, establishes an interdepartmental oversight board, establishes an external advisory committee with both federal and non-federal membership, establishes a quality assurance program, and delivers periodic reports to Congress. The Act largely, but not completely, separates the NCS from a single agency and formally gives the leadership function to the director of OSTP. However, a Central Operations Office, responsible for day-to-day administration of the proposed NCS, is to be placed in NOAA, but "operated as a cross agency priority by the Administrator." Special emphasis is placed on including regional centers of activity. Another draft National Climate Service Act proposes the establishment of a NCS within the Department of Commerce with NOAA as the lead agency. This proposed approach would include a national office and a network of "regional climate service enterprises" to produce climate information and products guided by an advisory council and coordinated through a climate services board. This proposal suggests that NOAA will be responsible for the delivery of climate observations, model results, and for overall coordination, but that the majority of the support to stakeholders will be provided by regional centers (selected through a competitive process) that will consist of collaborative arrangements between the NWS, other regional offices of federal agencies, RISAs, and other public and private sector climate service groups.

These proposals, together with significant discussion with the community, outline important elements of a national climate service, including a focus on user needs, interagency coordination, and support for regionally based activities. However, they focus primarily on the provision of information about climate change with less attention to climate modeling capability, which is needed to support key decisions about emission reductions, anticipate the onset of abrupt climate change, and understand interactions with other stresses.

Based on its analysis, the panel identifies three critical concerns regarding institutional requirements for successful climate services:

1. **Leadership at the highest level.** Making decisions related to climate change is daunting and will involve many people at various scales. This requires leadership at the highest level of government to coordinate agencies, manage risks associated with multiple spatial scales, confront the increased frequen-

cies and intensities of extreme events, and make recommendations on potential tradeoffs between emissions reductions and adaptation. With a credible
information stream from climate services, decision makers would be able to
make more balanced policy decisions even when faced with uncertainty. Good
leadership is critical and must be encouraged.

2. **Adequate funding and independent budget authority.** Because addressing climate risk requires building decision support infrastructure (e.g., training programs, data access systems, monitoring and assessment capacity, etc.),
   it does not lend itself well to an *ad hoc* funding source that is based on the
   goodwill of individual decision makers within the multiple federal science
   agencies. There needs to be significant, centralized coordination with budget
   authority to ensure that structural support is built and that outcomes are delivered. Priority setting should be based on risk and vulnerability (among other
   considerations) and be apolitical. Every sector and every region has needs, but
   not all will be met. Some user demands may be met successfully only after
   years of research. There should be clear milestones and periodic reviews to
   ensure that the work is progressing in the right direction and that there is a
   sustained commitment to the high priority elements.

3. **Coordination and engagement of federal agencies.** Although the roles of
   the various federal agencies in climate services have not been finalized, NOAA
   is likely to play a central role and has been identified as the potential lead
   agency in some reports. Building from previous reports, the panels' judgment
   is that NOAA cannot create an effective climate service on its own because it
   currently lacks comprehensive capacity and expertise in key functions (e.g.,
   vulnerability assessment and assessing user needs). To develop these functions in house would be costly and would duplicate some functions already
   available at the regional level and in other agencies. Incentives are necessary
   to encourage agencies to work together toward common climate service
   goals. There is no time and no money for turf battles over the components of
   this system. Making effective decisions related to climate change will require a
   variety of innovative partnerships with local and regional entities and universities, as well as functional partnerships between federal agencies. In addition, a
   climate service should serve as a clearinghouse for information produced and
   resolve any differences in information between agencies.

The case studies in Boxes 5.7 and 5.8 provide important examples of how elements of
climate services can be designed and implemented—the RISA program of the United
States and the U.K. Climate Impacts Programme (UKCIP) from Europe—and draw some
lessons from these cases.

GOALS FOR CLIMATE SERVICES OPERATION

Climate services need to have a clear set of principles to guide products and services and to ensure that they remain appropriately focused and are managed effectively. Any climate service should be an "honest broker" providing scientifically credible information with clarity and should be committed to a user-centric approach and scientific rigor. Its work and product development should be transparent and thoroughly vetted. All aspects of the observations, research, modeling, data management, and delivery need to be grounded in sound science and include sustained collaborations with various key partners (including non-federal governments, academia, and the private sector). It is important for information providers (scientists, federal agencies, etc.) and information users (farmers, resource managers, etc.) to build up mutual trust to balance information needs with the long lead time needed for research. This trust is essential for the team that actually delivers the service to stakeholders, thereby becoming a member of a community in which learning goes both ways. Time and collaborative work then merge to provide a valuable addition to the functional tool kit of the team delivering the service in the form of vetting information which may be suspect for a variety of reasons. A climate service can also demonstrate how science can be relevant to iterative decision processes by providing new information to incorporate into decisions (see Chapter 3).

The panel generally endorses the decision support principles set forth in previous NRC reports (1999, 2001, 2008, 2009a,d). The panel elaborates on four key principles for climate services:

1. **User-centric problem definition.** To provide the most effective services, there should be an ongoing effort and dialogue to identify the key decisions where climate information is needed by users and to frame at least some portion of the federal research program around those decisions and information needs. Basic understanding of the climate system and its interactions with humanity is still needed (i.e., social and economic science research), but increased emphasis on decision-relevant research questions is needed (see *Advancing the Science of Climate Chang*, NRC, 2010b) while maintaining research efforts that are likely to have future implications for the environment.
2. **Credibility of information.** Much is riding on the decisions associated with climate predictions, in some cases billions of dollars in infrastructure investments; in other cases, these decisions may make or break a family or a business. As discussed in Chapters 1 and 8, users need to trust the source of information. This, in turn, calls for testing the skill of tools that have been provided to users. In addition, rigorous scientific assessments at regular intervals are

---

BOX 5.7
Regional Climate Services: Lessons from the
Regional Integrated Sciences Assessments (RISA) Program

In February 1995, NOAA's Office of Global Programs (OGP), funded a pilot program in the Pacific Northwest, the Climate Impacts Group (CIG) based at the University of Washington, to link national and global developments in climate science to real decisions and decision makers at the regional spatial scale. This was to be accomplished by linking climate science, especially the advances in seasonal forecasting such as El Niño Southern Oscillation (ENSO) events up to a year ahead of time and the ability to downscale global climate model results from IPCC scenarios, to a specific place. The program was designed to focus on communities of stakeholders and their climate science and decision support needs. With modest funding, the CIG focused on regional hydrology and water resources management, forest ecosystems, aquatic ecosystems (both marine and terrestrial) including salmon, and the coastal zone. These were selected because they were among the most climate sensitive socioeconomic sectors in the Pacific Northwest, defined as encompassing Washington, Oregon, and Idaho (a large portion of the Columbia River Basin).

In 1997, building on the initial success of CIG, stakeholder-oriented research, and regional stakeholder workshops associated with the U.S. National Assessment, NOAA established the RISA program. Additional initiatives in the Southwest (Climate Assessment for the Southwest [CLIMAS], University of Arizona) and the Southeast (University of Florida and Florida State University) were established, focusing on ENSO impacts on agriculture, and subsequently expanded to include Georgia and Alabama as the South East Climate Consortium (SECC).

The focus was to derive societal benefits from the application of advances in climate science. By 2008, the RISA program had grown to nine regions and included a wide range of sectors. The RISA program, with its focus on *place-based, stakeholder-driven research, partnership,* and *services,* created an effective demonstration-scale climate service for parts of the nation. The experience of listening

---

necessary to ensure that user demands can be met by reliable and authoritative information, with a proper measure of scientific confidence, and characterization of uncertainty that is meaningful to decision makers. Trust also comes from partnership between information producers (e.g., scientists) and information users (e.g., policy makers).

3. **Adaptive management and performance evaluation.** Climate services need to encourage learning from past mistakes and successes and be responsive to new information. There is enormous value in continuous assessment and evaluation of climate services at national and regional levels, with regional evaluations providing a critical "finger on the pulse" of user needs and responses to climate services. New management infrastructure and information systems should be designed to incorporate changing climate conditions (both

to stakeholders, partnering with them on research, and developing decision support tools and other products is especially valuable. RISAs couple and integrate national efforts to provide global observations, research, and modeling with regional scale needs. They serve on the front lines in support of regionally based state and local agencies, NGOs, the private sector, and the public, all of whom must become climate literate and plan for adaptation and emissions reductions on a multi-decadal timescale. However, the RISAs are dependent on the larger national institutions, such as NOAA and NSF, who fund research and climate modeling and communicate climate information.

The RISA programs have shown that a critical element of the regional focus is the intense, sustained contact with users that is necessary to uncover, assess, and refine the ways in which climate services can best meet user needs. Because the research that is undertaken at this scale is largely, though not completely, determined in an interactive process with stakeholders, the activities of these units often break new ground and are, therefore, a continuing source of innovation (Kennel, 2009). The RISA process has, in several cases, driven new scientific discoveries through responding to stakeholder interests, including the discovery of the Pacific Decadal Oscillation (PDO) (Mantua et al., 1997). Successes have also been achieved in the application of seasonal forecasts to water supply, in drought prediction, planning, monitoring, and assessment. Collaboration among RISAs in the western United States has produced a major addition to U.S. capabilities to cope with drought hazards in the creation of NIDIS. RISAs have also been in the forefront of significant innovations in decision tools for managers of water supply systems, wildland fire management, and agriculture. For example, SECC and CLIMAS have made major advances in grafting a climate focus onto the traditional agricultural extension functions, including the development of decision tools and the creation of climate extension positions.

The RISAs are a relatively small and experimental program, but now there is a growing demand for the creation of such teams in regions where they do not currently exist. They have built up strong stakeholder constituencies and expertise in translating science, doing impact research, and working with regional offices of federal agencies such as USGS, USDA, and NOAA. The RISAs provide a model of the functions required for the regional component of a national climate service.

changes and variations in the physical climate and changes in the political climate), including using new communications techniques (e.g., cutting-edge informatics) that recognize non-stationarity in the climate system and the decision-making environment. It should also be responsive to changing user needs and socioeconomic contexts.

4. **Environmental justice and equal access to information.** The impacts of climate change are often unequally distributed, especially because of the differential vulnerabilities of regions, sectors, and social groups. Research has shown that the impacts of climate variability and change can fall disproportionately on poor, elderly, and minority populations (e.g., Arctic indigenous people and island nations) and that these groups may also lack access to climate information and adaptation options. Environmental justice considerations have

BOX 5.8
The U.K. Climate Impacts Programme

Several countries have established national climate services, including the United Kingdom, Australia, and Germany. The U.K. Climate Impacts Programme (UKCIP) was established in 1997 to help organizations assess how a changing climate will affect them and help them prepare to adapt to climate change by providing climate impacts information. UKCIP was initially funded by the U.K. government's Department of Environment (DEFRA) but is increasingly supported directly or in kind by other government agencies and NGOs.

The UKCIP operates as a grant funded entity located at a university but with deliverables strongly guided by detailed contracts with government. It relies heavily on the United Kingdom's climate analysis and modeling capability of the government funded Hadley Centre. More than a decade of experience with climate services in the United Kingdom provides several lessons of relevance to the United States, including the importance and cost-effectiveness of serious engagement with stakeholders (who co-produce many UKCIP reports), the challenges and rewards of reaching some sectors, the communicative value of a risk management approach, and the importance of sustained investment in both climate data and modeling, as well as in expertise beyond basic science for impact assessment, vulnerability analysis, communication, training, and adaptation planning within the climate service. Valuable decision support tools include climate and socioeconomic scenarios (including probabilistic climate ensembles); online tools for estimating costs, identifying adaptation options, and sharing best practices; and training experts to deliver information and tools to their local regions or organizations.

The UKCIP experience illustrates the value of climate services as a function that is seen as independent of the agency that collects climate information and runs climate models (the U.K. Meteorological Office). For example, UKCIP is, to some extent, able to distance itself from public skepticism about inaccurate weather forecasts and to work with the full range of national and local government agencies without being "owned" by any one agency. In particular, the funding base and partners for UKCIP broadens to include other government departments, regional and local government, the private sector, and NGOs. UKCIP has also taken a lead in adaptation planning (see *Adapting to the Impacts of Climate Change*, NRC, 2010a).

Other countries are developing climate services, taking into account the experiences of UKCIP. In Australia since 2007, climate information and tools for adaptation have been encompassed within a Department of Climate Change, which is charged to deliver information to decision makers for managing climate risks, especially through an adaptation and land management division. In Germany, the new Climate Service Center is intended to become the platform for inquiries and information about climate change in Germany and includes both natural and social science expertise and the goal of establishing networks. In all three international cases, the core staff is relatively small (20-50 people) and relies heavily on partners and on national climate observations, data, and modeling capability.

become a formal concern of agencies such as EPA, with specific programs and funding for disadvantaged groups and minority populations, avenues for legal actions, advocacy training, collaborative solution processes, and other actions. In developing a service, safeguards that ensure equal treatment of economically distressed and minority communities and that address the special concerns of tribes must be an overarching principle. There are also justice concerns regarding timing and access to information in relation to the role of climate information in futures markets, where a climate service must balance the private value of climate information with the public good (NRC, 1999).

METRICS FOR EVALUATING PERFORMANCE OF CLIMATE SERVICES

Among the most important metrics for evaluating the performance of climate services (Miles et al., 2006), the panel believes the most important are the following:

1. Responsiveness to user needs as measured by regular input from stakeholders and advisory boards, by feedback on the climate service portal, and by evidence that information and decision tools are actually being used in decision making and are improving climate literacy among users.
2. Use of the best available science as measured by timely integration of new observations, model results, and analysis of the climate system and associated social, ecological, and economic impacts and vulnerabilities.
3. Delivery of annual regional and sectoral assessments that provide user-relevant and scientifically based information on how the climate is changing, the latest projections for future change and vulnerabilities given policy alternatives, the current and potential impacts on regions and key sectors, and the progress and potential for adaptation and greenhouse gas emissions reductions.
4. Evidence of effective collaboration between agencies and other actors including (a) funding being appropriately balanced between national and regional activities, natural and social sciences, research, translation, and operations; (b) joint production of information, reports, and assessment; and (c) development of a single portal for stakeholders and the public.

CONCLUSIONS AND RECOMMENDATIONS

An informed response to climate change requires that the widest possible range of decision makers—public and private, national and local—have access to reliable information about current and future climate changes, the impacts of such changes, the vulnerability of different regions, the vulnerability of sectors and groups, and the options for reducing risks or adapting to them. Decision makers need information tailored to their particular needs, communicated clearly, and accompanied by decision support tools that allow the exploration of alternatives, emphasize local priorities, and encourage flexible responses.

Climate services can meet user needs by providing climate information to improve planning, risk management, resource allocation, impacts assessment, adaptation, and emission reduction strategies. In this chapter we have provided guidance on potential functions, institutional considerations, principles for operation, and performance metrics for climate services, taking into account previous reports and ongoing proposals. The panel's assessment is that current proposals include important elements of a service, but key functions may be overlooked in the attempt to base the system on existing federal capabilities. No single government agency or centralized unit can perform all the functions required by climate service. Therefore, coordination of agency roles and regional activities is a necessity for effective climate services.

A major barrier to providing climate services is the lack of clear federal roles which has stalled the implementation of information delivery systems. Federal roles should be clarified to recognize the respective missions, strengths, and limitations. Aligning the roles of federal departments and agencies for successful climate services will require coordination and very clear leadership.

A core service function is the timely delivery of user-relevant climate information. The development of new technologies will provide opportunities for rapid and cost-effective dissemination of climate information. Making effective decisions related to climate change will require a variety of innovative partnerships with local and regional entities and universities, as well as functional partnerships between federal agencies.

Current regional initiatives, such as RISAs, Sea Grant and Land Grant programs, and regional climate centers, provide important models illustrating how to interact with stakeholders and provide relevant climate information. NOAA, USDA, and other agencies with regional centers could advance the climate service idea by increasing operational support for existing regional centers and establishing partnerships with other federal agencies to implement the nationwide system.

Recommendation 6:

The nation needs to establish a coordinated system of climate services that involves multiple agencies and regional expertise, is responsive to user needs, has rigorous scientific underpinnings (in climate research, vulnerability analysis,

decision support, and communication), performs operational activities (timely delivery of relevant information and assessments), can be used for ongoing evaluation of climate change and climate decisions, and has an easily accessible information portal that facilitates coordination of data among agencies and a dialogue between information users and providers.

CHAPTER SIX

Informing Greenhouse Gas Management

S ome of the most important decisions on climate change concern the emissions of greenhouse gases (e.g., CO_2, CH_4, N_2O) and other greenhouse warming agents (e.g., tropospheric ozone, black carbon). High quality information on greenhouse gas emissions from multiple sources and at multiple scales is needed to detect trends, verify claims about reducing emissions, develop policies to manage carbon, and inform citizens. The importance of measurement, reporting, and verification of emissions (MRV) emerged as a key negotiating issue for the United States in the 2009 United Nations Climate Change Conference. Both public and private organizations report information on greenhouse gas emissions, often using standards and methods geared toward a specific application (e.g., regulation, carbon trading, or national obligations to the United Nations Framework Convention on Climate Change [UN-FCCC]). The resulting plethora of carbon information systems has created confusion for consumers, businesses, and policy makers and threatens to undermine the legitimacy of responses (Winkler, 2008). This chapter examines procedures used to measure and report emissions for greenhouse gas registries and energy efficiency and recommends ways to improve these procedures to better inform decisions to limit emissions in the United States. Limiting the magnitude of climate change, as noted in the *Limiting* and *Advancing* reports (NRC, 2010d;b), involves more than managing greenhouse gas emissions and can also include ecological strategies, such as changing albedo through land use, and may eventually involve geoengineering solutions to alter radiation or sequester carbon.

GREENHOUSE GAS ACCOUNTING SYSTEMS

Figure 6.1 provides a conceptual diagram of the processes and mechanisms needed to inform decisions on greenhouse gas management. Key elements of this information system are:

- The scientific underpinning
Processes:
- Monitoring
- Reporting protocols

- Verification

Mechanisms:

- Inventories and registries
- Carbon offsets

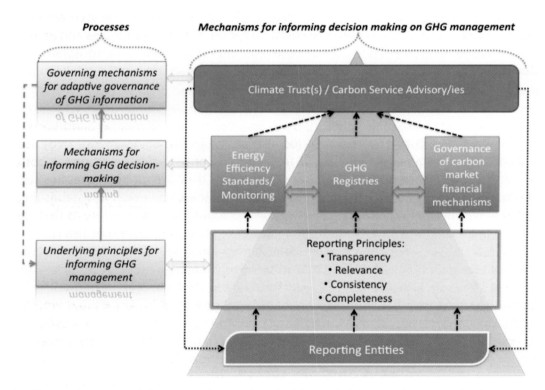

FIGURE 6.1 A conceptual diagram to illustrate the principles underlying greenhouse gas (GHG) reporting and accounting, mechanisms for reporting emissions information, and governance structures for GHG information systems. Black dashed arrows represent the transfer of emissions information collected by various entities (e.g., companies, local governments, and non-governmental organizations) for different purposes (energy efficiency, registries, and carbon markets) to overarching administrative bodies. The overarching bodies (created at state or federal levels) provide feedback and assistance on data collection (black dotted arrows). The blue dashed arrow illustrates the adaptive governance approach needed to respond to changing conditions and circumstances. The red triangle illustrates the increasing usefulness of GHG emissions information. This heuristic diagram is not meant to represent all the linkages between components of the GHG management chain.

Scientific Underpinning

An effective greenhouse gas (GHG) accounting system has a strong scientific basis for measuring, comparing, and verifying emissions. Most systems rely on methods developed by the Intergovernmental Panel on Climate Change (IPCC) and on reporting requirements of the United Nations Framework Convention on Climate Change (UNFCCC), which include six greenhouse gases (CO_2, CH_4, N_2O, SF_6, perfluorocarbons [PFCs], ad hydrofluorocarbons [HFCs]), converted into carbon dioxide equivalents using their global warming potentials. However, methods are continually evolving as more is learned about greenhouse gases. There are ongoing debates about whether other warming agents (e.g., SO_2 or black carbon) should be included, the accuracy of global warming potential estimates, accounting for sinks, and technologies and analytical tools for monitoring and estimating emissions. For example, the atmospheric lifetime of black carbon is short relative to other greenhouse gases and its global warming potential (along with other non-Kyoto greenhouse gases) is substantial. As a result, the IPCC regularly adjusts how it estimates GHG emissions. Thus, an ongoing U.S. research program on GHG monitoring—ranging from satellite and ground-based monitoring to analytical and modeling tools for estimating emissions from energy use and production data—is essential for an informed response to climate change at both national and international scales. The panel recognizes that greenhouse gases are not the only important feedback to climate change, and research is critical for establishing emission reductions targets and exploring policy options. For example, forests that are planted in northern temperate latitudes to facilitate carbon sequestration also reduce the albedo, thereby absorbing and transferring more energy to the atmosphere. Although the net effects of albedo and carbon sequestration are not yet certain, planting of forests at latitudes that support occasional snow cover is likely to cause climate warming, whereas a similar amount of forest growth at lower latitudes would have a clear cooling effect on climate. Changes in land use can therefore be a useful component of monitoring, reporting and verification in a system that assesses overall changes in the earth system that cause or limit climate change.

In the following three sections, monitoring, reporting protocols, and verification, collectively termed MRV, are discussed and the commonly accepted processes to manage greenhouse gas emissions are presented. In the next two sections, inventories and registries and carbon offsets are discussed, respectively, and example mechanisms are provided that serve the functions of MRV.

Monitoring

Most greenhouse gas emission estimates are based on calculations rather than direct measurements, although some large facilities have installed direct or continuous emission measurement systems (CEMS), especially for carbon dioxide and industrial gases with high global warming potential. Estimates are usually based on activity data (e.g., fossil fuel use) or other quantitative measures (e.g., waste volume) that can be converted into emissions using standard emission factors (DEFRA and DECC, 2009; IPCC, 2006). An example of such an estimate for global carbon emissions is shown in Figure 6.2.

Activity data are taken at a wide range of scales, from individual facilities (or even households for utilities) to entire states (e.g., gasoline use). The IPCC provides default emission factors for various source categories in four sectors (IPCC, 2006), but use of

FIGURE 6.2 Deutsche Bank, in collaboration with scientists at the Massachusetts Institute of Technology, created the Carbon Counter, a 20-meter billboard displayed in New York City's Madison Square Garden, to demonstrate to the public a running estimate of increasing global greenhouse gas emissions to the atmosphere. SOURCE: Deutsche Bank.

country-specific activity data and emission factors yields more accurate GHG estimates at the national level (Garg et al., 2006; Lowe et al., 2000).

The growing importance of carbon trading and regulation demands regional- and
facility-level estimates of greenhouse gas emissions, which in turn requires the development of local and regional emission factors. Protocols for estimating greenhouse
gases at smaller spatial scales are being developed. For example, the Environmental
Protection Agency's (EPA's) proposed rules for mandatory reporting of greenhouse
gases (EPA, 2009b) cover major emitters and facilities that are critical to monitor,
such as power plants. The International Local Government Greenhouse Gas Protocol
helps local governments quantify greenhouse gases emitted both from their internal
operations and from communities within their geopolitical boundaries. Sector specific
emissions standards are also being implemented by a number of industries and by
several state and local governments in the United States (see Table 6.1). Sector specific
methodologies are especially useful when emissions result from complex processes or
for industries that emit (e.g., aluminum, cement industries) or take up (sequester) high
levels of greenhouse gases.

Although direct measurement is expensive, the costs of facility-level reporting are
likely to come down as emissions monitoring becomes more automated and emissions management is incorporated into daily corporate practice (DEFRA and DECC,
2009). The reporting priority for any national system should be high intensity emission
sectors such as the electric power sector (which represents about 41 percent of total
U.S. emissions; EIA, 2009). In these sectors, the federal government can choose to mandate metering or continuous emissions monitoring systems for those facilities above a
minimum threshold.

Estimating greenhouse gas emissions carries a considerable degree of uncertainty
because of measurement error and model assumptions, and IPCC and other guidance recommend the analysis and reporting of this uncertainty. Sensitivity and error
analysis can be used to estimate variance resulting from uncertain evidence (such as
estimates on activity or emissions data) and poorly understood mechanisms (such
as particularly complex processes/relationships with emissions to the atmosphere;
Saltelli et al., 2000) and used as a quality assurance tool. EPA and other federal guidelines for documenting uncertainty are less comprehensive than needed for an effective national monitoring system. This is especially the case for complex sectors, such
as forestry, where clearly defined and consistently applied methodologies are still in
development, and different standards allow for different functional carbon reductions.
(For a review of U.S. state and voluntary initiatives, see Pearson et al. [2008]) Although
direct measurement of emissions is costly, it is likely to decrease as carbon trading and

TABLE 6.1 Examples of Top-Down, Sector Specific Methodologies for GHG Emission Calculations

Sector	Protocol/Supplier	Main Principles/Rationale for Sector Specific
Power	Power Generation/Electric Utility Reporting Protocol (CCAR)	Based on the California Climate Action Registry (CCAR) General Reporting Protocol (GRP). Complete (all gases, all facilities). Must be transparent, verified, and accurate on reporting. Power sector has large, specific GHG implications.
Cement	Cement Reporting Protocol based on GHG Protocol (CCAR, WRI GHG Protocol, Cement Sustainability Initiative [CSI])	Provides additional guidance on determining emissions from calcination in cement manufacturing process.
Forests	Forest Sector Protocol (FSP) (CCAR)	Includes non-biological (e.g., fossil fuels from forest machinery) and biological emissions. The GRP provides for non-biological emissions, while the FSP provides for forest biomass (i.e., biological) emissions.
Local government	Local Government Operations (LGO) protocol (CCAR, CARB, ICLEI)	Covers all operational aspects of local governments including transit and vehicle fleets, power generation, port and airport facilities, water and waste, buildings and fugitive emissions. Comprehensive approach to multifaceted institutional emissions.
Manufacturing refrigeration and air conditioning units	Direct HFC and PFC emissions resulting from commercial refrigeration and air conditioning (EPA Climate Leaders, based on WRI GHG Protocol)	Covers refrigeration, all gases not covered by the Clean Air Act (i.e., CFCs and HCFCs) including operating commercial equipment, service, disposal, and retrofit emissions for high global warming potential (GWP) gases.
Iron and steel production	Direct emissions from steel and iron production (EPA Climate Leaders, based on WRI GHG Protocol)	Identify and estimate GHG emissions (CO_2) from oxidation of reducing agent and flux in steel production and from removal of carbon from iron ore (separate from combustion).

BOX 6.1
Prices and Their Informational Content

The need to "put a price on carbon" arises from the fact that a market economy—if it is to work—has no way of dealing with essential commodities in the absence of a price signal to consumers. That is what a market economy is for—prices inform choices.

The *information content* of prices refers to the completeness and hence the correctness of market prices. Prices cannot provide appropriate signals to consumers if those prices do not convey information about the economic processes to which those prices pertain. As an illustration, there is now widespread agreement that the price of a gallon of gasoline does not reflect the full costs of highway congestion, exhaust pollution (smog), or the carbon emissions that contribute to global climate change. Similarly, the cost for a ton of coal does not account for either the environmental costs to mountains, waterways, and the atmosphere or the public health costs associated with inhaled pollutants.

In other words, if the full social costs of congestion delays, air pollution, and greenhouse gas emissions were correctly accounted for in the price of gasoline, consumers would almost certainly make different choices about automobile use. A price premium on energy reflecting the carbon content of petroleum products would help reflect these costs and would, as a result, convey more correct market signals to consumers.

---

regulation place higher prices on CO_2 and its equivalents and demand more rigorous reporting (Box 6.1).

Scientific observations and models provide a means to independently verify greenhouse gas emission estimates and thus to assess compliance with carbon management policy (Michalak, 2008; WRI, 2009). An analysis of the capabilities of these methods appears in NRC (2010e). For example, NASA's Orbiting Carbon Observatory, which failed on launch in February 2009, would have been able to monitor a sample of large local CO_2 sources, such as cities and power plants, over its 2-year mission lifetime (NRC, 2009c, 2010e).

As noted earlier, greenhouse gas emissions are only one component of earth system processes that maintain the planet at a livable climate. Policy making can benefit from integrated assessments that include, for example, all components of the carbon cycle or radiative processes, including land use, albedo, clouds, and aerosols, so that greenhouse gas emissions can be understood in the context of other important factors that affect the climate.

Reporting Protocols

There are a variety of protocols for reporting GHGs. Most protocols for reporting greenhouse gases are designed to achieve accuracy, transparency, completeness, and consistency (Penman et al., 2000). Accuracy is usually a function of the rigor of monitoring, estimation, and uncertainty. Transparency is an indication of the documentation, audit, and publication of information. Consistency is both internal to allow tracking over time and external to allow comparison with other reporting entities. Wide variations in where systems set their baseline, boundaries, scope, and thresholds for the reported emissions affect both consistency and completeness. There are advantages and disadvantages to a single scheme versus sector specific schemes. For example, the former allows for a better picture of the whole system, while the latter may be more useful for the specific needs of individual decision makers.

Baseline estimates of GHG emissions are usually established through a political negotiation in which countries and firms may attempt to secure baselines that provide favorable positions in a regulatory or trading system. A high baseline may provide more generous emissions allowances in a trading system with permit allocation based on historic emissions (as for some firms in the European trading scheme) or immediate emissions savings where current emissions have already fallen below the baseline (as was the case for Eastern Europe, the United Kingdom, and Germany entering Kyoto). In the United States, varying baselines partly reflect the patchwork of different reporting systems which have emerged in the absence of a mandatory national system (see Chapter 2).

The boundaries for reporting include the gases, sectors, reporting entity, and the geographic scale. Most GHG systems require the six UNFCCC gases, although some (e.g., WWF Climate Savers) include only CO_2 (WBCSD and WRI, 2005). Sectors that are difficult to measure are sometimes excluded from reporting systems, especially those associated with land use or landfills. Such exclusions can create significant gaps in the information needed for regional carbon management. Geographic boundaries are important to avoid double counting of emissions reporting and to ensure full coverage of emissions (Rypdal and Winiwarter, 2001), although emissions resulting from international activities (imports and exports, shipping, and aviation) can be difficult to assign. The boundaries are often chosen for the application—national for reporting under the UNFCCC or voluntary corporate accounting, facility, or state level for regional trading schemes.

For corporate level reporting, decisions must be made about whether to define responsibility by facility, national corporate entity, subsidiary, or even multinational scale and whether to define it by equity share, financial, or operational control (WRI, 2008, 2009). For example, the California Climate Action Registry requires reporting of CO_2 for the first 3 years (and thereafter the six UNFCCC gases) at the corporate level and allows participants to choose whether to report California or U.S. operations. The EU ETS reports CO_2 at the facility level within all member states. A flexible GHG reporting scheme for the United States is likely to require the collection of information on a wide range of greenhouse gases (including tropospheric ozone and black carbon) at multiple scales to enable crediting at the firm, state, and national scales. Many sectoral sources are covered under the EPA Mandatory Reporting of Greenhouse Gases proposed rule (EPA, 2009b). In principle, such methodologies can assist in cost-effective harmonization policies, create a more level playing field by providing standardized tools for all companies within a sector, build expertise on GHG calculation and reporting within a sector, and provide specific guidance on GHG-related issues in sectors with more complex industrial (e.g., cement), biological (e.g., forests), or institutional (e.g., local governments) emissions profiles.

The scope of emissions in a reporting system accounts not only for direct emissions (those from sources owned or controlled by the reporting entity) but also for indirect emissions (those that result from the activities of an entity but occur at sources owned or controlled by others; see Box 6.2). The California Climate Action registry requires reporting of direct emissions and indirect emissions associated with the generation of electricity, heat, and steam, whereas the EU ETS requires reporting of only direct emissions. Accounting for both direct and indirect emissions enables the most comprehensive carbon management.

The threshold for emissions reporting is chosen to balance cost considerations with the need to effectively manage the maximum amount of GHG emissions (Stolaroff et al., 2009). A threshold may allow reporting entities to exclude small sources that are expensive to monitor or difficult to estimate. For example, the California Climate Action Registry and the EU ETS allow entities to exclude 5 percent of emissions (WBCSD and WRI, 2005) and the proposed EPA mandatory system sets an economy-wide threshold for corporate reporting above 25,000 tons of CO_2. However, if there are a large number of small sources, a high threshold may create a significant material discrepancy in an overall emissions reporting system.

Verification

Assurance of the accuracy of emissions reporting is important within a carbon reduction system that will necessarily create winners and losers, the possibility for rule-

---

BOX 6.2

The World Resources Institute/World Business Council for Sustainable Development's Greenhouse Gas Protocol

The World Resources Institute/World Business Council for Sustainable Development's (WRI/WBCSD) GHG Protocol for Corporate Accounting and Reporting is the most commonly accepted global standard for the corporate accounting of greenhouse gases (Caponi et al., 2008). Developed internationally through a consultative process with over 500 stakeholders, and on its second revision, the GHG Protocol sets standards; lays out best practices for GHG accounting, reporting, and use at the organization or corporate level; and provides guidelines on boundary setting. The protocol also differentiates emissions into three categories: scope 1 (direct), scope 2 (indirect electricity and heat), and scope 3 (other indirect).

The GHG Protocol is used in a number of voluntary GHG reduction programs (e.g., U.S. EPA Climate Leaders, WWF Climate Savers, Business Leaders Initiative on Climate Change), GHG registries (e.g., California Climate Action Registry [CCAR], The Climate Registry, Regional Greenhouse Gas Initiative [RGGI], World Economic Forum Global GHG Registry), trading platforms (e.g., Chicago Climate Exchange, EU ETS), and sector specific protocols (e.g., International Aluminum Institute, International Council for Forest and Paper Associations, International Iron and Steel Institute, the WBCSD Cement Sustainability Initiative, and the International Petroleum Industry Environmental Conservation Association) (WRI, 2009). It is also the preferred protocol for the international Climate Disclosure Standards Board.

The main strengths of the GHG protocol are its focus on scopes 1 and 2 for high emitting sectors (e.g., power generation, cement manufacturing, and transportation) and its wide applicability. It is guided by principles central to the IPCC, flexible enough to encourage participation and incorporation into registries and standards, but standardized enough to allow comparative analysis over time and between organizations. Criticisms include its limited focus to date on scope 3 emissions (e.g., product use, waste disposal, storage, and logistics), which may comprise more than 90 percent of an entity's emissions profile (Matthews et al., 2008a), and on upstream and downstream emissions, which are needed for life cycle analysis and economic input-output models (e.g., Matthews et al., 2008b).[a] Greater flexibility in choosing financial or operational boundaries would allow wider corporate participation, although it would also add complexity in comparing entities that use different approaches.

---

[a]The WRI is currently undertaking concerted research on supply chain activities and product life cycle analysis to incorporate into the GHG Protocol (WRI, 2009a).

breaking (i.e., in underreporting of emissions sources) and high levels of political and public scrutiny. Third party verification can be costly and may not obviate internal conflicts of interest (Wara and Victor, 2008). Self-declaration, with obligations to provide supporting evidence, can be a useful and cost-effective way to ensure data quality, but it must be supported by other assurance methods, such as spot checks and desk re-view. National regulation or trading of GHGs demands the highest level of verification, including reviews of data systems and site visits, although batch verification can be developed for clusters of smaller organizations where the bulk of emissions are from electrical consumption or vehicles (CCAR, 2009).

Inventories and Registries

A greenhouse gas inventory is a quantitative accounting of greenhouse gases emitted or removed over a period of time for a particular country, region, firm, or other entity, whereas a GHG registry is usually defined as a collection of inventories from different groups, which can be used to collect, verify, and track emissions data from specific entities, such as facilities or companies (WRI, 2008).[1] A registry provides emissions information in standardized forms to enable comparison, trading, or regulatory over-sight. The reporting steps and protocols discussed above are required to ensure that inventories are complete and verifiable and that registries are high quality and per-ceived as legitimate. In Chapter 2 the panel described the various efforts to respond to climate change by local and state decision makers. Information on accurate emissions collected and accessible at the zip code level would allow decision makers to develop policies and programs specific for a region to reduce greenhouse gas emissions.

Greenhouse gas inventories are useful tools for carbon management because they can provide data on emissions by geographical area, administrative level, sector, or industry. For example, New York compiles an annual GHG inventory for the entire city, which has been instrumental for identifying key trends and sectors with high emis-sions (such as buildings), allowing policy to address specific emissions sources (Dickin-son, 2009). Registries can support a wide range of GHG reduction strategies by provid-ing a common framework to ensure accurate accounting in carbon market systems for issuing, holding, transferring, and canceling emission allocations or offset credits (Convery and Redmond, 2007).

The integrity of a registry is fundamental to the environmental effectiveness of a GHG reduction program and supplements other carbon markets by providing buyers and sellers with transparent, consistent information about legally verifiable allowances and offsets (Call and Hayes, 2007; Haites and Wang, 2006). However, they are not trad-ing platforms. They are data systems that quantify emissions attributable to specific entities and that protect the integrity of trading programs by ensuring that only the

---

[1] This is in contrast to an inventory, which aggregates, rather than establishes responsibility for, emissions.

fixed number of allowances embodied in an emissions cap are transferred and used for compliance. The ability of a registry to monitor and track emissions and offsets strongly affects its effectiveness for GHG reduction programs.

The United States has several GHG registries, based on voluntary reporting (e.g., DOE's 1605(b) program, Chicago Climate Exchange [CCX], EPA Climate Leaders, and CCAR) or on mandatory reporting (e.g., RGGI, California Air Resources Board). Some of these registries are linked to The Climate Registry, which is governed by 40 states, 12 Canadian provinces, 6 Mexican states, and 4 native sovereign nations. Table 6.2 summarizes characteristics of a number of these existing (and competing) registries. The principle differences between them include geographical remit, reporting standards, verification, and inclusion of other carbon management tools such as carbon offsets. However, little research has been done on the comparative benefits and performance of different carbon registries (but see Kollmuss et al., 2008; Pearson et al., 2008).

A global registry also exists. The Carbon Disclosure Project contains the largest amount of reported GHG data in the world. It has helped U.S. businesses better understand GHG reporting and registries. However, the use of standards is voluntary, making it difficult to compare data and limiting the usefulness of the Carbon Disclosure Project in supporting a global carbon management system. The Climate Disclosure Standards Board (CDSB) aims to standardize emissions reporting between entities in different countries governed by different carbon regimes in different countries. The CDSB approach allows companies to measure their overall GHGs and make them internationally comparable with other entities between and within sectors. This goes beyond the EU ETS because it can include any entity, not just the heaviest emitters.

A U.S. carbon management system will need to be compatible with international systems to enable carbon trading, to facilitate accounting for multinational corporations, and to support carbon management at the state, company, or facility levels. The national emissions inventory submitted to the UNFCCC is insufficient for these purposes. A U.S. system should, therefore, take into account the procedures and requirements of organizations such as the CDSB and, as a party to international agreements, negotiate for the most effective international reporting systems. It could also be available to the public in a single, easily accessible database.

Carbon Offsets

Carbon offsets present particular challenges to GHG information systems. An offset represents the reduction, removal, or avoidance of GHG emissions used to compensate for GHG emissions elsewhere (Quality Offset Initiative, 2009). Offsets can provide

TABLE 6.2 Examples of Current Registries and Informing Principles

Registry	Geographic Scope / Status	Level of Data Provided	Principles Standardization of Data/Guidance?	Mandatory Quality Assurance?	Reporting
California Climate Action Registry (Climate Action Reserve tracks emissions) California Air Resources Board (mandatory GHG reduction) indicated to work with CCAR for early action reward[a]	California, linked to Western Climate Initiative	All direct, and indirect emissions associated with heat and electricity Optional reporting on extra emissions All Kyoto gases	YES (Top-down consultative) Registry-specific General Reporting and Verification Protocols (based on WRI GHG Protocol; Guidance on sector specific protocols, new protocols in development)	YES All verified CCAR-accredited entities undertake verification	Provides emissions factors to improve standardization across data provision; uses Climate Action Registry Online Tool (CARROT) to assist.
The Climate Registry	U.S. states (over 40), Canadian Provinces (12), Mexican States (7), Native Sovereign Nations (4), Voluntary	100% all sources of GHGs Includes data from state level initiatives, such as Midwestern GHG Reduction Accord. Formed as by product of CCAR Requires corporate and facility-level reporting	YES (Top down consultative) Registry-specific Aimed as high integrity registry able to support various other reporting and reduction policies[b] General Reporting Protocol based on WRI GHG Protocol	YES All verified Proposed ISO 14065 Emissions Verification[b] / Should be ANSI or equivalent accredited body	Through Climate Registry website

continued

TABLE 6.2 Continued

Registry	Geographic Scope / Status	Level of Data Provided	Principles		
			Standardization of Data/Guidance?	Mandatory Quality Assurance?	Reporting
American Carbon Registry (Winrock International)	U.S.-wide	All scope 1 and 2 of GHG Protocol (scope 3 optional)	YES (Top-down consultative) Registry-specific Should observe GHG Protocol, ISO 16064, or equivalent standards. Sectoral protocols in development. Accepts input from VCS, CDM, EPA Climate Leaders	YES Registry-specific approved verifier	Online to ACR
Carbon Disclosure Project (CDP)	International, voluntary	Company/entity-based; variable, up to discretion of each reporting entity	NO No formal standard requirement; WRI GHG Protocol explicitly mentioned; some entities do not specify	NO Some third party, not a requirement	Through questionnaires administered by CDP
CDSB (Carbon Disclosure Standards Board)	International, voluntary (in preparation)	Company/entity based	NO Standards to be used in consultation at present	NO Likely	Online reporting, most likely similar to CDP

			YES	YES	
World Economic Forum Register	Global	Global consolidated emissions data	YES Should observe WRI/WBCSD GHG Protocol	YES Independently verified or allow spot checks by the Registry	Annual, online tools—not entirely apparent at the moment; Functional?
EU ETS (Community Registry/Community Independent Transaction Log)	EU Member States, mandatory	Facility-based Includes reporting from high-emitting sources	YES EU guidelines on reporting linked to IPCC guidelines on national inventories under the UNFCCC	YES Third-party verification required, specialized guidelines for verification	Annual submission of national inventories to Community Registry; active submission of allowances and trading credits
National Inventories under the UNFCCC [inventories] And International Transaction Log (responsible for carbon trading/CDM/JI)	Multi-national, mandatory for all signatories to the UNFCCC	National but sectorally differentiated (no attribution to specific actors) All emissions (sources) and reductions (sinks) All Kyoto gases	YES IPCC guidance on reporting national inventories	YES Provides guidance on third-party verification of reporting	Reported to the UNFCCC in countries national communications
Voluntary Carbon Standard Registry (VCS Registry) [offsets]	International, voluntary (just for offsets and voluntary carbon units using VCS)	Emission reduction data taken from offset registration	YES VCS methodologies for projects; also accepts CDM and CCAR methodologies	YES Projects must be third-party verified	Online registry managed by APX corporation

[a] Schmall and Sanders (2008).
[b] Delaney et al. (2008).
SOURCE: Winrock International.

219

a cost-effective way of reducing GHGs (Böhringer, 2003), and they are becoming a key strategy for governments, companies, organizations, and individuals to manage their emissions profiles (Bumpus and Liverman, 2008). They are an integral component of several voluntary, state-led schemes in the United States, including the Western Climate Initiative (WCI), RGGI, CCX, and of proposed national schemes (Waxman and Markey, 2009). The federal government, some states and cities, and a large number of businesses and non-governmental organizations (NGOs) are using offsets as part of their carbon management strategy. However, their legitimacy for both regulatory and voluntary emission reduction schemes has been questioned.

The process of creating an offset includes the development of an emission reduction project, the issuance of emission reduction credits, and the sale of the credits to those seeking to compensate for their emissions. The NRC report *Limiting the Magnitude of Future Climate Change* (NRC, 2010d) discusses offsets and their role in overall climate and technology policy; here we focus on the need to establish effective information systems and standards for offset reporting.

A wide range of approaches and technologies can be used to create a carbon offset, including investments in industrial and household energy efficiency, industrial and waste facility gas capture or destruction, fuel conversion, renewable energy, forestation and forest protection, and soil management. Within the international climate regime, the Clean Development Mechanism (CDM) provides a flexible option for countries and firms to offset emissions and gain credits toward reduction obligations through investments in emissions reductions in the developing world.

Some approaches to creating carbon offsets are easier to validate than others. For example, it is easier to assert carbon reductions in industrial gas destruction projects, where smokestack scrubber processes can be quantified and monitored relatively easily, than in decentralized projects, such as the distribution of improved stoves or light bulbs (Bumpus, 2009), where there are significant differences in uptake and monitoring between sites. Credit for offsets depends on quantifying the GHG saved or sequestered compared to baseline and business as usual scenarios and relies on the development of accurate and verifiable accounting methods. Offsets have requirements beyond standard GHG monitoring and estimating procedures (Steenhof, 2009). For an offset to be valid, it should demonstrate additionality[2] and should be real, permanent, verified, and unambiguously owned (WCI, 2009).

---

[2] Reduction in emissions by sources or enhancement of removals by sinks that is additional to any that would occur in the absence of a Joint Implementation or a CDM project activity as defined in the Kyoto Protocol Articles on Joint Implementation and the Clean Development Mechanism (IPCC, 2007d).

Additionality requires that the offset should materially reduce emissions beyond what would have ordinarily happened and/or should demonstrate that carbon finance was critical to the viability of a project (Greiner and Michaelowa, 2003; Michaelowa, 2005; Müller, 2009). High quality offsets should also take account of the risks of project failure and leakage (e.g., failure of a forest offset because of fire; forest protection leads to deforestation to other places) and link the monitoring and permanence of reductions to the flow of credits to purchasers. The UNFCCC has established methods for such accounting for the CDM specific to different types of projects and technologies. Incorporating offset projects into a standardized registry that allows emission reduction credits to be issued, tracked, and retired would avoid "double-counting" of reductions.[3] These principles are essential if the commodity of a carbon credit is to be both commensurable with money in a market (the credit) and to have a functional effect on the atmosphere (the reduction; see Bumpus and Liverman, 2008).

Several studies have questioned the independence and accuracy of offset verification (DNV suspension, 2008; Kollmuss et al., 2008; Wara and Victor, 2008). As a result of criticisms—especially around additionality, leakage, omission of key offset technologies, and the small scale of projects—both the compliance markets (CDM, ETS) and voluntary markets are considering reforms and new standards for offset accounting. Possible CDM reforms include minimizing leakage in sectoral emission reduction programs or increasing effectiveness by targeting the entire electricity or cement sector of a country. Credit for GHG emission reductions through protecting forests (not currently allowed under Kyoto) is the basis for proposals for carbon finance to countries agreeing to Reduced Emissions from Deforestation and Degradation (REDD), although this program also faces tremendous accounting challenges. Other proposals include allowing offsets for forest protection (not currently allowed under the Kyoto Protocol) or for investments in nuclear power generation or carbon capture and storage. New standards are also emerging within regulatory regimes. For example, the U.K. government, concerned about public confusion and criticism of voluntary offsets, has established guidelines for approved offsets that include the need for an independent registry and verification. Voluntary markets are proposing new voluntary standards, including identification of the basic criteria for effective carbon offsets and standards for carbon reduction and technologies (Boyd et al., 2007; Bumpus, 2009; Bumpus and

---

[3] For example, "green" certificates, such as Renewable Energy Certificates (RECs), can assist states in tracking investments in renewable energy, and they can effectively be sold as credited commodities. However, they are difficult to include as offsets because they can result in double counting of ownership (i.e., indirect emissions reductions from RECs make it difficult to assign reductions to one individual or entity) and may not demonstrate additionality (i.e., investment in renewables may or may not have happened as a result of the REC; Quality Offset Initiative, 2009).

Liverman, 2008; Gillenwater et al., 2007; Kollmuss et al., 2008; Lovell, 2009; Lovell and Liverman, 2010). A careful and objective assessment of current methods and standards, especially those of the CDM, would help the U.S. government decide whether to establish such guidelines or standards for voluntary schemes and how to treat offset accounting within national and international trading systems.

INFORMATION ON EMISSIONS AND ENERGY USE AND THE PUBLIC RESPONSE TO CLIMATE CHANGE

Citizens can play a role in responding to climate change by choosing to reduce emissions within their households and travel and by selecting lower carbon products and services. The United States has considerable experience in using standards, labels, and other information to increase energy efficiency, but little experience in using similar approaches to reduce greenhouse gas emissions. While government and industry standards can contribute to emissions reductions, easy access to that information, as well as other economic incentives (e.g., tax incentives, utility incentives, etc.) is just as important for the consumer to make informed decisions.

Household Emission Reduction

Households account for 27 to 38 percent of U.S. carbon dioxide emissions through direct energy use in homes and in non-business travel (Bin and Dowlatabadi, 2005; EIA, 2008; Gardner and Stern, 2008). Household activities also indirectly drive most of the remaining emissions, which come from producing, distributing, and disposing of the goods and services that households purchase (Bin and Dowlatabadi, 2005). A tremendous untapped potential exists to reduce direct emissions from households through the acquisition and use of energy-efficient equipment that is economically attractive to the household and that does not change household lifestyle (Dietz et al., 2009; Gardner and Stern, 2008; Granade et al., 2009; Vandenbergh et al., 2008). Household actions that would reduce emissions include replacing household equipment (e.g., vehicles, appliances), improving the efficiency of home heating and cooling systems (e.g., with insulation and more efficient furnaces), improving maintenance of vehicles, changing energy-using behaviors to reduce unnecessary use (e.g., turning off standby electric power, accelerating cars more slowly), purchasing carbon offsets, investing in distributed renewable energy (e.g., solar photovoltaics) to replace fossil fuel powered electricity, and downsizing.

The potential for carbon emissions reductions from the first five of these classes of action has been estimated at 27 to 37 percent of household direct emissions (Dietz et al., 2009; Granade et al., 2009). Equipment choices usually have greater potential for savings than alterations in the use of equipment once in place (Gardner and Stern, 2008; Stern and Gardner, 1981), and the savings are more easily maintained over time.

Despite the potential savings, households do not always act in their self interest (e.g., Brown et al., 2008; NRC, 1984; Parformak et al., 2009). One reason is that households do not know where the cost-effective opportunities lie, in part because useful information is unavailable or difficult to obtain or interpret. For example, information on the energy cost of home ownership is not generally available for comparison shoppers (Box 6.3). For homeowners, trustworthy information on expected savings from adding insulation or replacing leaky windows is obtainable, but only at a cost and with considerable difficulty. Multiple standards and guidelines for consumer and public education can confuse and discourage action on emissions reduction. Research has also shown that, although people want information on energy efficiency, they interpret it according to local contexts and personal choices (Gram-Hanssen et al., 2007).

It is also the case that people sometimes act on incorrect or misleading information. People tend to overestimate savings from curtailing highly visible energy uses, such as televisions and lights, and to underestimate savings from less visible energy uses, such as improving efficiency of furnaces and water heaters (Stern, 1986). People also err by using accurate information inappropriately, such as when they compare vehicles' fuel economy by subtracting miles-per-gallon (mpg) ratings, which are in fact not a linear measure of fuel use per service provided. In such cases, better information could help close the behavioral gap.

Available research shows that the effect of information on behavior depends on how specifically it is matched to households' situations and choices, how easy it is to understand and use, and the extent to which it comes from trustworthy sources (Abrahamse et al., 2005; Gardner and Stern, 2002; NRC, 2002b). However, information is only one of many barriers to behavioral change. For example, the initial cost is a significant barrier to replacing household equipment with more efficient models. Other barriers include infrastructure limits (e.g., unavailability of convenient alternatives to private cars for transport) and institutional problems (e.g., inability of renters to invest in building shells, and regulatory impediments to household investment in solar electricity production).

Some key areas where information provided to households and consumers could lead to greenhouse gas emission reductions are described below.

BOX 6.3
Homes and Building Efficiency Information

Home energy ratings provide summary information on the energy efficiency of an entire home, normally focused on the building shell and sometimes also the heating and cooling systems, in simple, understandable form. They are intended to be used to enable buyers, renters, appraisers, real estate agents, mortgage lenders, builders, and others to assess the energy cost of operation of homes and to make comparisons among them. They can create an incentive for owners to upgrade the energy efficiency of their properties in advance of transfer to make them more attractive to buyers. However, home energy audits have not been effective in improving the energy efficiency of homes (Hirst et al., 1981; McDougall et al.,1983), except when combined with techniques to overcome other barriers, such as free or reduced-cost installation of recommended improvements (NRC, 1985; see also Chapter 4).

A National Research Council (NRC, 1985) report found that over 40 home energy rating systems were in operation in the United States by 1982, but that the effects of these programs were largely unknown (see also Chapter 5). EnergyStar houses in Texas are an example where people are financially rewarded, in addition to energy savings, for efficient energy use (Entergy Texas, 2009). Work in Europe has found that information from experts is more likely to influence behavior than energy ratings on houses (Gram-Hanssen et al., 2007), and that dynamic instant information on household use that can be seen on PCs or websites can yield 8.5 percent savings (Benders et al., 2006). Utilizing on-demand and solar water heating systems instead of the more typical American systems that heat water 24 hours per day from electricity or natural gas is another example of how home energy consumption could be substantially reduced with existing technologies. In the United Kingdom, the introduction of Energy Performance Certificates, which must be included in house sale information, has started to drive modest investments in home energy efficiency.

Future research in this area should take an interdisciplinary approach to understanding household energy strategies (for review, see Steg, 2008). The federal government may also choose to provide a more reliable source of information by setting guidelines or standards for informing consumers about home energy or emissions audits. The challenge is to encourage emission reductions in existing housing to complement the standards set for new homes by industry and local government.

Feedback Information on Energy Use

Information on the actual consumption of electricity or other energy sources over time enables people to learn ways to reduce usage. Used effectively (i.e., daily), it can reduce energy use by 5 to 12 percent (Abrahamse et al., 2005; Fischer, 2008) and more if combined with comparisons to the energy use of other consumers. Information can be provided to households through their utility bills, which can be designed to facilitate feedback on energy (and water) use, but greater reductions occur when people receive more frequent information. "Smart meters" and in-line power consumption meters have the potential to provide detailed, tailored information to consumers about their energy use. However, these technologies have typically been designed for other purposes, such as load control by utilities. Human factors research will likely be required to optimize these technologies for end-use emissions reduction.

Technology for providing fuel economy feedback in motor vehicles is already installed on some newer models and could be useful for illustrating the effects of changes in driving techniques (e.g., slower acceleration). The effectiveness of fuel economy feedback has received little research attention, although lessons can be drawn from the growing literature on the effect of instantaneous feedback on energy consumption (see Darby, 2006; McCalley and Midden, 2002; van Houwelingen and van Raaij, 1989).

The government could choose to require that utilities provide standardized feedback on billing or install smart metering systems and that automobile manufacturers provide improved feedback systems in new models. Industry could also create standardized reporting systems that provide consistent and clear feedback to consumers.

Information on Energy Efficiency

Information about the energy efficiency of homes, vehicles, and appliances is commonly available in the form of certifications, ratings (e.g., EPA vehicle fuel economy ratings), and labels (e.g., EnergyStar and EnergyGuide labels, as illustrated in Figure 6.3). Some ratings include multiple types of information, such as the European system that rates appliance models according a color-coded 7-point (A-G) scale (EST, 2009) and also provides information on consumption per unit of service for various appliances (Boardman, 2004). Another example is Australia's appliance energy efficiency program,[4] which rates all consumer and industrial products based on energy consumption and estimated operational costs over a specified time period. Such a system of informing consumers about the energy consumption and costs associated with a wide variety of goods, could be a means of driving behavioral changes in consumption patterns.

Energy efficiency information is currently provided by government agencies, such as the EPA and DOE EnergyStar certification program, and by private sector networks, such as the widely adopted Leadership in Energy and Environmental Design (LEED) certification program for buildings. Figure 6.4 illustrates the various levels of LEED certification that a given building can attain depending on the level of sustainability

---

[4] See *http://www.energyrating.gov.au.*

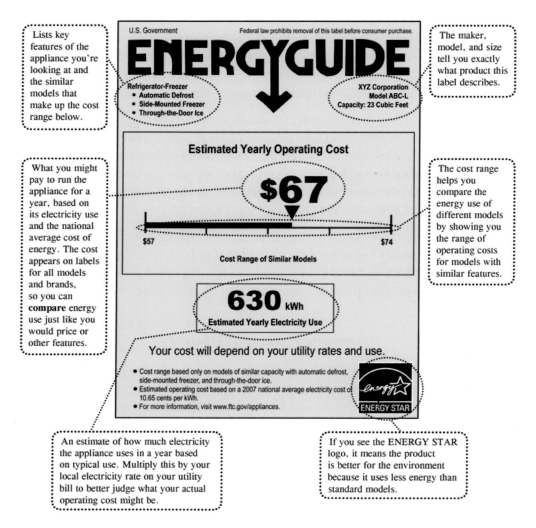

FIGURE 6.3 EPA and the DOE developed the EnergyGuide and EnergyStar labeling program for various home appliances to inform consumers about a product's energy consumption, cost of operation, and energy efficiency. SOURCE: EPA and DOE.

and efficiency measures built into its design. The DOE produces a building energy data book providing statistics on residential and commercial building energy consumption. The Data Book is evolving and could be developed as a useful tool for decision makers. EnergyStar leverages bottom-up approaches to managing energy efficiency by providing information and incentives to consumers to take action that is in their self interest and that meets wider energy efficiency goals. The greatest savings have been

FIGURE 6.4 The various levels of LEED certification attesting to the sustainability and efficiency measures built into a building's design. The program is administered by the U.S. Green Building Council. SOURCE: USGBC (2010).

in the areas of office equipment and computers, with an estimated reduction in emissions of up to 107 TgC between 1993 and 2006 and up to 278 TgC from 2007 to 2015 (Sanchez et al., 2008).

Several new systems are being developed to rate the GHG and energy performance of companies (Horne, 2009). The "GreenStar" system, to be launched in 2010, will assign a star rating to the top half of each sector based on brand level corporate emissions accounting, then leverage market forces and consumer choice to drive down emissions through yearly competitive rankings (GreenStar, 2009). This approach has some similarities to Japan's successful "Top Runner" program, in which the government set standards for products based on the current highest efficiency with the demanding standard becoming mandatory for all by a target year. Many products reached the standards before the target year, companies found themselves more competitive as government publicized the program, and efficiency improved more than the standard for some products (Jänicke, 2008).

Carbon Calculators

Carbon calculators have recently begun appearing on the web pages of NGOs, private companies, research groups, and government agencies. These calculators are used to estimate emissions from everyday activities. For example, the Nature Conservancy calculator estimates emissions from home energy, driving and flying, food and diet, and recycling and waste.[5] Although in principle such calculators can inform household

---

[5] See *http://www.nature.org/initiatives/climatechange/calculator/*.

decision making, different calculators give different results from the same input, and most are not transparent enough to allow an analyst to understand the discrepancies (Padgett et al., 2007). One of the criticisms of offset companies in Europe is the wide range of emission estimates given and the resulting effect on the costs of carbon offsets when the same flight is entered into different calculators. This has contributed to a decline in public confidence in carbon calculations. The EPA has developed a household emissions calculator, which has a good scientific basis but does not include airline travel or food—two of the most important sources of individual GHG emissions where Americans could make informed choices.[6]

An informed response to climate change in the United States might benefit from a wider diffusion of an improved EPA/DOE calculator and attention to the risks to public confidence from the proliferation of other calculators and their inconsistencies. Development of a standard by non-federal actors could also be helpful.

Carbon Labeling of Products

Carbon labeling of products offers a more nuanced analysis of product (or company) carbon intensities or ratings than the binary labeling schemes discussed above. Based on life cycle assessment (LCA), carbon labeling aims to provide consumers with information on the embedded carbon footprint of certain products, presumably so that they are better able to make climate friendly choices. The U.K.-based supermarket chain, Tesco, has announced its intention to develop carbon labels for all products, and carbon labeling schemes are being developed for biofuels in Europe (Rutz et al., 2007).

Carbon labeling is attractive because it uses a simple metric (CO_2e emissions), but there are real problems in defining boundaries for the footprint, and the LCA approaches used miss important environmental variables (Weidema et al., 2008). No single protocol exists, and variable carbon footprinting assumptions lead to a wide variety of outcomes (White, 2007). In addition, different countries may manufacture the same products using different energy mixes. As a result, carbon labels may need to convey complex information to make individual choices relevant and contextual (Horne, 2009; Schmidt, 2009). Creating easy-to-understand comparative labels for complex products is difficult. An alternative approach is to label the emissions of companies, rather than the individual products they produce. Such an approach would inform consumers of emissions attributable to certain brands and also prevent emitting companies from "hiding behind" some low-carbon products.

---

[6] See *http://www.epa.gov/climatechange/emissions/ind_calculator.html*.

INSTITUTIONAL OPTIONS FOR INFORMED GREENHOUSE GAS MANAGEMENT

Governments can support informed greenhouse gas management decisions by providing information directly to the public and/or by creating or supporting organizations that help the public and private sector reach emission reduction targets. Examples of support organizations include the United Kingdom's Carbon Trust and the Oregon Climate Trust. The Oregon Climate Trust is a non-profit organization that provides offsets and advisory services to government, utilities, and large business (The Oregon Climate Trust, 2009). California and Florida have announced their intention to create similar organizations in their states to assist in reducing emissions.[7] The U.K. Carbon Trust is partly funded by a U.K. government levy on electricity, gas, and coal. Its functions include information, education, and advisory services (e.g., carbon audits), loans to business for low-carbon technology and energy efficiency, and development of techniques (e.g., for carbon labeling) and standards.

The United States lacks a single point of contact for comprehensive information on GHG emissions reductions or best practices for moving toward a low-carbon economy. The DOE, EPA, and other agencies provide information on energy efficiency and emissions reductions. The DOE's Energy Information Administration provides some information on greenhouse gases, but it is not tailored to methods for emissions reductions. In general, information on methods for emissions reductions is either not available or is difficult to find on U.S. federal agency websites.

The clarity and accessibility of information available could be improved by upgrading agency websites or relying more heavily on state and local government or the private sector to provide greenhouse gas information and management services. A more ambitious option is to create a structure within the government to provide greenhouse gas management services, perhaps as part of a Climate Service. The arguments for such a service include the growing public and private need for credible information and guidelines on emissions reductions and the evidence that information can—when coupled with incentives, regulation, and technology—foster changes in behavior. The functions might include:

- Assistance to entities (e.g., firms, government offices) in greenhouse gas monitoring and reporting;
- Work with state and private sector climate trusts and carbon management services to ensure consistent reporting and information, thereby providing a level playing field for business and reducing confusion among consumers;

---

[7] See *http://www.myflorida.com*, Governor Crist Signs Agreement With United Kingdom's Carbon Trust

- Conduct periodic independent reviews and report to the administration and Congress on national progress on emissions reductions;
- Encourage GHG reductions in different sectors;
- Ensure climate justice objectives by empowering local and community activities;
- Fund technology demonstration and research in areas where private investment is lacking;
- Demonstrate the need and methods for GHG reductions and assess competing low-carbon strategies;
- Provide information and guidelines on complex GHG management strategies and policies, such as cap-and-trade and offsets;
- Provide carbon audit services or guidelines.

A greenhouse gas management service that understands the regulation of emissions (policy), the mechanisms for counting and reporting emissions (emissions protocols and registries), and the practical implications for companies would be both business-friendly and effective for emissions reductions. Given current agency responsibilities and expertise, such a service might be best placed in or supported by EPA and DOE working in close coordination.

Another important institutional issue relates to the governance of carbon markets and finance. A carbon market advisory could be established to ensure standardization and transparency in carbon accounting (similar to those in financial systems) and to establish ground rules and rigor in carbon markets. With accurate carbon data and a carbon price in place, the carbon market and associated trading should be governed by the Securities and Exchange Commission to ensure quality in carbon commodities and actual carbon reductions. Consultation with international entities working on this area, such as the Climate Disclosure Standards Board (CDSB), would help ensure that a U.S. system is compatible with carbon reporting and trading mechanisms elsewhere in the world (see Chapter 3).

COMPETITION AND EQUITY CONSIDERATIONS

Overarching issues in the implementation of greenhouse gas information systems include those relating to competition and equity. Because greenhouse gas emissions relate to the design of products and other information that can influence private sector competitive advantage, some firms and even governments may be reluctant to disclose detailed information. On the other hand, disclosure and labeling can promote more effective actions in a competitive market or social context with firms, local governments, and households competing to claim the largest emission reductions or for higher positions on tables that rank commitment to climate response and emission cuts. As in the case of climate services (Chapter 5), there are also substantial concerns about access to information and greenhouse gas information systems, including labels and standards should be easily accessible and understandable to the full range of U.S. citizens. Many people may find it difficult to afford products or energy that produces lower emissions, and the less well off may be unable to respond to information unless it is accompanied by programs that support specific decisions to install new energy systems or upgrade appliances. And, as with climate services, greenhouse gas information systems must engage with outreach to the public and private sectors to understand their needs and ensure that people understand the information and find it useful.

CONCLUSIONS AND RECOMMENDATIONS

Informed decisions on greenhouse gas reductions and trading require information systems for the reporting of emissions by a variety of actors (e.g., governments, companies, and organizations) and at multiple levels (e.g., city, state, regional, and national). These diverse data sets will have to be developed according to standard accounting principles for internal consistency in order to generate a comprehensive, transparent system that is capable of supporting a wide range of carbon management decisions. Developing such a system will require

- Research on greenhouse gas science, monitoring, and the effectiveness of accounting systems;
- Agreement on a national accounting system and standards to report the full range of greenhouse gas emissions using consistent methods, boundaries, baselines, and acceptable thresholds;
- High-quality verification schemes, including those for offsets;
- Methods to facilitate carbon management in supply chains and to control emissions at the most effective stage in the production-consumption chain; and
- A national greenhouse gas registry to track emissions from specific entities.

The development of a national GHG system should be informed by existing systems at international, regional, and state scales, operated by governments, consortia, or the private sector. In adopting existing systems at the federal level, care is required to ensure that national systems are fair, cost-effective, and designed to a high standard with options to adapt to new science and monitoring technologies and to link to international systems that might benefit American firms and citizens.

Four main conclusions can be drawn from this chapter. First, no uniform approach to managing greenhouse gas emissions exists. Harmonization of different approaches is important to ensure that GHG emissions reporting is fair and accountable and that it provides information needed for a broader carbon reduction regime. Non-federal actors, such as cities, states, companies, and NGOs, have already taken important steps in standardizing emissions reporting. Information for a comprehensive GHG accounting system must be accurate, transparent, relevant, consistent, and complete. These principles are fundamental for informing the design and use of protocols to measure GHG emissions. However, at present they are applied inconsistently between different carbon standards and therefore do not support a comprehensive or commensurable understanding of emissions information. A GHG regime should develop mechanisms built on these principles.

A national climate registry should stipulate standard methodologies and expectations and include regulated entities with the capability to add voluntary reduction and disclosure (i.e., the system should be scalable). The registry should complement international GHG reporting systems and should be extendable to create emissions inventories for city, state, and sectoral jurisdictions. A nationwide cap-and-trade system will require a harmonized registry built on the principles listed above to be atmospherically legitimate and to allow the incorporation of scientifically based programs into GHG reporting. A national registry needs to be "policy neutral" and to provide a flexible architecture for the incorporation of other programs, such as a tiered system that can account for reporting at state and federal levels, with various GHG reduction programs.

Finally, a federal carbon or GHG management service may assist the public and organizations in understanding their GHG emissions and potential reduction strategies and in providing accurate data to formal registries and national assessment activities. A "climate trust" type of organization(s) may be the best positioned to achieve this and create an effective long-term adaptive governance arrangement that can continually improve upon reporting protocols, increase efficiency, and assist in the evaluation of the provision of useful emissions information.

We conclude that there is a strong need for consistent methodologies for both emissions accounting and the development of energy efficiency information. Information needs to be accessible and reportable through harmonized accounting and registry systems. Consumers can be encouraged to limit emissions through feedback on energy use and credible labeling, especially when supported by federal or industry-wide

standards. We also judge that the United States could benefit from a federally supported, high profile, single organizational contact and structure for greenhouse gas
management and information, which would include informational, advisory, standard
setting, assessment, and research functions.

Recommendation 7:

**The nation should establish a federally supported system for greenhouse gas
monitoring, reporting, verification, and management that builds on existing
expertise in the EPA and the DOE but could have some independence. The system should include the establishment of a unified (or regionally and nationally
harmonized) greenhouse gas emission accounting protocol and registry. Such
an information system should be supported and verified through high quality
scientific research and monitoring systems and designed to support evaluations
of policies implemented to limit greenhouse gas emissions.**

Recommendation 8:

**The federal government should review and promote credible and easily under-
stood standards and labels for energy efficiency and carbon/greenhouse gas
information that build public trust, enable effective consumer choice, identify
business best practices, and can adapt to new science and new emission reduction goals as needed. The federal government should also consider the establishment of a carbon or greenhouse gas advisory service targeted at the public
and small and medium enterprises. Core functions could include information
provision, assessment of user needs and national progress in limiting emissions,
carbon auditing guidelines and reporting standards, carbon calculators, and support for research.**

CHAPTER SEVEN

International Information Needs

I nformation from other countries is essential to plan for and respond to climate change for a number of reasons: (1) the economic and market couplings of the United States with the rest of the world, particularly in agriculture; (2) shared water and other natural resources; (3) disease spread and human health; (4) humanitarian relief efforts; and (5) national security. This chapter highlights that the United States needs to be an active participant in improved sharing of global data, increased monitoring and surveillance of climate variability and climate change, and developing institutions that can be flexible and respond to changing circumstances.

ECONOMIC AND MARKET COUPLINGS

The world is connected by the flow of goods, materials, food products, and more. This means that activities and events that affect one region halfway across the world, or across our borders, can affect the U.S. economy. The reverse is also true. In particular, agricultural linkages are very important, especially when considered in the context of climate change. In general, countries that currently import most of their food may have to increase their net imports as climate changes negatively affect their crops and domestic agriculture. Developing countries, where climate change will make presently dry areas even drier, will become increasingly reliant on food imports (World Bank, 2010). The trade markets in food are very dependent on key regions, with just the United States, Australia, and Russia being major net exporters (Figure 7.1) (FAO, 2008). Countries that import most of their food will become increasingly dependent on the agriculture and food production from those major net exporters and thus will become more vulnerable to shifts in production and price due to extreme weather events associated with climate change (World Bank, 2010). Severe weather events, such as droughts, floods, and typhoons, have reverberations around the world as the recent food crises have shown. United States farmers and commodity markets are sensitive to changes in climate and markets across the world and can benefit from timely and accurate information about crop conditions elsewhere. Such information is currently provided by the USDA Foreign Agricultural Service, U.S. embassies, the USAID Famine Early Warning systems, and initiatives such as NOAAs ENSO forecasts and benefits

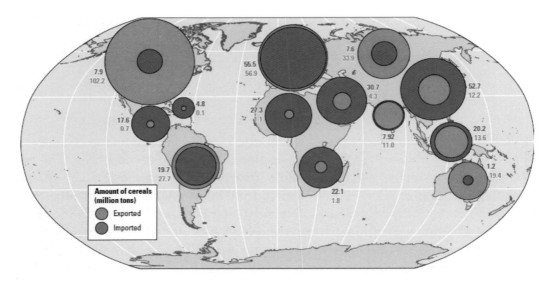

FIGURE 7.1 This world map illustrates how global cereal production and trade depends on very few countries. SOURCE: FAO (2008), as redrawn in World Bank (2010).

from innovations in remote sensing. United States food and fiber processing industries also require information on conditions in key countries exporting to the United States, with some companies maintaining in house climate expertise in order to gain comparative advantage in climate-sensitive global markets. Climate change will make it harder to produce enough food for the world's growing population, and the global rate of agricultural productivity growth will need to almost double while minimizing the associated environmental damage (IPCC, 2007; Rosegrant et al., 2009). This will require dedicated efforts to identify crop varieties able to withstand climate shocks, as identified by the reports *Advancing the Science of Climate Change* (NRC, 2010b) and *Adapting to the Impacts of Climate Change* (NRC, 2010a), as well as improved early warning information about extreme weather events. Global events such as the 1997-1998 strong El Niño and the weaker El Niños in 2002, 2004, and 2006 have shown how timely and effective climate forecasts and assessment information lead to enhanced resilience in domestic and international sectors such as disaster management and agriculture (NRC, 1999). Better forecasting and management of continental drought, with an emphasis on risk management rather than crisis management, can help in coping with more extensive climate change in the future. Two models of such approaches are the Australian Drought Policy (Wilhite et al., 2005) and the U.S. Western Water Assessment (WWA).[1] *The United States will increasingly need information about seasonal,*

[1] Boulder, CO, *http://wwa.colorado.edu/index.html.*

interannual and decadal extreme events to effectively prevent and respond to domestic and international food crises.

SHARED RESOURCES AND ECOLOGICAL SERVICES

The United States shares not only water across our borders (Figure 7.2), but also fisheries, migrating species (birds, whales, waterfowl, etc.), and many ecosystem services (such as flood control by buffer areas and wetlands filtration of pollutants). As climate change makes resources harder to manage, and growing populations increase demand, countries will need to cooperate more intensively to manage international waters, forests, wildlife, and fisheries (SEG, 2007). Thus, the United States must provide and procure information across borders, making sharing of real-time data more important. For example, changes in the timing and availability of water will affect agricultural production, sanitation, drinking-water quality and cost, water supply reliability, ecosys-

FIGURE 7.2 The San Pedro River watershed is an example of shared water resources across our borders. SOURCE: Dale Turner and The Nature Conservancy; TNC (2010).

tem services, and hydropower generation in neighboring Canada and Mexico. Impacts in more distant countries will also have broader economic and trade effects on the United States.

Climate change impacts will also strain existing domestic and international institutions and resource-management capabilities. Given the universal and transnational character of climate impacts on natural resources, there is a special role for integration of international information and agencies to facilitate helping the United States and other nations to cope with climate change. Few international organizations were designed with climate change in mind but will increasingly need to incorporate information about changing trends. For example, the International Joint Commission was set up between the United States and Canada to protect water quality and quantity in the Great Lakes regions (GLWQ, 2003). Future lake water levels are expected to drop as climate changes with concomitant increase in evaporation and decreases in ice cover. This will require managing outside of historic norms and both countries will need to set new operating rules (Bierbaum et al., 2008). Similarly, the United States is currently a party to (or considering becoming a party of) a number of environmental treaties designed to tackle problems other than climate change, but which will interact with efforts to respond to climate change. The established goals and roles of different countries in these treaties should be revisited. Among the pertinent treaties are those on biodiversity (Convention on Biological Diversity) and desertification (United Nations Convention to Combat Desertification), the Ramsar Convention on Wetlands, the Convention on Shared International Watercourses, and the International Treaty on Plant Genetic Resources for Food and Agriculture. The United States and other nations need to jointly evaluate how to sustain international ecological services from lands and waters in the face of climate change. Decisions will have to be with other countries about how to allocate harvesting of marine resources and (already overexploited) fisheries, and how to manage species as they relocate into new areas, such as the Arctic Ocean, to find suitable conditions. This will also be true for terrestrial species, as climate change is proceeding at rates 2 to 10 times the normal rates of migration (IPCC, 2007a); many ecosystems are being disrupted already and may not be able to persist without nations facilitating migration toward the poles or active preservation of individual species.

State and local governments along borders also have strong interest in the management of shared waters and ecosystems, as do the large number of U.S. conservation groups that operate internationally in attempts to protect internationally significant ecosystems. *Non-federal governments and conservation groups are generating useful data and information at the local level, can supplement information that is missing at national and international levels, and are major users of international data generated by the U.S. Federal government and international organizations.*

HUMAN HEALTH

The H1N1 outbreak has reinforced how easily infectious agents cross borders. Historically, the Centers for Disease Control and Prevention in the United States focused mainly on health effects within our borders and on efforts to keep these under control. Many threats to human health will clearly increase as the U.S. climate becomes warmer and leads to more heat stress in vulnerable populations. As well, warmer temperatures provide suitable habitat to more disease vectors. Increasingly, U.S. doctors will need to be trained in tropical medicine in order to recognize new diseases. For example, dengue fever (a viral disease) has been expanding its geographic range northward to the United States, and climate change is already accelerating the comeback of dengue to the Americas. In the United States, the incidence has already doubled over the past decade (PAHO, 2009). To detect and monitor the spread of diseases and prevent them from reaching epidemic proportions, national health systems will need to upgrade surveillance and enhancement of early warning systems and include information about conditions along and beyond the U.S. border (WHO, 2008).

The United States is also dependent on receiving good information from international sources to protect its own citizens from disease spread as they travel the world. Today, surveillance in many parts of the world fails to anticipate new disease pressure, for example, in Africa, where malaria is increasing rapidly (Keiser et al., 2004). Satellite remote-sensing and biosensors can improve the accuracy and precision of surveillance systems and prevent disease outbreaks through early detection of changes in climate factors (Rogers et al., 2002). Advanced seasonal climate forecast models can now predict peak times for malaria transmission and give regional authorities in Africa information to operate an early warning system and longer lead times to respond more effectively. Improving such capabilities will also safeguard American travelers (Frumkin and McMichael, 2008). *The United States benefits from close cooperation and information sharing with other countries and with international organizations such as the World Health Organization as well as international foundations (e.g., The Gates Foundation).*

HUMANITARIAN REASONS

The United States responds to global disasters with increases in foreign aid following floods, droughts, cyclones, earthquakes, and tsunamis for humanitarian and strategic reasons. Even in the absence of climate change, the United Nations University's Institute for Environment and Human Security cites predictions that by 2010 the world will need to cope with as many as 50 million people fleeing environmental degradation (SEG, 2007). Climate change could further augment the number of environmental refugees given that the frequency and intensity of floods, droughts, and fires will increase. Flooding of low-lying areas as sea level rises will displace additional tens of millions of people across the globe. As well, hurricanes and cyclones that form are expected to be more intense as climate changes, with higher wind speeds and increased rainfall (IPCC, 2007b). Environmental refugees due to climate may already being appearing as the climate warms and the environment deteriorates (Black, 2001). Natural droughts, compounded by poor agricultural practices and land-tenure policies, have contributed to severe famines in recent decades, which in turn led to the displacement of large numbers of people worldwide. Closer to home, Hurricane Mitch created so much local devastation in Central America that it drove thousands of displaced people to relocate, causing significant refugee pressures in nearby countries, including the United States (Glantz and Jamieson, 2000). Worldwide, weather-related disasters in 2005 exceeded $300 billion in insured and uninsured losses.

No nation is immune to the impacts that these global disruptions cause. The United States has experience in responding to climate-driven disasters such as hurricanes and droughts and such recovery efforts would benefit from good information systems to evaluate the success of the responses, and to reduce vulnerability in the future. The United States needs better information on the frequency and intensity of extreme events, which requires improved global monitoring systems. Also, federal and international organizations need to prepare to respond to a refugee and resettlement problem of significantly greater magnitude and longer duration than that for which they have currently planned.

Humanitarian NGOs have considerable interest in international disaster response because responding to international humanitarian emergencies is at the core of their missions. For example, following Hurricane Mitch in 1998 (Figure 7.3), emergency response teams from aid organizations such as USAID and Oxfam were sent to care for the injured and to provide water, food, and shelter. *Humanitarian NGOs are an important source of information on what is happening at the local level within other countries (e.g., Oxfam's climate witness program and Red Cross/Red Crescent's Climate Centre) and use information generated by the U.S. government and international organizations to target their efforts, inform the public, and assess the successes and failures of their programs.* In addition, the U.S. military has the capability to respond quickly and effectively to disasters and to promote stability in the affected regions; however, the expected in-

FIGURE 7.3 Hurricane Mitch. SOURCE: NOAA (1998).

crease in the frequency and intensity of natural disasters may negatively affect military readiness in other parts of the world (CNA, 2007).

NATIONAL SECURITY

Many of our overseas deployments reflect the strategic decision to ensure the free flow of oil to the United States and to our allies (CNA, 2009). But climate disruption and consequent social upheaval anywhere in the world can also lead to conflicts that will adversely affect U.S. interests of one kind or another, potentially necessitating a U.S. diplomatic and/or military response. Climate change itself can have impacts on the types of missions security forces must perform as well as the ways in which the military carries out those missions. For example, the effectiveness of some military operations could be influenced by the durability of equipment in extreme weather, the location and vulnerability of military bases to issues such as sea level rise and hurricanes, and a lack of reliable infrastructure to support transportation and energy needs (CNA, 2007). Impacts such as floods, droughts, wildfires, powerful storms, and pest outbreaks can increase "civil defense" demands. An ice-free Arctic will lead to shipping traffic, which will increase patrol requirements (see Box 5.1).

Climate change could interact with other international tensions and increase the chance of conflict (CNA, 2007). Water shortages in international basins, contention over ownership and access to ice-free arctic resources, and disputes and tensions over responsibility and compensation for climate-change damages are likely to ensue. If the United States struggles to relocate its own citizens as sea level rises (e.g., some of the land south of New Orleans will lose up to a meter this century from subsidence alone; sea level rise of up to a meter will be on top of that), migration to our borders of millions of people from Latin America and the Caribbean could tax humanitarian assistance (Woolsey, 2008).

Remedies chosen to address climate change could either reduce or enhance the potential for conflict. Reducing dependencies on Middle East oil could lessen conflict. Efforts to increase natural gas use (because its CO_2-to-energy ratio is better than that of oil or coal) may increase dependence on Russian gas and the potential for conflict. Expanding nuclear energy enough for a big impact on CO_2 emissions means thousands of reactors globally; controlling enrichment (highly enriched uranium) and reprocessing (plutonium) at this scale is daunting and the potential for proliferation would greatly increase.

In the face of these challenges to national security, several military experts recommended that the United States adjust its national security and national defense strategies to account for the possible consequences of climate change. For example, the Department of Defense could conduct an impact assessment of how rising sea levels, extreme weather events, and other effects of climate change might affect U.S. military installations over the next three to four decades. Beyond the direct military dimension, enhancing the resilience of the international community in the face of climate-related threats by strengthening the governance, health care, and disaster prevention and relief capabilities of foreign countries will also indirectly help the United States (Campbell and Parthemore, 2008).

WAYS FORWARD

Creating a Global System of Observations

Demand for sustained and reliable data and information on trends, unusual events, and long-range predictions at international scales has never been greater than it is today. Each of the international information needs sections above highlights the importance of observations for monitoring short- and long-term climate trends, warning of ecological or socioeconomic tipping points, characterizing regional vulnerability and impacts, and evaluating the efficacy of mitigation and adaptation responses. Climate information from both domestic and international sources must be increasingly incorporated into planning by sectors as diverse as agriculture, transportation, energy, insurance, water, and fisheries.

Improved seasonal outlooks for both U.S. and overseas locations can be used to assist farmers in the choice and timing of crops they plant and animal stocking levels, and to help water resource managers plan storage levels in dams to avoid floods and retain water needed for irrigation. Improved observations of the atmosphere will lead to more accurate and more extended weather forecasts, which will become particularly important as climate change leads to more intense convective rainfall events, more powerful and flood-inducing tropical cyclones, and more intense and frequent heat waves. Increased warning times will allow more effective protection of human life and property, better prediction of storm tracks and flood potential, and improved forecasts of air pollution levels. Advanced observations of the land surface will allow careful monitoring of the state of forests, grasslands, and other ecosystems, which will become particularly important as climate change shifts their natural ranges. This information should help in identifying those regions most susceptible to fire, in managing wildlife as snow cover and sea-ice extent change, and in irrigating crops and reducing pest damage through periods of drought or excess moisture. Coupled to a broader network of observations to be assembled into a Global Earth Observation System of Systems (GEOSS), these measurement systems and others—along with special measurements of the state of the polar ice sheets, the melting of which could more rapidly raise sea level around the world—have the potential to reduce the vulnerability to adverse impacts of climate change by enhancing warning times and resilience (see Box 7.1).

A global climate observing system could provide the climate-relevant information that society needs to better plan for and anticipate climate conditions on timescales from months to decades. Such an enterprise would build on existing observation systems but must go far beyond them. It must provide information to help farmers decide on appropriate crops and water management during droughts, on appropriate infrastructure to cope with the new 100-year extreme precipitation and storm surge events, monitor changing carbon stocks and flows in forests and soils, and evaluate efficacy of disaster response strategies under changing climate conditions. The original scientific objectives of NASA's Orbiting Carbon Observatory (OCO), which failed on launch in February 2009, was to study natural CO_2 sources and sinks. However, OCO would also have provided proof of concept for spaceborne technologies to monitor greenhouse

BOX 7.1
TOGA-COARE and Seasonal Climate Prediction

The Tropical Ocean Global Atmosphere Coupled Ocean Atmosphere Response Experiment (TOGA-COARE) has made major advances in the understanding of the strongest climate variation on seasonal-to-interannual time scales, El Niño and Southern Oscillation (ENSO). ENSO events, which occur every 3 to 8 years, involve a periodic change in the tropical Pacific atmosphere and ocean. These events are associated with extreme weather and climate anomalies, such as droughts in Australia, India, and Africa; floods in South America; and severe winter storms in the United States. Climate variations such as ENSO have serious impacts on human affairs, including loss of life, crop failures, and depletion of fisheries. In the past, these events could not be predicted in advance, which inhibited decision makers and governments from take effective actions to respond to their impacts. The 1982-1983 ENSO, which was not predicted or even detected, resulted in economic damages in the North and South Americas, including the United States, and also in Africa and Asia. This event spurred concrete actions by decision makers.

To better understand the implications of the ENSO and predict climate phenomena on time scales of months to years, the TOGA program was initiated in 1985 by the World Meteorological Organization (WMO) with contributions from 16 nations including the United States. Goals were achieved through long-term monitoring of the upper ocean and the atmosphere, specific process-oriented studies, and modeling (WCRP, 1985), which were implemented by a series of national, multinational, and international efforts (e.g., NRC, 1986b; WCRP, 1986a). In-situ observation programs such as the array of moored buoys in the Pacific known as the Tropical Atmosphere-Ocean (TAO) project were developed to provide these oceanographic data sets along with satellite observations and other sources of data. Seasonal forecasting of the ENSO cycle has dramatically influenced operational decisions made within industries such as agriculture and utilities. Research has estimated that ENSO forecasting may benefit U.S. agriculture decision making, resulting in a net economic value between $507 and $959 million/year (Chen et al., 2002).

According to the 1986 NRC report, "TOGA opened the way to the future of seasonal-to-interannual climate predictions. The follow-on programs will further develop the means of predicting the climate for the ultimate benefit of humankind." The TOGA program is widely regarded as an epitome of success, bringing together observation and monitoring, research and modeling, and service elements, and resolving issues of practical significance that were traceable to climate variations. This entails resources of all kinds (intellectual, monitoring, computational, etc.) and highlights the importance of relevant milestones of progress. Nevertheless, the success of the TOGA program brought together a multiplicity of countries, along with climate and social scientists, and economic and resource planners, and it represents a stark success of the goals that can be achieved provided the problem and the pathway to solving it are framed in a substantive manner and executed with due diligence.

gas emissions, as well as baseline emissions data. Such monitoring and verification of emissions reductions to support a greenhouse gas reduction treaty will also be needed (NRC, 2009c, 2010e). A successor to OCO will be decided on by the U.S. government in the next year.

Some of the additional necessary information to develop a Global System of Observations is being provided by United States National Meteorological and Hydrologic Service Centers and increasingly by Global Climate Observing System contributions through various government agencies and nongovernmental institutions. When the UNFCCC was negotiated, provision for the establishment of a Global Climate Observing System (GCOS) was initiated. Subsequently, GCOS was established by the WMO, UNEP, ICSU, and the Intergovernmental Oceanographic Commission (IOC) of UNESCO to ensure that all the observations required for climate monitoring, research, prediction, services, assessment, and climate change mitigation and adaptation are obtained, archived, and made broadly accessible to address multiple societal needs.) Also, a number of other institutions, such as the World Data Centers and the International Research Institute, regularly provide climate-related data and products including forecasts on monthly to annual timescales.

There are also a few examples of fledgling regional but international climate services. One such example is the Pacific Climate Information System (PaCIS), which provides a regional framework to integrate ongoing and future climate observations, operational forecasting services, and climate projections. PaCIS facilitates the pooling of resources and expertise, and the identification of regional priorities. One of the highest priorities for this effort is the creation of a web-based portal that will facilitate access to climate data, products, and services developed by the NOAA and its partners across the Pacific region.

Another example is the formation of regional climate centers, which the WMO has formally sought to define and establish since 1999. The WMO has been sensitive to the idea that the responsibilities of regional centers should not duplicate or replace those of existing agencies but instead support five key areas: (1) operational activities, including the interpretation of output from global prediction centers; (2) coordination efforts that strengthen collaboration on observing, communication, and computing networks; (3) data services involving providing data, archiving it, and ensuring its quality; (4) training and capacity building; and (5) research on climate variability, predictability, and impacts in a region.

Building a comprehensive and integrated system to monitor environmental changes across the planet is beyond the means of any single country, as is analyzing the wealth of data it would generate. That is why the Group on Earth Observation (GEO), a voluntary partnership of governments and international organizations, developed the concept of a Global Earth Observation System of Systems (GEOSS). Providing the institutional mechanisms to ensure the coordination, strengthening, and supplementation of existing global Earth observation systems, GEOSS supports policy makers, resource managers, scientific researchers, and a broad spectrum of decision makers in nine areas: disaster risk mitigation, adaptation to climate change, integrated water resource management, management of marine resources, biodiversity conservation, sustainable agriculture and forestry, public health, distribution of energy resources, and weather monitoring. Information is combined from oceanic buoys, hydrological and meteorological stations, remote sensing satellites, and internet-based Earth-monitoring portals. Some early progress includes the following:

- In 2007, China and Brazil jointly launched a land-imaging satellite and committed to distribute their Earth observation data to Africa.
- The United States recently made freely available 40 years of data from the world's most extensive archive of remotely sensed imagery.
- A regional visualization and monitoring system for Mesoamerica, SERVIR, is the largest open-access repository of environmental data, satellite imagery, documents, metadata, and online mapping applications. SERVIR's regional node for Africa in Nairobi is predicting floods in high-risk areas and outbreaks of Rift Valley fever.
- GEO is beginning to measure forest-related carbon stocks and emissions through integrated models, in situ monitoring, and remote sensing.

Improving Analytical Capabilities, Sharing of Best Practices, and "Learning by Doing"

Increasing observational capabilities is only the first step in a longer process of improving decision making in the context of climate change. The data collected by satellites, gauges, and instruments, and the information produced by modeling and forecasts, are only helpful if useable. Acting upon this information requires not only distilling and packaging the information into formats familiar to decision makers, but also domestic and international managers, consultants, and practitioners to acquire and employ new techniques, methods, and skill sets. Throughout this report the panel has described various information needs for different decision makers such as a hydro-electric dam manager, transportation official, and a fisheries manager (see Boxes 2.3, 5.2, and 5.3, respectively). Box 7.2 also illustrates analytic and information needs for decision makers in the health and land management sectors.

While it is certainly helpful to be able to observe changes and shifts in climate (and

---

BOX 7.2
Examples of Information Needs

Human Health

A public health official is concerned with long-term climate impacts associated with major disease vectors and vulnerabilities, poor air quality, malnutrition, and extreme weather events to assess the various effects of climate change on human health to determine the portion of the population that is most at risk. However, the limited amount of research and data available hinders the official's understanding of the magnitude and distribution of current and future health risks. The impacts need to be understood in order to address appropriate preventative measures and response systems to manage the risks. As with other aspects of climate change research, extensive monitoring programs are needed to track the health impacts in order to improve responses. Climate change forecasts, including meteorological and air quality predictions, are needed to characterize the specific vulnerabilities of populations in relation to the environmental or societal stressors that are already in place. Similarly, research and testing should take advantage of local knowledge and perspectives to identify patterns of severe health impacts and to develop effective adaptation methods.

Federal Land Manager

The ecosystems on federal land supply a wide range of services, including energy production, recreation, mining, and agriculture, as well as "non-market goods" such as carbon storage, air purification, and flood control. A land and resource manager must consider a variety of impacts related to climate change, including rising temperatures and the increase of extreme events such as drought, floods, and wildfires. For example, the land resource manager needs to understand how multiple stresses such as invasive species, forest fires, changes in the hydrologic cycle, and other impacts associated with shifts in temperature and precipitation will affect specific regions and populations and how to make decisions on land use changes. Moreover, synthesizing this information into proactive planning documents that identify priorities, thresholds, and decision triggers will require vulnerability and adaptation assessments (World Bank, 2010). This requires extensive monitoring to anticipate changes in the distribution and abundance of various plant and animal species (GAO, 2007). The land resource manager will also have to determine to what extent the demand for exploration and expansion of energy resources, either biomass or fossil fuel, will compete with the demand to maintain carbon reservoirs. These decisions made by the federal resource manager will have implications across various space and time scales and could hinder or empower similar decisions made by private, municipal, or state entities.

the subsequent impacts from these changes), anticipating the implications of these impacts in advance will likely reduce economic, resource, health, humanitarian, and security costs.

To understand the range of potential impacts, there is clearly a need for conducting detailed integrated vulnerability and adaptation assessments at local, regional, and national scales. Support for such assessments and the synthesizing of changing vulnerabilities and response capabilities is important to determine the level of risk that the United States faces from both internal and external pressures. Understanding how other countries will be affected by climate change will in turn help the United States understand potential impacts on trade, international aid needs, and spread of disease vectors.

Shared data sets of "best practices" can help identify decisions that have proven to be robust across a range of possible outcomes. In particular, compiling and sharing best practices across a variety of sectors and socioeconomic development patterns will be helpful for jumpstarting domestic and international mitigation and adaptation as cities, regions, and nations "learn by doing." Establishing and managing a "clearinghouse" that processes and makes available success stories and options from around the world will help communities design appropriate strategies. The *America's Climate Choices* panel report *Adapting to the Impacts of Climate Change* (NRC, 2010a) concluded such a clearinghouse should be "built on a series of consistent metrics and deliver information, training, and capacity-building services for climate change adaptation and mitigation that are broadly available to government, NGOs, and private sector interests."

Verifying efforts to reduce emissions, reduce vulnerability, and enhance adaptive capacity will require frequently updated international socioeconomic data, such as population density, changing land use and land cover, and infrastructure development (Bowen and Ranger, 2009). Changing carbon stocks and flows and changing demands on water and land can then be tracked in real time. Satellite and geographic information technology provide powerful means to generate physical and socioeconomic information rapidly and cost-effectively in a changing climate (NRC, 2007b,c). In order to monitor and verify treaty commitments, such rapid assessment tools will be necessary.

Recurrent extreme climate events—storms, floods, droughts, and wildfires—characterize many parts of the world. To cope with climate-related disasters, the United States and the world need to expedite development of enhanced forecast models for the likelihood of occurrence of these extreme events and for increasing the effectiveness of capabilities for coping with them. The U.S. forecast modeling centers should also consult with several international organizations such as the WMO, the World Health Organization, and UNESCO, in addition to other modeling centers around the world.

Early warning and surveillance systems can harness information technology and communication systems to provide advance warnings of extreme events. For such information to save lives, disaster management agencies need mechanisms in place to receive and communicate information to communities well ahead of the event. This requires systematic preparedness training; capacity building and awareness raising; and coordination between national, regional, and local entities. Taking swift and targeted action after a disaster is equally important, including social protection for the most vulnerable and a strategy for recovery and reconstruction. Climate change will change patterns of extreme events, but negative impacts can be reduced through systematic risk management—assessing risk, reducing risk, and mitigating risk (World Bank, 2009).

Finally, facilitating knowledge infrastructure in developing countries is key in order to avoid unsustainable development and promote clean energy and adaptation options. Supporting the development of institutions such as universities, schools, training institutes, research and development institutions, and laboratories, and such technological services as agricultural extension and business incubation, can support the private and public capacity to utilize mitigation and adaptation technologies.[2] Research institutes can then partner with government agencies and private contractors to identify and design appropriate coastal adaptation technologies and to implement, operate, and maintain them. They can help devise adaptation strategies for farmers by combining local knowledge with scientific testing of alternative agroforestry systems or support forestry management by combining indigenous peoples' knowledge of forest conservation with genetically superior planting material. If developing countries can better prepare for the consequences of climate change, it will reduce the chances for catastrophic losses, reduce the need for international aid, and promote adaptation and sustainable development.

CONCLUSIONS AND RECOMMENDATIONS

Informing the response to climate change in the United States can benefit from information about climate impacts and responses in other regions of the world. International information needs to be integrated into U.S. decisions to integrate information

---

[2] Bangladesh, which is particularly prone to hurricanes and sea-level rise, is an extreme example: university students enrolled in engineering represented barely 0.04 percent of the population (World Bank, 2010).

to gain the best understating of how climate is changing and how national and international polices interact with each other. A wide range of users— farmers, business, humanitarian NGOs, transboundary resources managers, and security agencies—can benefit from access to international information on climate change and thus from U.S. investment in international information systems. Information on impacts, greenhouse gas emissions, and response strategies internationally is essential for effective United States decisions because of the effect that international conditions have on United States climate, competitiveness, carbon prices, security, standards, and protocols for business. Valuable information is provided through federal agencies that collect, monitor, and disseminate international information such as the U.S. Department of Agriculture, U.S. Fish and Wildlife Service, USAID, and the U.S. military.

Recommendation 9:

The federal government should support the collection and analysis of international information, including (a) climate observations, model forecasts, and projections; (b) the state and trends in biophysical and socioeconomic systems; (c) research on international climate policies, response options, and their effectiveness; and (d) climate impacts and policies in other countries of relevance to U.S. decision makers.

CHAPTER EIGHT

Education and Communication

limate change is difficult to communicate by its very nature. Greenhouse gases are invisible, and their accumulating effects (e.g., global warming, precipitation changes, and extreme weather events) can take years before they are felt. Worldwide warming trends are hard for the average person to detect amidst the variability of everyday weather and the causes are far removed, in both time and space, from the impacts. Climate change is thus an example of "hidden hazards"—risks that, despite potentially serious consequences for society, generally pass unnoticed or unheeded until they reach disaster proportions (Kasperson and Kasperson, 1991).

Education and communication are among the most powerful tools the nation has to bring hidden hazards to public attention, understanding, and action. Citizens, governments, and the private sector cannot factor climate change into their decisions without a reasonably accurate understanding of the problem. To make informed decisions, people must have at least a basic knowledge of the causes, likelihood, and severity of the impacts, and the range, cost, and efficacy of different options to limit or adapt to climate impacts.

There are a variety of ways to empower decision makers and citizens with knowledge, ranging from formal educational curricula to public service announcements. Maps, graphs, and model-based projections can be especially useful and effective for presenting complex information clearly and understandably. It is critical both to provide new knowledge and to correct common misconceptions. For example, as people typically use fairly simple mental models to understand complex phenomena, it is not surprising that many people currently hold fundamentally incorrect mental models of climate change (Bostrom, 1994; Kempton, 1997; Kempton et al., 1995; Leiserowitz, 2006; O'Connor et al., 1998). Many people believe greenhouse gases are like smog and other kinds of air pollution that dissipate in a matter of days. However, the major greenhouse gases, such as carbon dioxide, will stay in the atmosphere and continue to alter climate for centuries to millennia (Kempton, 1997). Climate change will not stop the moment we limit emissions of greenhouse gases. This basic misconception may thus lead some people to underestimate the risks of delaying action to limit of the magnitude of climate change.

To improve public understanding, natural and social scientists must play an active role in the dissemination of their findings about climate change. At the same time, both formal and informal educators must develop new ways to translate this information. The steps we describe here can help empower the nation's present and future decision makers with the basic knowledge required to make informed choices. Although there are many other important approaches, we focus here on three specific areas: climate change education in the classroom, for the general public, and for decision makers.

K-12, HIGHER EDUCATION, AND INFORMAL SCIENCE EDUCATION

A student today may become tomorrow's business leader setting strategic priorities amidst changing energy markets, a mayor considering a seaport plan, a farmer adapting to new weather patterns, a designing policies to limit greenhouse gas emissions, or a citizen reducing his or her own carbon footprint. Today's students will need the knowledge and skills that will enable them to make informed decisions and actions, whether as engaged citizens or as future leaders.

The underlying science, while increasingly clear and compelling, involves complex concepts and processes, global perspectives, decades-long planning, and some irreducible uncertainties. Furthermore, decisions involve many other factors besides climate science, including economics, social values, competing priorities, and the risk and inherent messiness involved in virtually all complex decisions. This complexity, coupled with the long-term dynamics of the climate system, makes climate education challenging. Yet this richness and complexity provide an interdisciplinary context for deep learning, grounded in real-world challenges, and a content domain that will help schools implement required standards in science, mathematics, and social studies. For example, K-12 students can learn about the economic, political, and moral dimensions of climate change in addition to the basics of climate science (Figure 8.1). At the college and university level, the issue of climate change provides opportunities to engage students in both basic science and professions such as law and business and is especially useful in training students in the sort of interdisciplinary thinking needed for real world careers and decisions.

Current State of Climate Education

Climate change education confronts many of the same challenges as the broader effort to improve scientific literacy in schools and colleges, including the difficulties many teachers have in keeping up to date with rapidly evolving science and related issues. Professional societies such as the American Association for the Advancement

FIGURE 8.1 Summer science adventure programs, such as this one co-sponsored by the National Renewable Energy Laboratory, Denver Public Schools, and the Keystone Science School, expose schoolchildren to environmental science concepts and sustainable living practices. SOURCE: National Renewable Energy Laboratory.

of Science (AAAS) have published detailed definitions and learning progression maps of what citizens should know in order to be science literate. Federal agencies are just beginning to undertake climate change education and training initiatives, either as services they directly provide or as competitive grant programs to fund research, development, and implementation by experts in the field. These initiatives include the following:

- NASA: *Global Climate Change Education.* A grant program to develop K-12 education materials, provide teacher training, and conduct research on effective methods (and Space Grant).
- NOAA: *Environmental Literacy.* Grants support educators and scientists to develop and implement climate and environmental literacy programs. *Sea Grant.* A national network of university-based programs that provide workforce development and public education that now include climate as a theme.

- NSF: *Informal Science Education.* Grants support museums, after school programs and other informal venues; several current grants focus on climate.
- U.S. Forest Service: *Educator Resources.* A web site providing educational resources and highlighting the climate/forest connection.
- Department of Energy: *Global Change Education Program.* Provides summer undergraduate fellowships and graduate fellowships for global change.

In 2006, NOAA, in partnership with the AAAS Project 2061,[1] funded a workshop to discuss the need for a common set of curriculum guidelines specifically for climate education to be used at the local, state, and national levels. This workshop resulted in a broader interagency effort to coordinate and produce *Climate Literacy: The Essential Principles of Climate Science,*[2] a guide for the integration of climate education into national and state education standards. The document established a peer-reviewed overview of key concepts for climate education and has been used to support teacher workshops to ensure that educators are proficient in teaching climate science concepts.[3] Their essential principles of literacy in climate science included the following:

1. The Sun is the primary source of energy for Earth's climate system.
2. Climate is regulated by complex interactions among components of the Earth system.
3. Life on Earth depends on, is shaped by, and affects climate.
4. Climate varies over space and time through both natural and man-made processes.
5. Our understanding of the climate system is improved through observations, theoretical studies, and modeling.
6. Human activities are impacting the climate system.
7. Climate change will have consequences for the Earth system and human lives.
8. Humans can take actions to reduce climate change and its impacts.

The recognized need for more coordination and support for climate education across the country led to the formation of the Climate Literacy Network in 2007, a group of non-profit organizations, universities, and government agencies that identifies critical needs and opportunities and shares ideas, materials, and resources, both within the network and through its website. The Climate Literacy Network also provides a forum for strategic thinking among the member organizations and develops a common framework for their collective efforts.

---

[1] *http://www.project2061.org/*
[2] *http://www.climate.noaa.gov/education.*
[3] Climate Change Workshops for K-8 Teachers, Monmouth, Oregon, August 26-28, 2009.

In 2002, the Technical Education Research Center (TERC) Center for Earth and Space Science Education reviewed the science standards of all 50 U.S. states and rated the depth and breadth of their treatment of several key concepts in Earth science. The study found that climate change is taught in 30 states directly and 10 states indirectly, the Earth as a dynamic system is taught in 35 states directly and 15 states indirectly, and environmental literacy was taught in 20 states directly and 14 states indirectly. Another national study by the University Corporation for Atmospheric Research (UCAR) found that the backgrounds of high school Earth science teachers (using Earth science as a proxy for climate) also vary widely: only 60 percent of teachers had a teaching certification in earth science and only 27 percent had an undergraduate degree in earth science. Moreover, only 25 percent of high school graduates had taken a course in Earth science. In middle schools, only 13 percent of students had studied some aspect of Earth science. While earth science and climate change are ever more important, these studies demonstrate that climate and earth science education in the K-12 curriculum have been patchy and inadequate.

Research, however, on the best way to integrate climate science into core curricula is limited. Climate education efforts have traditionally been organized as a branch of Earth or physical science and often do not include the human dimensions of climate change or the science of climate response that is now emerging as a critical future need (see further discussion in *Advancing the Science of Climate Change*, NRC, 2010b).

It is also essential to understand what teachers need to better instruct students about climate change and to involve them in the development of educational tools. The University Corporation for Atmospheric Research (UCAR) has a number of teacher development programs offered through online courses and workshops and has learned several key lessons about integrating climate change in the classroom:

1. Teachers do not want to teach only the science, but also the solutions. They want to use climate change to improve problem solving skills and creative thinking.
2. Teachers do not want to scare their students when teaching climate change. They want to raise students' awareness and their self-efficacy to address the problems posed by climate change.
3. Students need to understand climate change at the local as well as the global scale, and place current changes in the context of longer time scales.

Some schools do not teach Earth or climate science simply because of a lack of textbooks or standards-based education materials. On average, textbooks are revised only once every 6 years and older editions often have little content on climate change and variability. Schools that lack the resources to buy new textbooks are thus forced to turn to alternative sources of teaching materials. Moreover, climate science often advances more rapidly than textbook revisions, so even the latest textbooks are soon out of date. Although there is now a wide range of information about climate change on the internet, teachers and students may find it difficult to judge the reliability of different sources.

Fortunately, a growing number of ancillary materials have appeared as supplements to mainstream high school textbooks. Some specific examples include *Oceans' Effect on Weather and Climate: Changing Climate*, *Climate Change From Pole to Pole: Biology Investigations*, and *Earth's Changing Surface: Humans as Agents of Change.*[4]

In addition to printed materials, educators today can use a wide range of media and learning materials, including text, graphics, maps, simulations, websites, television programs, movies, field trips, experiments, and citizen science projects. An NSF-funded cognitive research project, Visualizing Earth, found that images and animations of Earth from space help students understand the complex interplay of Earth's atmosphere, land, oceans, and life, but that students sometimes needed help in the transition from local to global scales, and in understanding the complexities of change over multiple time scales (Barstow et al., 1999). The Global Learning and Observations to Benefit the Environment (GLOBE) program, established in 1995 through NASA, NSF, the U.S. Department of State, and UCAR, has students around the world collecting environmental data, which they submit to a central database for use by students and scientists to monitor environmental change over time. This program helps elementary and middle school students learn core concepts of Earth system science and climate change and develop scientific thinking and data analysis skills.

To help students better understand the impacts of climate change, teachers may want to use images to which people can relate. For example, some people are more likely to respond to depictions of the impacts of climate change on local areas to which they have emotional connections (O'Neill and Hulme, 2009). Describing the impacts on a time scale of less than 50 years also makes the icons more relatable (Nicholson-Cole, 2005; O'Neill and Hulme, 2009). Unlike icons drawn from the scientific literature (such as the Antarctic ice sheet), to which most people cannot relate, people find it easier to imagine the effects on local places based on personal experience (O'Neill and Hulme, 2009). In addition to impacts, seeing images of "things people could do" may not increase participants' interest in climate change, but it can raise their perceived ability to change their behavior (O'Neill and Nicholson-Cole, 2009). Catastrophic imagery, however, may not be particularly effective for education (Nicholson-Cole, 2005). Fear-based

---

[4] National Science Teachers Association Learning Center, *http://learningcenter.nsta.org/default.aspx*.

images can help people see the issue as more important, but they can also inspire helplessness or apathy (O'Neill and Nicholson-Cole, 2009).

Using different media and venues also helps teachers reach a wider range of students with diverse learning styles. Students are often more interested in learning about climate change (or any topic) when an immediate and concrete decision needs to be made. The large and growing array of climate-related decisions can be used as compelling contexts for learning and provide an opportunity for students to make direct connections between what they learn and the world around them.

Many colleges and universities are now teaching about climate change, usually as part of environmental science and studies programs, and a growing number of graduate students are studying both the physical and human dimensions of climate change and seeking careers as climate change researchers, educators, and decision makers. However, recent financial cuts at many higher education institutions often make it difficult to add new faculty with climate change expertise and to fund graduate students who wish to study the topic. New NSF and NASA funded programs are seeking to transform climate change education by including a focus on solutions and developing new curricula through consortia and collaborations of institutions.[5] Broader curricula in the principles of sustainability and environmental education would be even more effective in equipping students for the issues they will face later in their lives and have already been adopted as university requirements at several institutions, such as the University of Georgia.

Educational institutions can model what they teach and turn their campuses into "living laboratories and classrooms" by making their facilities more sustainable, including efforts to conserve energy and limit greenhouse gas emissions. Michael M. Crow, President of Arizona State University, has stated the challenge clearly: "More than ever, universities must take leadership roles to address the grand challenges of the twenty-first century, and climate change is paramount amongst these." The American College & University Presidents' Climate Commitment[6] (ACUPCC) brings this recognition into focus. Participating universities and colleges are submitting Climate Action Plans that list specific steps schools are taking to dramatically reduce their emissions toward the goal of climate neutrality. Students are often heavily involved in the design and implementation of these plans, which provide invaluable opportunities to teach the science of solutions and analytical skills and provide hands-on training. The ACUPCC provides several important recommendations about climate change education in higher education, including student involvement in greening campuses; elective, major, minor,

---

[5] See *http://www.nsceonline.org.*
[6] See *http://www.presidentsclimatecommitment.org/.*

freshman, capstone, and general education classes on sustainability; and the full range of strategies for teaching, including learning through projects with local communities and internships, case studies, and internet-based activities (ACUPCC, 2009).

Federal support to facilitate the greening of schools, museums, and universities could help catalyze a transformation of both education and the nation toward sustainability. Depending on their institutional context, such initiatives can provide living laboratories for students, teachers, parents, and the broader public to explore, learn, and understand what sustainability means and how this relates to reducing the risk of climate change. To give just a very few examples of such initiatives among many:

- Ball State University will cut its carbon dioxide emissions by about 50 percent through the installation of a campus-wide geothermal district heating and cooling system.
- The University of New Hampshire will generate up to 85 percent of the energy used by the campus from the EcoLine™ project, a landfill methane gas-to-energy initiative.
- At Green Mountain College, a new heat and power biomass facility is predicted to shift 85 percent of current fuel oil usage to sustainable-sourced biomass.

Informal learning in museums, on the web, in after-school programs, in the Girl and Boy Scouts, and in community activities can also provide numerous opportunities to extend and deepen learning for students as well as the general public (see Figure 8.2 and Box 8.1). In fact, the role of informal learning in education has grown considerably over the years, and in some ways it has become as important as formal school learning (NRC, 2009b). While schools provide a formal structure and organized developmental sequence of learning, informal learning environments can provide other experiences of discovery, relevance, and adventure. Increasingly, there are also innovative informal learning settings that provide climate change education through special exhibits, presentations and discussions in town hall meetings, museums, science centers, zoos, aquariums, botanical gardens, planetariums, and other venues that help students and adults learn about climate change and make informed decisions about how to respond.

THE GENERAL PUBLIC

Communication about the risks posed by climate change requires messages that motivate constructive engagement and support wise policy choices, rather than engendering indifference, fear or despair. (Frumkin and McMichael, 2008)

FIGURE 8.2 Museum visitors learn about climate change and its impacts through interactive media and eye-catching displays. SOURCE: U.S. Climate Action Report (2010).

Although nearly all Americans have now heard of global warming, the greenhouse effect, and climate change, many have yet to understand the full implications of the problem, the alternatives for a national response, and the opportunities that lie in the solutions. The field of climate change communication is still relatively young but has identified a number of key knowledge and information needs, roadblocks to understanding, guiding principles, and potential models for improved education and communication that can help advance public understanding of climate change, inform individual and collective choices, and support public deliberation about potential responses.

While no formal national assessment has yet been conducted to determine the full state of public understanding of climate change causes, consequences, and potential solutions, several nationally representative scientific studies, as well as numerous public opinion polls, do provide important insights.

A study of American climate change beliefs, risk perceptions, policy preferences, and

BOX 8.1
Informal Science Education

"You learn—it's amazing" (a 73-year-old visitor to the Huntington Botanical Gardens, quoted in Jones, 2005:6).

Informal science education (ISE) venues include museums, zoos, aquariums, botanical gardens, cultural centers, summer camps, public lecture series, media programs, science magazines, and even backyards and dinner conversations. ISE thus provides a wide variety of learning opportunities and experiences that speak to multiple audiences and continue through a lifetime. Museums and science centers draw hundreds of millions of visitors every year (NRC, 2009b), while zoos and aquariums attracted 143 million visitors in 2007 alone (Falk et al., 2007). ISE also includes "science- and math-based television and radio programs [that] reach some 100 million children and adults each year" (NRC, 2009b). Memorable science *experiences*, including encounters with knowledge (e.g., through interactive exhibits, simulations, and movies) are provided through ISE: "Associating scientific thinking with engaging and enjoyable events and real-world outcomes can create important connections on a personal level" (NRC, 2009b). In addition, many scientists trace the origins of their interest and love for science to childhood visits to natural history museums, zoos, aquariums, or botanical gardens, or to television programs like Carl Sagan's *Cosmos*.

ISE can also inform environmental decisions and actions. For example, in 1997 the Monterey Bay Aquarium established the Seafood Watch Program to raise awareness about overfishing and the importance of purchasing sustainable seafood. In less than a decade, the program has helped individual consumers to change their eating habits and caused seafood restaurants around the world to change their standards (Quadra Planning Consultants Ltd. and Galiano Institute, 2004). Participation in citizen science projects has also been shown to develop scientific thinking and skills among the general public (Bonney et al., 2009).

Informal science educators have also been addressing climate change for years. Projects and programs have ranged from large exhibitions like the Association of Science-Technology Centers' (ASTC) *Greenhouse Earth*, which opened in 1992, to public television specials such as NOVA's "What's Up with the Weather?" More recently, ASTC has initiated a citizen science project called "Communicating Climate Change." This project is helping science centers across the nation identify local indicators of climate change—such as pine bark beetle infestations in the pine forests of Arizona, spreading disease on coral reefs in Hawaii, and changing patterns of bird migration in Philadelphia—and training "citizen scientists" to observe, monitor, and track these changes over time, thereby helping communities understand global climate change as also a local issue.

Despite these innovative examples, there is a demand for climate change materials, including accurate explanations of how the climate system works; how climate affects and is affected by fundamental human systems like food, water, and energy; and the causes, likely impacts, and potential solutions to climate change both locally and globally. Because ISE reaches audiences numbering in the hundreds of millions, connects professional educators through national networks, and partners with diverse media, this field is well situated to help improve public climate change awareness, understanding, and informed decision making.

behaviors completed in January 2010 found that 57 percent of Americans said that global warming is happening[7] and 53 percent believed that human activities are a contributing cause (Leiserowitz et al., 2010). Likewise, 50 percent were personally worried about global warming (Newport, 2008), while 65 percent considered it a serious threat (Pew Research Center for the People and the Press, 2009). Less than half the public (41 percent), however, said global warming is a near-term threat having dangerous impacts on people around the world either now or within the next 10 years. In line with several national polls, this study found significant decreases since 2008 in levels of public belief that global warming is happening (–14 percentage points), that it is human caused (–9), personal worry (–13), and perceptions of climate change as a serious threat (–8) (see Associated Press/Stanford/GfK Roper, 2009; Leiserowitz et al., 2009; Pew Research Center for the People and the Press, 2009; Saad-Gallup, 2009).

A 2008 national survey found that a majority of Americans believed that if nothing is done to address it, global warming will cause more droughts and water shortages, severe heat waves, intense hurricanes, the extinction of plant and animal species, intense rainstorms, famines and food shortages, forest fires, and the abandonment of some large coastal cities due to rising sea levels within the next 20 years. However, most Americans perceived it as a geographically distant problem that will primarily impact people, places, and species far away. Most also had little to no understanding of the potential health impacts resulting from increased climate change, while several studies have documented poor public understanding of some of the fundamental properties of climate change itself (Bostrom, 1994; Kempton et al., 1995; Leiserowitz, 2006; O'Connor et al., 1998; Read et al., 1994).

Importantly, however, a large majority of Americans desire more information about climate change. In 2008, only 12 percent said that they were very well informed about the different causes, consequences, or solutions to global warming,[8] while a year later 69 percent of Americans said that they would like more information. Furthermore, 70 percent said "schools should teach our children about the causes, consequences, and potential solutions to global warming" while 60 percent said the "government should establish programs to teach Americans about global warming" (Leiserowitz et al., 2010).

---

[7] Also see recent national studies by the Pew Research Center (2009) and the Miller Research Center for Public Policy at the University of Virginia (Rabe and Borick, 2008).

[8] Also see Washington Post/ABC/Stanford (2007).

Information Sources

Television remains the primary source to which most Americans turn to keep up with current news and world events (59 percent), followed by the internet (22 percent), print newspapers (10 percent), radio (8 percent), and magazines (1 percent). Most Americans sometimes or often watch their local TV news (75 percent)—many more than the national nightly network news broadcasts on CBS, ABC, or NBC (55 percent), the Weather Channel (47 percent), Fox News (42 percent), CNN (38 percent), or MSNBC (31 percent). Regarding topics, 70 percent of Americans somewhat or very closely follow the local weather forecast—many more than national politics (53 percent), world affairs (47 percent), health (43 percent), the environment (35 percent), sports (33 percent), and science and technology (32 percent) (Maibach et al., 2009). Thus, while the internet and social media are growing rapidly in use, local television remains the primary source of news and information for most Americans.

The perceived trustworthiness and credibility of different messengers also plays a key role in public responses to hazards, especially for scientifically complex problems like climate change (Slovic, 1999). A 2008 study asked respondents who they trusted to tell them the truth about global warming and found that 82 percent of Americans trusted scientists, followed by family and friends (77 percent), environmental organizations (66 percent), and television weather reporters (66 percent). Nearly half of Americans trusted religious leaders (48 percent) or the mainstream news media (47 percent), while only 19 percent of Americans trusted corporations as a source of information (Leiserowitz et al., 2009).

While scientists are still the most trusted source of information, a 2010 survey found that trust in scientists had dropped to 74 percent. Moreover, only 34 percent of Americans believed that most scientists agree that global warming is happening, while 40 percent believed that there is a lot of disagreement among scientists (Leiserowitz et al., 2009). Public perception of whether or not there is widespread agreement among scientists is particularly important.

A number of groups opposed to climate change legislation have attempted to reinforce the perception of scientific uncertainty as a means to delay action (McCright and Dunlap, 2000, 2003). Journalistic norms have also played a role in this process, often through articles that pair quotes from climate scientists with quotes from a small minority of climate change deniers who cast doubt on either the reality of or human contribution to climate change, thereby supporting an inference that the scientific community is equally divided (Boykoff, 2007; Boykoff and Boykoff, 2004). More recently, the unauthorized release of a set of emails from climate scientists at the University of

East Anglia in the United Kingdom, as well as the discovery of errors in the Fourth Assessment Report of the Intergovernmental Panel on Climate Change (IPCC), have been widely reported, used by critics to allege scientific misconduct, and may have contributed to lower public trust in scientists and the perception of greater disagreement among scientists about the reality of climate change.

Policy Support

In a democratic system, public support or opposition to proposed policies plays a vital role in shaping choices about what climate change and energy policies are adopted at all levels of government. While any survey represents a snapshot in time, many public opinion polls have found that the American public supports a wide variety of policies currently being considered by decision makers in local, state, and national governments (Nisbet and Myers, 2007). For example, a survey completed in January 2010 found strong public support, across political lines, for policies to develop renewable energy, improve energy efficiency, establish new regulations, and sign an international treaty to reduce greenhouse gas emissions (see Figure 8.3).

Eighty-five percent of Americans supported funding more research into renewable energy sources, such as wind and solar power, while 82 percent supported providing "tax rebates for people who purchase energy-efficient vehicles or solar panels." Seventy-one percent supported the "regulation of carbon dioxide (the primary greenhouse gas) as a pollutant," while 61 percent supported signing "an international treaty that requires the United States to cut its emissions of carbon dioxide 90 percent by the year 2050."

Fifty-eight percent supported policies to "require electric utilities to produce at least 20% of their electricity from wind, solar, or other renewable energy sources, even if it cost the average household an extra $100 a year." Likewise, 55 percent supported a government subsidy to replace old water heaters, air conditioners, light bulbs, and insulation, even if the subsidy cost the average household $5 a month in higher taxes. Finally, at the local level, majorities of Americans have also supported changing their own city or town's zoning rules to decrease suburban sprawl and concentrate new development near the city center (68 percent) and to require that neighborhoods have a mix of housing, offices, industry, schools, and stores close together (65 percent; Leiserowitz and Feinberg, 2007).

Thus, the American public supports a variety of local, state, national, and international policies to reduce greenhouse gas emissions. Surveys have also found, however, that Americans strongly oppose higher taxes on energy. For example, sixty-four (64

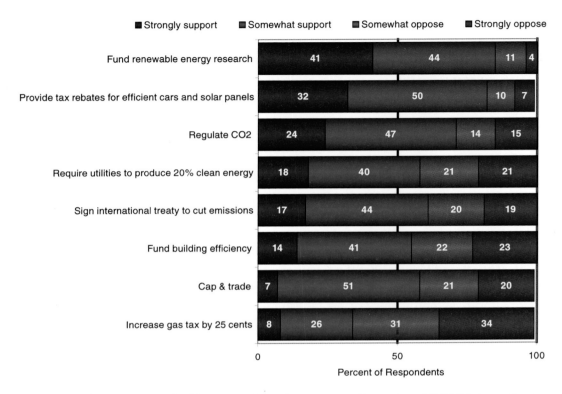

FIGURE 8.3 The percentage of public support for various energy policy options. SOURCE: Leiserowitz et al. (2010).

percent) opposed "a 10-cent city or local fee added to each gallon of gas you buy, to encourage people to use less gasoline" (Leiserowitz and Feinberg, 2007), while 67 percent of Americans opposed a policy to "increase taxes on gasoline by 25 cents per gallon and return the revenues to taxpayers by reducing the federal income tax" (Leiserowitz et al., 2009). Likewise, a 2009 AP-Stanford poll found that 78 percent of Americans opposed a policy to "increase taxes on electricity so people use less of it" (Associated Press/Stanford/GfK Roper Poll, 2009). Moreover, climate change remains a relatively low national priority compared to other national challenges. For example, in January of 2009, a poll conducted by the Pew Research Center for the People and the Press found that global warming was ranked last of 20 national priorities, well below the economy, jobs, terrorism, social security, education, Medicare, health care, the deficit, and others. Energy, however, was the fifth highest national priority (Pew Research Center for the People and the Press, 2009).

Global Warming's "Six Americas"

Behind these aggregate data, however, lies a diversity of views. In the United States there is considerable variation in how much people know about climate change, the extent to which they have taken personal action to reduce their own carbon footprints, and in deeper-level values, which strongly shape public interpretations of and preferred responses to climate change (see Leiserowitz, 2006). Researchers have identified six distinct segments of the public that conceptualize and respond to the issue in very different ways (Figure 8.4) (Leiserowitz and Feinberg, 2005, 2007, Leiserowitz et al., 2010; Maibach et al., 2009).

For example, one segment of the public is very worried about global warming (the "Alarmed"), strongly supportive of national policies, and beginning to change some of their own climate-related behaviors. By contrast, a different segment of the public does not believe that global warming is happening or human caused (the "Dismissive"). Some within this group simply deny it is happening, others say the science has not been proven yet, while others believe it might be happening but is just part of a natural cycle. Some argue it is just media hype, while a few believe global warming is a hoax (Leiserowitz and Feinberg, 2005, 2007). Between these two extremes lie four other groups: the Concerned, Cautious, Disengaged, and Doubtful.

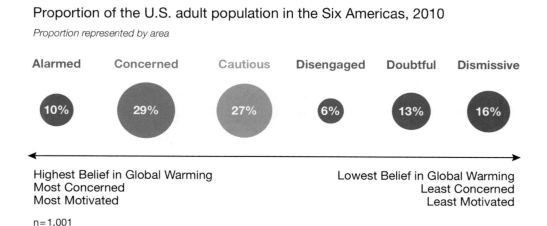

Proportion of the U.S. adult population in the Six Americas, 2010

Proportion represented by area

Alarmed · Concerned · Cautious · Disengaged · Doubtful · Dismissive

10% · 29% · 27% · 6% · 13% · 16%

Highest Belief in Global Warming
Most Concerned
Most Motivated

Lowest Belief in Global Warming
Least Concerned
Least Motivated

n=1,001
Source: Yale Project on Climate Change

FIGURE 8.4 Adults in the Six Americas are grouped into six categories (Alarmed, Concerned, Cautious, Disengaged, Doubtful, and Dismissive) based on their responses to global warming SOURCE: Leiserowitz et al. (2010).

Overall, each group responds to the issue in its own way. For example, 92 percent of the Alarmed are either very or extremely sure that global warming is happening versus 64 percent of the Concerned, 17 percent of the Cautious, 12 percent of the Disengaged, 6 percent of the Doubtful, and only 3 percent of the Dismissive. Similarly, 85 percent of the Alarmed believe global warming is caused mostly by human activities versus 77 percent of the Concerned, 46 percent of the Cautious, 41 percent of the Disengaged, 14 percent of the Doubtful, and only 2 percent of the Dismissive (Leiserowitz et al., 2010). Moreover, each of these groups has unique demographic characteristics, attitudes, values, political orientations, media preferences, and other attributes.

Each of these audiences is also likely to trust different messengers. For example, 46 percent of the Alarmed strongly trust former Vice President Al Gore as a source of information on global warming, while 89 percent of the Dismissive strongly distrust him. By contrast, 67 percent of the Dismissive somewhat to strongly trust their own family and friends as a source of information on global warming, while 44 percent trust their religious leaders (Maibach et al., 2009). Likewise, messages that highlight the potential impacts of climate change on polar bears and other species may resonate with the strong environmental values of the Alarmed and Concerned, but fall flat with the Doubtful or Dismissive. By contrast, messages that point out the national security implications or a religious-based stewardship ethic may help some Americans understand this issue from the perspective of their own deeply held values. In recent years, many military leaders and analysts have argued that climate change is a national security threat (CNA, 2007; Gilgoff, 2009; Goodstein, 2005; Pumphrey, 2008). In addition, many religious leaders contend that it is a moral imperative. Research has also found that minorities are often more likely to believe climate change is happening, human caused, and a serious threat than white Americans (Leiserowitz and Akerloff, 2010). Climate change is also an issue of environmental justice, as poor and minority communities are often the most vulnerable to the impacts of climate change, despite contributing relatively little to the problem (Morello-Frosch et al., 2009). Thus, efforts to educate the public about climate change must recognize that there are in fact multiple, different audiences within the American public—each requiring tailored information, outreach, and engagement efforts.

Communication Roadblocks and Other Influences on Attitudes

Obstacles to the effective communication of information about climate change include the nature of climate change itself, limitations in individual perception and decision making, structural barriers, and ineffective communication strategies. For example, the causes of climate change are largely invisible to the human eye. Every day, all around us, carbon dioxide, methane, nitrous oxide, and other greenhouse gases pour from automobiles, buildings, airplanes, feedlots, factories, and power plants. Yet these emissions are invisible and thus remain, at best, abstract knowledge, out of sight and out of mind. These emissions also have distant effects. The full impact of rising greenhouse gas concentrations in the atmosphere will not be felt for years into the future. Moreover, thus far the primary impacts of rising emissions have been felt far away from the cities and factories that produced them (e.g., in the polar regions). Unlike typical pollutants, many of which can be seen and smelled and have a direct impact on the area immediately surrounding the source, greenhouse gases are quickly mixed and diffused throughout the global atmosphere. Finally, climate science itself is highly complex, requiring the integration of diverse and complicated disciplines, and full of inherent uncertainties about the timing, likelihood, and severity of the impacts—all of which make this issue difficult for non-specialists to understand. Some barriers to understanding lie in the individuals themselves. An extensive range of cognitive biases and heuristics influence human decision making (Gilovich et al., 2002), including the reliance on inappropriate mental models to understand climate change. For example, many Americans do not understand the difference between weather and climate, leading some to interpret unusual cold weather as proof that global warming is not happening (Bostrom and Lashoff, 2007). Many others conflate climate change with the ozone hole, leading them to the (erroneous) conclusion that the best solution to climate change is to ban aerosol spray cans (Leiserowitz, 2007).

The first step to improving understanding may be for scientists and educators to recognize that they often understand key climate change concepts differently than the general public. In general, climate scientists investigate complex interactions within the earth system, a way of thinking that many non-scientists struggle with. Climate science itself is often described in abstract, statistical, or technical terms that have little resonance for most Americans (Hassol, 2008). Common examples include debates over how to limit global warming to 2° Celsius (made worse by the fact most Americans use Fahrenheit), atmospheric concentrations of 350, 450, or 550 parts per million, annual emissions of "gigatons of carbon," and "megawatts of energy" produced by different power sources. Scientists understand that the climate system has non-linear interactions, multiple stable states, and inherent unpredictability. In contrast, non-scientists sometimes struggle with the idea of positive and negative feedback loops, and may even misunderstand the term "positive feedback" as a good thing (see Table 8.1).

Members of the public often lack the spatial and geographic skills that earth scientists have developed. As climate change involves complex interactions and feedbacks as well as delayed impacts at multiple spatial scales, non-scientists often have difficulty understanding how their personal actions influence climate and how geographically

TABLE 8.1 Commonly Misunderstood Terms Relating to Climate Change.

What scientists say	Words More Easily Understood by the Public
Anthropogenic	Human caused
Spatial and temporal	Space and time
Positive feedback	Self-reinforcing cycle, vicious circle
Theory	Understanding of how something works
Degrees Centigrade	Degrees Fahrenheit
Solar radiation	Solar energy
Positive trend	Upward trend
Climate sensitivity	How the climate responds to more greenhouse gases in the atmosphere
Uncertainty	Range of possibilities
Enhanced greenhouse effect	Stronger greenhouse effect

SOURCE: Hassol (2008).

distant impacts relate to their current lives. Similarly, climate scientists often think about geologic time scales. Some investigate events that occurred millions of years ago, and envision an earth of the past that has little resemblance to the current planet. Because of their familiarity with long time scales, most geoscientists are comfortable imagining climate change's impacts centuries into the future. By contrast, most people's ability to conceptualize the future goes "dark" around 15 to 20 years, with 50 years as an upper limit (Nicholson-Cole, 2005; Tonn et al., 2006). At other times, some of those with little knowledge of climate change imagine future impacts that are often unscientific and unrealistically catastrophic (Nicholson-Cole, 2005). Many people have a mental image of climate change as very far away in time and space, unrelated to their actual experience (Leiserowitz and Feinberg, 2005; Nicholson-Cole, 2005). Some research has found, however, that those with a higher level of climate change understanding described impacts that were on a closer time scale, were more realistic, and related to their lives (Nicholson-Cole, 2005). For many Americans, climate change is often lost among the myriad of other day-to-day risks, including unemployment, terrorism, the H1N1 flu virus, nanotechnology, genetic engineering, nuclear power, toxic waste, obesity, tobacco, illegal drugs, violent crime, and health care, among many others. Most Americans have little time to focus on climate change, due to the day-to-day needs of parenting, working, paying bills, socializing, and the distractions of entertainment. Even individuals who are motivated to act often confront other barriers, such as the lack of capital to make household energy efficiency improvements; the lack of clean, safe, and affordable public transportation options; the lack of knowledge about effective actions; the ruts of personal habit and routine; lifestyle expectations; peer pressure; and pervasive advertising and marketing campaigns to increase consumption. Finally, some Americans view climate change through a political and ideological lens, leading them to be deeply suspicious of both climate science and proposed responses (Dunlap and McCright, 2008).

Although past communication efforts have successfully placed climate change on the national and international agenda, there has been limited change in public knowledge, attitudes, or behavior. Past communication efforts have often implicitly assumed that more information leads automatically to greater public concern and action, but this is often not the case. Studies in related fields demonstrate that although scientific information is often necessary for such change, it is often insufficient, especially for changing behavior (Chess and Johnson, 2007). Communication efforts that have attempted to evoke fear, guilt, or shame to motivate public action have often proven unsuccessful or have even backfired (Moser, 2009).

Another important barrier to improved public understanding in the United States has been the well-organized and well-financed campaigns by special interests, contrarian scientists, and defenders of the status quo to create public perceptions of scientific disagreement and to promote the idea that climate change is not happening, and that if it is happening, it is caused by natural factors or cycles, or will actually be a good thing (McCright and Dunlap, 2003; Menestrel et al., 2002; Oreskes, 2004).

Finally, media—the primary way most Americans learn about climate change—is currently undergoing a massive and historic structural transformation, with many newspapers and broadcast networks cutting science and environment reporters and news desks (Jones, 2009; Meyer, 2004). At the same time, new online social media are emerging as alternative sources of climate and scientific information. However, the promise and perils of this shift to less professionalized, more decentralized, fragmented, and often more biased information sources are still evolving, with uncertain consequences for public understanding (Brown, 2005; Jones, 2009).

These many barriers make communication about climate change challenging. Yet studies of public opinion demonstrate that, despite all of these constraints, the American public does have some understanding (albeit limited) of climate change, its causes, potential impacts, and solutions. A majority of Americans view it as a serious threat, support a wide variety of public policies, and express a willingness to change some of their own behaviors, especially those around saving energy at home and on the road, and as consumers.

Guidelines for Effective Communication

Information campaigns are often used to help achieve different social goals, including raising awareness; improving public understanding of the causes, consequences, and potential responses to hazards; informing consumer choices (e.g., EnergyStar labels); and motivating behavioral changes, such as promoting energy efficiency and conservation, developing emergency plans, taking proactive protective action (e.g., vaccines), and evacuating to avoid extreme weather events (e.g., hurricanes). Different goals require different strategies; thus, any communication effort must start by clearly articulating what it is trying to achieve.

Likewise, different actors in American society often pursue different communication goals: public health officials work to reduce risky health behaviors (e.g., smoking, obesity, and drug use); scientists seek to raise public and policy maker awareness and understanding of new hazards; government agencies seek to provide concrete and usable information to specific stakeholders (e.g., investors, farmers, and natural resource managers); while advocacy groups may attempt to raise public support for or opposition to proposed public policies.

Some information campaigns are essentially one-way, such as from government agencies to the public at large (e.g., public service announcements). These may be most effective at raising public awareness of a risk (such as an impending hurricane). Some information campaigns use formal procedures, such as the development of school curricula, while others work through informal processes, such as changes in social norms, activation of social networks, and the influence of opinion leaders. Other information campaigns are participatory, with experts and different stakeholders within the public engaging in extended dialogue and structured decision making about an issue or hazard (e.g., siting new facilities or changing zoning rules).

Making communications interactive and including citizens as participants in the decision-making process are key components of participative and deliberative decision making and can improve the effectiveness of decision making and action. Deliberative processes provide citizens with basic information about a problem, allow them to process and discuss it through structured deliberations, and reach considered conclusions. The focus is on active rather than passive involvement and on process rather than results (Blowers et al., 2005; Edwards et al., 2008). Events are often described as conversations (rather than one-way information provision) and begin by encouraging participants to ask questions. The recent National Research Council report on public participation in environmental decision making (NRC, 2008b) offers additional recommendations on effective participation, such as inclusive and transparent involvement, agreement on procedural rules, in-person interaction to build trust, attention to both facts and values, being explicit about assumptions, and iteration to adapt to new information, evaluations, and participants.

Many governmental, non-governmental, and academic organizations use social media outlets to convey information to the general public. For example, NOAA uses tools such as Twitter, Facebook, YouTube, and podcasts to reach broad audiences. Non-governmental organizations (NGOs) like the World Wildlife Fund are currently utilizing tools such as those mentioned above and additional social networking websites to interact with the public and encourage supporters to recruit others with similar interests. Universities are also increasingly using new social media as an integral part of instruction and to keep their current and prospective students informed of current events and campus news. These and other online communities can diffuse information rapidly through a population. The information is often relatively short and easy to understand, which facilitates sharing and can help those with little background on a subject to understand basic concepts and take meaningful action.

An effective national response to climate change will involve many different goals, actors, and approaches. Although the goals are varied and different actors will use different strategies, certain principles are relevant in each of these contexts (see Table 8.2 and Box 8.2).

One of the first rules of effective communication is to "know thy audience." Formative research to identify the target audience and their needs is thus a critical first step. Communicators of all types need to know exactly who their audiences are (demographically, geographically, politically, and what communication channels they rely upon), including what they currently know (and misunderstand) about the issue; how they perceive the risks, costs, and benefits of acting (or not acting); whether they understand how they might choose, decide, or act differently; what specific barriers they confront; and whether they believe their actions would be effective. Using local people who are already familiar with an audience may help non-local communicators better express their message (see Box 8.3), especially if they are trusted leaders of community groups such as churches or other social organizations.

This initial understanding of the audience and its specific needs should then guide the production and dissemination of information to meet these needs. In general, the information to be provided needs to be understandable, memorable, useful, and timely. It should be communicated through the channels that the target audience pays attention to, using credible messengers that the audience trusts. For optimal effectiveness, information campaigns should incorporate evaluation and feedback mechanisms to assess what works and what does not, using an iterative, adaptive learning model.

TABLE 8.2 Guidelines for Effective Climate Change Communication.

Principle	Example
Know your audience	There are different audiences among the public. Learn what people (mis)understand before you deliver information and tailor information for each group.
Understand social identities and affiliations	Effective communicators often share an identity and values with the audience (e.g., a fellow CEO or mayor, parent, co-worker, religious belief, or outdoor enthusiast).
Get the audience's attention	Use appropriate framing (e.g., climate as an energy, environmental, security, or economic issue) to make the information more relevant to different groups.
Use the best available, peer-reviewed science	Use recent and locally relevant research results. Be prepared to respond to the latest debates about the science.
Translate scientific understanding and data into concrete experience	Use imagery, analogies, and personal experiences including observations of changes in people's local environments. Make the link between global and local changes. Discuss longer time scales, but link to present choices.
Address scientific and climate uncertainties	Specify what is known with high confidence and what is less certain. Set climate choices in the context of other important decisions made despite uncertainty (e.g., financial, insurance, security, etc.). Discuss how uncertainty may be a reason for action rather than a reason for inaction.
Avoid scientific jargon and use everyday words	Degrees F rather than Degrees C "Human caused" rather than "anthropogenic" "Self-reinforcing" rather than "positive feedback" "Range of possibilities" rather than "uncertainty" "Likelihood" or "chance" rather than "probability" "Billion tons" rather than "gigatons"
Maintain respectful discourse	Climate change decisions involve diverse perspectives and values.
Provide choices and solutions	Present the full range of options (including the choice of business as usual) and encourage discussion of alternative choices.
Encourage participation	Do not overuse slides and one-way lecture delivery. Leave time for discussion or use small groups. Let people discuss and draw their own conclusions from the facts.
Use popular communication channels	Understand how to use new social media and the internet.

TABLE 8.2 Continued

Principle	Example
Evaluate communications	Assess the effectiveness of communications, identify lessons learned, and adapt.

SOURCES: CRED (2009), Nerlich et al. (2010).

BOX 8.2
Public Health Communication

Campaigns to support improved public health decision making and behavior have a long history in the United States, dating back at least to 1721, when Cotton Mather used pamphlets and personal appeals to encourage the citizens of Boston to get inoculated against smallpox (Gross and Sepkowitz, 1998; Paisley, 2001). In the 19th and early 20th centuries, public information campaigns were used to fight slavery, child labor, and tuberculosis. Other well-known national examples include the Smokey the Bear campaign against forest fires, heart disease prevention, anti-tobacco and anti-drug campaigns, skin cancer prevention, seat belt use, mammography screening, and campaigns for improved diets and physical exercise. The U.S. Congress directly mandated two large national campaigns: the $1 billion national youth anti-drug campaign and the Youth Physical Activity Campaign (the VERB Campaign), while recent tobacco court settlements mandated support for the anti-teen-smoking "Truth" campaign (Snyder, 2007).

Meta-analysis evaluations have generally found that overall, health campaigns can have a positive, if limited, impact on public health knowledge, attitudes, and behaviors. For example, Snyder and Hamilton (2002:375) found that, on average, campaigns led to "9% more people performing the behavior after the campaign than before" across a variety of health issues. Other researchers note, however, that many prior campaigns have not followed effective design principles, while well-designed campaigns can have significantly larger effects (Noar, 2006) (see Table 8.2 for a summary of effective campaign design principles).

In addition, health communication researchers have found that the more information is tailored to the needs of a particular audience, the more effective it is; trusted opinion leaders are especially effective messengers; and public participation, discussion, and dialogue can greatly increase the reach and impact of an information campaign. There is, however, often a tradeoff between reach and impact. A mass media campaign "with a small-to-moderate effect size that reaches thousands [or millions] of people will have a greater impact [than] an individual or group-level intervention with a larger effect size that only reaches a small number of people" (Noar, 2006:36). Each approach is valuable but may be better suited for different goals. For example, mass media campaigns may be better at raising national awareness and basic knowledge about a risk, while behavior change may be more effectively encouraged through grassroots campaigns involving friends, families, co-workers, and local communities. Finally, information campaigns are often more effective when they are combined with new incentives, supporting infrastructure, or the enforcement of laws and regulations (Maibach et al., 2008; Witte and Allen, 2000).

BOX 8.3
Using Local Opinion Leaders

Personal interaction can be an important component of climate change communication to the general public. Institutions, such as universities or non-profit organizations, can provide the data and other resources that individuals may not be able to obtain on their own. They can also provide training that can teach important communication or hands-on skills, such as public speaking or conducting energy audits. Drawing upon these resources, communicators who understand the local geographic and social landscape can target their messages to match their audiences' values and interests. Providing an ongoing "shared space" for these volunteer communicators to learn from one another can reinforce their enthusiasm and improve their messaging.

One such program is the Oxfordshire ClimateXChange, run by the Environmental Change Institute at Oxford University. This project connects local people interested in climate change who want to communicate about it to others in their schools, churches, and towns. Events have included weatherization training, skill sharing within the group, and showcases of environmental and energy government agencies. The program's website has social networking features, offering the opportunity for people to continue the conversation between events. The program also provides resources to the "Climate Explorers," including informational articles, a film/DVD library, "pub quiz" questions, electronic presentations, and an energy monitor lending program. These events and resources equip ClimateXChange members to speak accurately and effectively to people they know in their communities. Meeting similar people in the same geographic area who are doing related work allows for collaborations. One networking meeting attracted representatives from 30 different local groups. Overall, the project has reached over 13,000 people through more than 120 events.

The informational needs of American society in response to climate change range from basic awareness and understanding of the problem itself to extremely technical information used only by specialists in specific fields. Communicators at all levels of government and across all sectors of society will thus need to provide a wide variety of different information types for different audiences, from individual households to the nation as a whole. Both scientists and decision makers should seek ways to effectively communicate the complex issues of climate change to others.

COMMUNICATION AND EDUCATION FOR DECISION MAKERS

As citizens, policy and decision makers can also benefit from improved climate change communications to the public. Those communicating with policy makers should keep in mind the principles outlined in the previous section and in Table 8.2. But given their

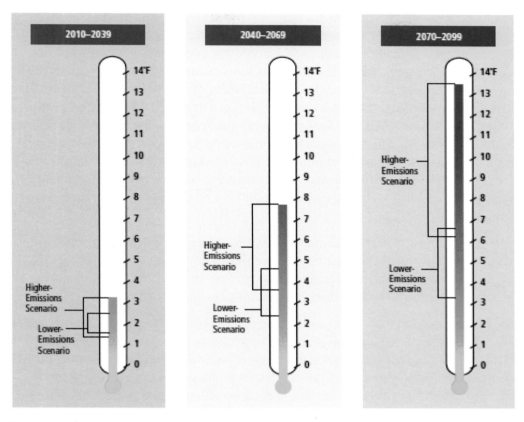

FIGURE 8.5 Changes in Northeast regional summertime temperatures. SOURCE: Union of Concerned Scientists (2006).

unique powers and responsibilities, policy makers require specific efforts to build their capacity to make effective choices. Decision makers are often faced with a variety of near-term climate-related choices that will have implications for generations to come. Providing them with the climate information required to make those decisions is crucial (see Figure 8.5). As just one example, local, state, and national policy makers are constantly making decisions about public infrastructure investments (e.g., roads, bridges, hospitals, and schools) that should incorporate considerations of climate change.

Previous chapters discuss a range of information systems and decision support tools that can assist policy makers in making climate decisions—such as climate and greenhouse gas information services and models for evaluating choices—but these will be ineffective if those making decisions have an incomplete or inaccurate understanding of climate change and the options for reducing the risks. It is important to recognize that many decision makers are extremely busy, balancing competing priorities, and as leaders may find it difficult to admit to an inadequate understanding of the science or economics of climate change. For Congress, long-established services such as the Congressional Research Service or Government Accountability Office reports provide concise summaries of key issues (as does the National Research Council) and a number of think tanks and NGOs also seek to provide summaries of climate change science and options for decision makers including those in the private sector. As with citizens, decision makers in the private and public sectors are most likely to desire information that is directly relevant to their localities and jurisdictions, uses a minimum of jargon, respects their expertise, is clear about uncertainties, and takes account of their context, interests, and needs.

As with other groups, climate information may be best communicated by someone who is familiar and trusted by a particular group of policy makers—by a fellow CEO or mayor, a former political colleague, a senior staff person or a constituent—and in a venue where questions can be asked and differences aired in confidence.

For those professionals who are asked to take on new responsibilities for reporting or responding on climate change within their organization, there is a need for capacity building and training to help them in their new roles, and to build expertise in appropriate areas of climate science and policy. This might involve short courses or professional certification such as those emerging in carbon finance and project development, although there is a need to ensure that such courses are of high quality.

While the scientific community has achieved great progress in the identification, description, and projection of the risks of climate change, the informational needs of the nation—to either limit future warming or adapt to climate impacts—now extend far beyond the relatively narrow boundaries of climate science. As described in Chapter 7, meeting the rising demand for information will require a significant investment by governments and other organizations in developing, validating, and providing high quality information about the causes, consequences, and potential responses to climate change.

CONCLUSIONS AND RECOMMENDATIONS

Current and future students, the broader public, and policy makers need to understand the causes, consequences, and potential solutions to climate change; develop scientific thinking and problem-solving skills; and improve their ability to make informed decisions. To achieve these goals, the United States needs to make considerably more progress in national, state, and local climate education standards, climate curriculum development, teacher professional development, and production of supportive print and web materials. Hands-on or experiential approaches are particularly effective ways to promote learning among students. The United States also needs a national strategy and supporting network to coordinate climate change education and communication activities for policy makers and the general public, including the identification of essential informational needs; development of relevant, timely, and effective information products and services; construction and integration of information dissemination and sharing networks; and continuous evaluation and feedback systems to establish which approaches work best in what circumstances.

The panel judges the following 5 elements as important guidelines for all climate education and communication programs to help people think deliberately, responsibly, and respectfully about climate change and the many related decisions they will face. All such programs should

1. **Be based on the best available, peer-reviewed science.** Accurate science, based on the latest data and analysis, must lie at the core of any education activity. Educational content should be derived from respected scientific sources such as the IPCC and reports from the U.S. Global Change Research Program such as the *Global Climate Change Impacts in the United States* (USGCRP, 2009). Education activities should be careful to avoid exaggerations or misrepresentation of the science. Climate change education, like environmental education generally, is much more than just natural science. In addition to the physical climate system, climate change education must also include other critical dimensions of the issue, including the human drivers of greenhouse gas emissions; energy efficiency and conservation; renewable energy, carbon capture and storage, and other options for limiting climate change; issues of social vulnerability to climate change; options for adaptation; and the economic, political, psychological, social, cultural, and moral dimensions of the issue. It should also help students understand risk management and learn how to use this framework in climate-related decision making.

2. **Use examples, images, language, and units of measure that are accessible and relevant to the American public and decision makers.** Scientists must translate their information and findings into the language and units of everyday life. For example, use degrees Fahrenheit instead of Celsius, talk about the possible range of results rather than uncertainty, and use examples that relate to food, health, water, and familiar ecosystems.

3. **Provide linkages between global and local activities.** Climate change affects people from the local to the global scale, and at different places in differ-

ent ways at different times. All localities in the United States produce greenhouse gas emissions and all will experience impacts in one form or another; therefore, all need to be part of climate solutions. As students learn about climate change, they should understand both the local and the global contexts, that climate change involves the entire Earth, with interactions among the atmosphere, oceans, land, and life as well as human systems, including agriculture, industry, transportation, consumer markets, and social values.

4. **Include a focus on longer-term time scales, but connect to the present.** Decisions made today will have very important and long-term consequences for the climate of the future. Decisions that make sense from the perspective of the short term may not make sense from the perspective of a longer time frame. For young people, this often involves a fundamental shift from thinking merely days and months into the future to thinking about years, decades, and beyond. Yet climate change will affect them, as they become adults, in profound and far-reaching ways, and thus can provide powerful connections to their own life scales and time frames.

5. **Maintain respectful discourse.** Climate change decisions involve a wide range of perspectives, including not just the complexities of natural and social science but also divergent social, political, environmental, religious, and ethical values and views of the proper role of individuals, the private sector, and government in responding to climate change. Thus, climate educators and communicators at all levels of society should set a tone of respect for diverse perspectives and an open and honest consideration of the implications of various responses to climate change. When discussion moves from core scientific concepts to more complex issues of societal values, students should learn how to engage in responsible and respectful discourse and debate as well as critical thinking and analysis skills.

At the federal level, support for climate education is scattered across several federal agencies and programs, notably NOAA, NASA, and NSF. While there are nascent efforts among these agencies to collaborate around climate education, this collaboration needs a more formal structure and a clear mandate to contribute to an overarching set of national goals for climate science education, with clear objectives and measures of success. *Climate Literacy: The Essential Principles of Climate Science*, cited above, provides an early example of the benefits of such federal coordination. A national education and communication network would help support, integrate, and synergize diverse efforts by sharing best practices and educational resources; building collaborative partnerships; and leveraging existing education, communication, and training networks across the country.

The challenge to science and education has been seen and met before. The National Defense Education Act of 1958, in response to Sputnik, fundamentally strengthened our nation's science, mathematics, engineering, and technology education. Thirty years later, the Global Change Research Program, including NASA's Mission to Planet Earth, created a fundamental step-change in graduate education and research in America's universities and colleges. It established an ambitious target: to increase our understanding of the environment and improve our ability to predict changes on a global scale. This broad initiative needs to be both focused and revitalized: if the nation desires to develop a national strategy and resources to support climate change education and communication, a national climate education act could serve as a powerful response to the educational challenges of climate change. It would have the advantage of a single focal point of congressional action and would provide an integrated federal strategy and funding. It could also include integrated support for informal science education, university-level initiatives, and workforce development in climate-related fields.

Since states define their own educational standards, state-based reform is critical. State education agencies have already begun revising their educational standards to include climate and energy literacy, as well as the "21st-century thinking skills" of engineering, problem solving, systems thinking, teamwork, and communications (Hoffman and Barstow, 2007).

The United States needs a better base of knowledge and expertise in climate education. These needs include research to establish priority learning goals, development of effective methods in climate education and innovative approaches to assessment, and conduct of national surveys of current practice at state and local levels. Research is also needed to understand students' correct and incorrect mental models about climate change, barriers to learning and understanding, and learning pathways to adequately address climate change.

For the broader public, many barriers to public understanding and engagement with climate change science and responses exist, including the nature of climate change itself, limitations in individual perception and decision making, structural barriers, and ineffective communication strategies. Despite these barriers, however, majorities now believe it is real, happening, human caused, and a serious threat. Likewise, majorities want their elected officials at all levels to take more action and support a variety of policies to reduce national greenhouse gas emissions. Many Americans are interested in making individual changes to save energy and reduce their own contributions to climate change, but still confront critical obstacles such as up-front capital costs and lack of knowledge about what actions to take (Leiserowitz, 2007).

Perhaps most important, Americans express a clear desire for more information about climate change, including how it might affect their local communities and how they as individuals and the nation as a whole can act to reduce greenhouse gas emissions and prepare for the impacts. Thus, today's adults evidence a critical need for more and improved information about climate change at all levels and across all sectors of American society. Local, state, and national governments must play an important role, but so too will the private sector, civil society, the mass media, and individual Americans within their own social networks. Climate communicators can learn from information campaigns in other domains, such as public health. For example, communicators should start by clearly defining the goals of the campaign: to merely inform and educate individuals about climate change, to encourage societal action and behavioral change, or to encourage deeper changes in social norms and cultural values? Successful campaigns must also identify the specific target audience, the message frame (see Chapter 1), the core message itself, the best messengers, and most effective communication channels (Moser, 2009) (see Table 8.2). It is also critical that all communication campaigns include evaluations and metrics to assess campaign effectiveness.

The panel judges that the nation needs a national strategy and supporting network to coordinate climate change education and communication activities. If the nation so desires, a task force could be convened to assess the current state of formal and informal climate change formal and informal education and communication in the United States, identify knowledge gaps and opportunities, and evaluate the advantages and disadvantages of different national organizational structures to promote climate change education and communication. This will require coordination between relevant organizations (e.g., federal, state and local agencies, and public and private sector organizations involved in kindergarten through adult education) and increased federal funding for research on education and communication. The federal agencies that manage research activities mandated under the U.S. Global Change Research Act could choose to establish a research program to

- Establish baseline levels of public understanding and responses to climate change and monitor changes in American climate literacy, including knowledge, risk perceptions, and behavior;
- Assess the effectiveness of different climate change education and communication strategies and programs; and
- Provide federal support to increase the capacity of educational institutions, scientists, and students to collaborate with diverse groups and stakeholders needing climate change information.

The federal government could also choose to

- Promote teacher training programs for climate education;
- Develop climate change–related educational tools, materials, and technologies, including web-based materials;
- Set national climate education goals and provide support to states to design and implement climate education standards; and
- Provide guidelines and support for climate change education in informal environments such as museums, zoos, and aquariums.

States could choose to integrate principles of climate literacy into educational standards, such as:

- Expand the definition of climate education beyond the physical science of climate to the interdisciplinary sciences, including the social sciences, needed to respond to climate change;
- Share their expertise and experience, through such groups as the National Coalition for State Science Supervisors;
- Develop and share methods for teacher professional development, and for assessing student learning; and
- Provide guidelines and resources to local schools to implement climate education standards.

Recommendation 10:

The federal government should establish a national task force that includes formal and informal educators, government agencies, policymakers, business leaders, and scientists, among others, to set national goals and objectives, and to develop a coordinated strategy to improve climate change education and communication.

References

Abrahamse, W., L. Steg, C. Vlek, and T. Rothengatter. 2005. A review of intervention studies aimed at household energy conservation. *Journal of Environmental Psychology* 25:273-291.

Abrahamson, E. 2009. Necessary conditions for the study of fads and fashions in science. *Scandanavian Journal of Management* 25(2):235-239.

ACIA (Arctic Climate Impact Assessment). 2004. *Impacts of a Warming Arctic: Arctic Climate Impact Assessment.* Cambridge, UK: Cambridge University Press.

ACUPCC (American College and University President's Climate Committment). 2009. *Education for Climate Neutrality and Sustainability: Guidance for ACUPCC Institutions.* Boston, MA: ACUPCC.

AGO (Australian Greenhouse Office). 2006. *Climate Change Impacts and Risk Management: A Guide for Business and Government.* Canberra: AGO.

Allan, C., and G. H. Stankey. 2009. Adaptive environmental management: A practitioner's guide. Dordrecht, The Netherlands: Springer Science & Business Media.

Allen, M. 2003. Liability for climate change. *Nature* 421(6926):891-892.

Allen, M. R., D. J. Frame, C. Huntingford, C. D. Jones, J. A. Lowe, M. Meinshausen, and N. Meinshause. 2009. Warming caused by cumulative carbon emissions towards the trillionth tonne. *Nature* 458:1163-1166.

Ambrose, S. E. 2000. *Nothing Like It in the World: The Men Who Built the Transcontinental Railroad, 1863-1869.* New York: Simon & Schuster.

Associated Press/Stanford/GfK Roper Poll. 2009. *The Associated Press-Stanford University Environment Poll.* Available at *http://www.ap-gfkpoll.com/pdf/AP-Stanford_University_Environment_Poll_Topline.pdf.*

Ayres, R. U. 1995. Life cycle analysis: A critique. *Resources, Conservation and Recycling* 14:199-223.

Bain, D. H. 1999. *Empire Express: Building the First Transcontinental Railroad.* New York: Viking. Available at *http://lcweb.loc.gov/catdir/toc/99033375.html.*

Barstow, D., E. Frost, L. Liben, S. Ride, and R. Souviney. 1999. *Final Report to the National Science Foundation: Visualizing Earth.* Arlington, Virginia. National Science Foundation.

Bartels, D. 2001. Wartime mobilization to counter severe global climate change. *Human Ecology* (Special Issue 10):229-232.

Beckerman, W., and C. Hepburn. 2007. Ethics of the discount rate in the Stern Review on the Economics of Climate Change. *World Economics* 8(1).

Benders, R. M. J., R. Kok, H. C. Moll, G. Wiersma, and K. J. Noorman. 2006. New approaches for household energy conservation— In search of personal household energy budgets and energy reduction options. *Energy Policy* 34 (18):3612-3622.

Bierbaum, R. M., D. G. Brown, and J. L. McAlpine, eds. 2008. *Coping with Climate Change: National Summit Proceedings.* Ann Arbor, MI: The Regents of the University of Michigan.

Bin, S., and H. Dowlatabadi. 2005. Consumer lifestyle approach to US energy use and the related CO_2 emissions. *Energy Policy* 33(2):197-201.

Birkland, T. A. 1997. *After Disaster: Agenda Setting, Public Policy and Focusing Events.* Washington, DC: Georgetown University Press.

Black, R. 2001. *Environmental Refugees: Myth or Reality?* Working Paper 34. Falmer, Brighton: University of Sussex.

Blakes. 2009. *Laying the Foundation for Cap-and-Trade: WCI Releases Final Essential Requirements of Mandatory Reporting.* Available at *http://www.blakes.com/english/view.asp?ID=3300.*

Blowers, A., J. Boersema, and A. Martin. 2005. Experts, decision making and deliberative democracy. *Environmental Sciences* 2(1):1-3.

Boardman, A. E., D. H. Greenberg, A. R. Vining, and D. L. Weimer. 2001. *Cost-Benefit Analysis: Concepts and Practice*, 2nd edition. Upper Saddle River, NJ: Prentice Hall.

Boardman, B. 2004. Achieving energy efficiency through product policy: The UK experience. *Environmental Science and Policy* 7(3):165-176.

Bodansky, D. 2001. The history of the global climate change regime. *International Relations and Global Climate Change* 23-40.

Boden, T.A., G. Marland, and R.J. Andres. 2009. Global, Regional, and National Fossil-Fuel CO2 Emissions. Carbon Dioxide Information Analysis Center, Oak Ridge National Laboratory, U.S. Department of Energy, Oak Ridge, Tenn., U.S.A. doi 10.3334/CDIAC/00001.

Böhringer, C. 2003. The Kyoto Protocol: A review and perspectives. *Oxford Review of Economic Policy* 19(3):451-466.

Bonney, R., C. B. Cooper, J. Dickinson, S. Kelling, T. Phillips, K. V. Rosenberg, and J. Shirk. 2009. Citizen science: A developing tool for expanding science knowledge and scientific literacy. *BioScience* 59(11):977-984.

Bostrom, A. 1994. What do people know about Global Climate Change 1 & 2: Mental models. *Risk Analysis (An International Journal)* 14(6):959-982.

Bostrom, A., and D. Lashof. 2007. Weather or climate change? In *Creating a Climate for Change: Communicating Climate Change and Facilitating Social Change*, S. C. Moser and L. Dilling, eds. New York: Cambridge University Press.

Bowen, A. and N. Ranger. 2009. Mitigating climate change through reductions in greenhouse gas emission: the science and economics of future paths for global annual emissions. *Centre for Climate Change Economics and Policy.* Policy Brief December 2009.

Boyd, E., N. E. Hultman, J. T. Roberts, E. Corbera, J. Ebeling, D. M. Liverman, K. Brown, R. Tippman, J. C. Cole, P. Mann, M. Kaiser, M. Robbins, A. G. Bumpus, A. Shaw, E. Ferreira, A. Bozmoski, C. Villiers, and J. Avis. 2007. The Clean Development Mechanism: An assessment of current practice and future approaches for policy. In *Tyndall Centre Working Paper 114*. Norwich: Tyndall Centre.

Boykoff, J., and M. Boykoff. 2004. Journalistic Balance as Global Warming Bias. FAIR: Fairness and Accuracy in Reporting. Global Warming and the U.S. prestige press. Global Environmental Change 14 (2):125-136.

Boykoff, M. T. 2007. Flogging a dead norm? Newspaper coverage of anthropogenic climate change in the United States and United Kingdom from 2003 to 2006. *Area* 39(4):471-481.

Brewer, G. 1975. *An Analysis View of the Uses and Abuses of Modeling of Decision Making.* Santa Monica: Rand Corporation.

Brooks, N., and W. N. Adger. 2005. Assessing and enhancing adaptive capacity. In *Adaptation Policy Frameworks for Climate Change*, B. Lim, E. Spanger-Siegfried, I. Burton, E. L. Malone, and S. Huq, eds. New York: Cambridge University Press.

Brown, L. R. 2009. A wartime mobilization. *Earth Policy Institute* 2008.

Brown, M. 2005. Abandoning the news. *Carnegie Reporter* 3(2).

Brown, M., J. Chandler, M. Lapsa, and B. Sovacool. 2008. *Carbon Lock-In: Barriers to Deploying Climate Change Mitigation Technologies.* Oak Ridge, TN: Oak Ridge National Laboratory.

Bulkeley, H., and M. M. Betsill. 2003. *Cities and Climate Change: Urban Sustainability and Global Environmental Governance.* London: Routledge.

Bumpus, A. G. 2009. *Carbon Development? A Political Ecology Analysis of Carbon Offsets: Governance, Materialities and Local Development in Honduras.* Oxford: University of Oxford.

Bumpus, A. G., and D. M. Liverman. 2008. Accumulation by decarbonization and the governance of carbon offsets. *Economic Geography* 84(2):127-155.

California State. 2006. California, New York Agree to Explore Linking Greenhouse Gas Emission Credit Trading Markets; Gov. Schwarzenegger Tours Carbon Trading Floor, Press Release, Sacramento: Office of the Governor. October 16, 2006.

Call, J., and J. Hayes. 2007. A description and comparison of selected forest carbon registries: A guide for states considering the development of a forest carbon registry. Washington D.C: USDA Forest Service.

Calvin, K., J. Edmonds, B. Bond-Lamberty, L. Clarke, S. H. Kim, P. Kyle, S. J. Smith, A. Thomson, and M. Wise. 2009. Limiting climate change to 450 ppm CO_2 equivalent in the 21st century. *Energy Economics* 31:S107-S120.

Campbell, K. M., and C. Parthemore. 2008. National security and climate change in perspective. In *Climatic Cataclysm: The Foreign Policy and National Security Implications of Climate Change*, K. M. Campbell, ed. Washington, DC: Brookings Institution Press.

Caponi, F., S. Cheng, G. Daigger, M. Hartman, J. Sandoval, R. Schmidt, and J. Witherspoon. 2008. The impacts of the California Global Warming Solutions Act of 2006 (AB 32) on POTWs. *Proceedings of the Water Environment Federation* 2008(4):752-762.

Cash, D. W. 2001. In order to aid in diffusing useful and practical information: Agricultural extension and boundary organizations. *Science, Technology and Human Values* 26(4):431-453.

Cash, D. W., W. C. Clark, F. Alcock, N. M. Dickson, N. Eckley, D. H. Guston, J. Jager, and R. B. Mitchell. 2003. Knoledge systems for sustainable development. *Proceedings of the National Academy of Sciences of the United States of America* 100(8):8086-8091.

CCAP (Chicago Climate Action Plan). 2008. *Corporate Risk Case Study: City of Chicago Climate Change Task Force*, O. Wyman, ed. Chicago: City of Chicago Climate Change Task Force.

CCAR (California Climate Action Registry). 2009. Industry Specific Protocols. *http://www.climateregistry.org/tools/protocols/industry-specific-protocols.html*. Accessed August 23, 2009.

CCSP (Climate Change Science Program). 2007. *Scenarios of Greenhouse Gas Emissions and Atmospheric Concentrations, Sub-report 2.1A of Synthesis and Assessment Product 2.1 by the U.S. Climate Change Science Program and the Subcommittee on Global Change Research*, L. Clarke, J. Edmonds, H. Jacoby, H. Pitcher, J. Reilly, and R. Richels, eds. Washington, DC: Department of Energy, Office of Biological and Environmental Research.

———. 2009. *Best Practice Approaches for Characterizing, Communicating, and Incorporating Scientific Uncertainty in Decisionmaking*. Synthesis and Assessment Product 5.2. Washington, DC: U.S. CCSP.

CCSP and The Subcommittee on Global Change Research 2008. *Our Changing Planet: The U.S. Climate Change Science Program for Fiscal Year 2009*. Washington, DC: CCSP Office.

CDP (Carbon Disclosure Project). 2008. *Carbon Disclosure Project Cities Pilot Project 2008*. London: CDP.

———. 2009a. *Carbon Disclosure Project Report 2009: Global 500*. London: CDP.

———. 2009b. Carbon Disclosure Project: Walmart instructs suppliers to report through Carbon Disclosure Project. London: CDP.

CDSB (Climate Disclosure Standards Board). 2009. *Climate Disclosure Standards Board (CDSB)—Home*. Available at *http://www.cdsb-global.org/*.

Center for Health and the Global Environment. 2006. *Climate Change Futures: Health, Ecological and Economic Dimensions*. Cambridge: Harvard Medical School.

Chakravarty, S., A. Chikkatur, H. de Coninck, S. Pacala, R. Socolow, and M. Tavoni. 2009. Sharing global CO_2 emission reductions among one billion high emitters. *Proceedings of the National Academy of Sciences of the United States of America* 106(29):11884-11888.

Chen, C. C., B. A. McCarl, and H. Hill. 2002. Agricultural value of ENSO information under alternative phase definition. *Climate Change* 54(3):305-325.

Chess, C & Johnson, B. 2007. Information is not enough. In S. Moser, & L. Dillon (Eds.), *Creating a climate for change: Communicating climate change and facilitating social change* (223-237). New York and Oxford, UK: Oxford University Press

City of Chicago. 2008. J. Parzen, ed. *Lessons Learned: Creating the Chicago Climate Action Plan*. Chicago: City of Chicago Department of Environment.

ClimateInteractive. C-ROADS-CP Flier. Available online at *http://climateinteractive.org/simulations/C-ROADS/c-roads-cp/croads%20twopager%201029v6.pdf/view*.

Cline, W. R. 1992. *The Economics of Global Warming*. Washington, DC: Peterson Institute for International Economics.

CNA Corporation. 2007. *National Security and the Threat of Climate Change*. Alexandria, VA: CNA Corporation.

———. 2009. *Powering America's Defense: Energy and the Risks to National Security*. Alexandria, VA: CNA Corporation.

Cogan, D. G. 2008. Corporate Governance and Climate Change: The Banking Sector. Boston, MA: CERES.

Collins, D. 2000. *Management Fads and Buzzwords: A Critical-Practical Perspective*. London: Routledge.

Collins, M. 2007. Ensembles and probabilities: A new era in the prediction of climate change. *Philosophical Transactions of the Royal Society A* 365(1857):1957-1970.

Convery, F. J., and L. Redmond. 2007. Market and price developments in the European Union emissions trading scheme. *Review of Environmental Economics and Policy* 1(1):88-88.

Corfee-Morlot, J. 2009. *California in the Greenhouse: Regional Climate Change Policies and the Global Environment.* Geography Department, University College of London, London.

Correll, R. 1990. The United States Global Change Research Program: An overview and perspectives on the FY 1991 program. *Bulletin of the American Meteorological Society* 71(4):507-511.

Cowell, S. J., R. Fairman, and R. E. Lofstedt. 2002. Use of risk assessment and life cycle assessment in decision making: A common policy research agenda. *Risk Analysis* 22:879-894.

CRDI (Climate Risk Disclosure Initiative). 2007. *Global Framework for Climate Risk Disclosure: A Statement of Investor Expectations for Comprehensive Corporate Disclosure.* Boston, MA: Ceres, Inc.

CRED (Center for Research on Environmental Decisions). 2009. *The Psychology of Climate Change Communication: A Guide for Scientists, Journalists, Educators, Political Aides and the Interested Public.* New York: Columbia University..

CRS (Congressional Research Service). 1990. *Mission to Planet Earth and the U.S. Global Change Research Program.* CRS Report for Congress, 90-972.

———. 2000. *Global Climate Change: A Survey of Scientific Research and Policy Reports,* W. A. Morrissey, ed. Report RL30522. Washington, DC: CRS.

Cyert, R. M., and J. G. March. 1992. *A Behavioral Theory of the Firm.* New York: Wiley.

Darby, S. 2006. *The Effectiveness of Feedback on Energy Consumption: A Review for DEFRA of the Literature on Metering, Billing and Direct Displays.* Oxford: Environmental Change Institute.

Dasgupta, P. 2007. The Stern Review's Economics of Climate Change. *National Institue Economic Review* 199(1).

DEFRA (Department of Environment, Food, and Rural Affairs) and DECC (Department of Energy and Climate Change). 2009. *Draft Guidance on How to Measure and Report Your Greenhouse Gas Emissions.* DEFRA/DECC, U.K.

Delaney, B. T., B. Pe, and B. Previdi. 2008. Voluntary greenhouse gas reporting—not an option for much longer. *Environmental Claims Journal* 20(2):110-123.

Den Elzen, M. G. J., and P. Lucas. 2005. The FAIR model: A tool to analyse environmental and costs implications of climate regimes. *Environmental Modeling & Assessment* 10(2):115-134.

DeSombre, E. R. 2000. The experience of the Montreal Protocol: Particularly remarkable, and remarkably particular. *UCLA Journal of Environmental Law and Policy* 19:57-62.

Dickinson, J., ed. 2009. *Inventory of New York City Greenhouse Gas Emissions: A Greener, Greater New York.* New York: Mayor's Office of Long Term Planning and Sustainability.

Dietz, T., and P. C. Stern. 1998. Science, values, and biodiversity. *BioScience* 48:441-444.

Dietz, T., G. T. Gardner, J. Gilligan, P. C. Stern, and M. P. Vandenbergh. 2009. Household actions can provide a behavioral wedge to rapidly reduce US carbon emissions. *Proceedings of the National Academy of Sciences of the United States of America* 106(44):18452-18456.

Dilling, L. 2007. Towards science in support of decision making: Characterizing the supply of carbon cycle science. *Environmental Science and Policy* 10:47-61.

DiMaggio, P. J., and W. W. Powell. 1983. The iron cage revisited: Institutional isomorphism and collective rationality in organizational fields. *American Sociological Review* 48:147-160.

———. 1991. "Introduction." In *The New Institutionalism in Organizational Analysis.* Chicago: University of Chicago Press.

DNV suspension another jab at battered CO_2 scheme. 2008. *Reuters.* Dec. 2, 2008,

DOE (Department of Energy). 1985. *Carbon Dioxide Research: State-of-the-Art Report Series.* Washington, DC: DOE.

Doran, P. T., and M. K. Zimmerman. 2009. Direct examination of the scientific consensus on climate change. *EOS* 90(3):22.

Drori, G. S., J. W. Meyer and H. Hwang. 2006. *Globalization and Organization: World Society and Organizational Change.* New York: Oxford University Press.

Dunlap, R. E., and A. M. McCright. 2008. A widening gap: Republican and Democratic views on climate change. *Environment* 50(5):26-35.

Economics of Climate Adaptation Working Group. 2009. Shaping Climate-Resilient Development: A Framework for Decision Making. A report of the Economics of Climate Adaptation Working Group. A Report of the Economics of Climate Adaptation Working Group, Climate Works Foundation, European Commission, Global Environmental Facility, McKinsey and Company, Rockefeller Foundation, Swiss Re, UNEP.

Edwards, P., R. Hindmarsh, H. Mercer, M. Bond, and A. Rowland. 2008. A three-stage evaluation of a deliberative event on climate change and transforming energy. *Journal of Public Deliberation* 4(1):Article 6. Available at *http://services. bepress.com/jpd/vol4/iss1/art6*.

EIA (Energy Information Administration). 2008. *EIA—Emissions of Greenhouse Gases in the U.S. 2007—Carbon Dioxide Emissions 2008*. Available at *http://www.eia.doe.gov/oiaf/1605/ggrpt/carbon.html#total*.

———. 2009. *EIA—Press Releases—U.S. Energy-Related Carbon Dioxide Emissions Declined by 2.8 Percent in 2008*. Available at *http://www.eia.doe.gov/neic/press/press318.html*.

Engel, K. H. 2006. Harnessing the benefits of dynamic federalism in environmental law. *Emory Law Journal* 56:159-159.

Enquist, C., and D. Gori. 2008. Implications of Recent Climate Change on Conservation Priorities in New Mexico. The Nature Conservancy, New Mexico Science Program, Sante Fe, New Mexico.

Entergy Texas. 2009. *ENERGY STAR Qualified Homes*. Available at *http://www.entergy-texas.com/energy_efficiency/es_ overview.aspx*.

EPA (Environmental Protection Agency). 1997. *The Benefits and Costs of the Clean Air Act, 1970 to 1990: U.S. Environmental Protection Agency, Office of Air and Radiation, Office of Policy, Planning, and Evaluation*. *http://purl.access.gpo. gov/GPO/LPS3436*.

———. 2008. *State and Local Governments—State and Regional Climate Policy Maps*. Available at *http://www.epa. gov/ climatechange/wycd/stateandlocalgov/state_planning. html#two*. Accessed May 5, 2010.

———. 2009a. *Partnership Directory: Climate Leaders Setting the Standard in Greenhouse Gas Management*. Washington, DC: EPA.

———. 2009b. *Greenhouse Gas Reporting Rule/Climate Change Greenhouse Gas Emissions*. Washington, DC: EPA.

EST (Energy Saving Trust). 2009. *Energy Labels—Energy Efficiency Logos—Energy Saving Trust*. Available at *http://www. energysavingtrust.org.uk/Energy-saving-products/About-Energy-Saving-Recommended-products/Other-energy-labels*.

Falk, J. H., E. M. Reinhard, C. L. Vernon, K. Bronnenkant, N. L. Deans, and J. E. Heimlich. 2007. *Why Zoos & Aquariums Matter: Assessing the Impact of a Visit*. Silver Spring, MD: Association of Zoos and Aquariums.

Fankhauser, S. 1995. *Valuing Climate Change: The Economics of the Greenhouse*. London: Earthscan Publications.

FAO (Food and Agriculture Organization). 2008. *Food Outlook: Global Market Analysis*. Rome: FAO.

Farman, J. C., R. J. Murgatroyd, A. M. Silnickas, and B. A. Thrush. 1985. Ozone photochemistry in the antarctic stratosphere in summer. *Quarterly Journal of the Royal Meteorological Society* 111(470).

Fennell, M. L., and J. A. Alexander. 1987. Organizational boundary spanning in institutionalized environments. *Academy of Management Journal* 30(3):456-476.

Fenner, F. 1993. Smallpox—emergence, global spread, and eradication. *History and Philosophy of the Life Sciences* 15 (3):397-420.

Ferree, M. M., W. Gamson, J. Gerhards, and D. Rucht. 2002. *Shaping Abortion Discourse: Democracy and the Public Sphere in Germany and the United States*. Cambridge, UK: Cambridge University Press.

Fingar, T. 2009. *National Intelligence Assessment on the National Security Implications of Global Climate Change to 2030*. Testimony Submitted to the House Permanent Select Committee on Intelligence and the House Select Committee on Energy Independence and Global Warming. 25 June.

Finucane, M. L. 2008. Emotion, affect and risk communication with older adults: Challenges and opportunities. *Journal of Risk Research* 11:983-997.

Fischer, C. 2008. Feedback on household electricity consumption: A tool for saving energy? *Energy Efficiency* 1(1):79-104.

Fishel, M. C. 2006. *Fifth Survey of Climate Change Disclosure in SEC Filings of Automobile, Insurance, Oil & Gas, Petrochemical, and Utilities Companies*. Friends of the Earth. Available at *http://www.foe.org/camps/intl/SECFinalReportand Appendices.pdf*.

Fortun, M., and H. J. Bernstein. 1998. *Muddling Through: Pursuing Science and Truths in the 21st Century*. Washington, DC: Counterpoint.

Franco, G. D., A. Cayan, M. Luers, M. Hanemann, and B. Croes. 2008. Linking climate change science with policy in California. *Climate Change* 87:7-20.

Frumhoff, P. C., J. J. McCarthy, J. M. Melillo, S. C. Moser, and D. J. Wuebbles. 2007. *Confronting Climate Change in the U.S. Northeast: Science, Impacts and Solutions*. Synthesis report of the Northeast Climate Impacts Assessment. Cambridge, MA: Union of Concerned Scientists.

Frumkin, H., and A. J. McMichael. 2008. Climate change and public health: Thinking, communicating, acting. *American Journal of Preventive Medicine* 35(5):403-410.

Gamson, W. A., and A. Modigliani. 1989. Media discourse and public opinion on nuclear power: A constructionist approach. *American Journal of Sociology* 95(1):1-37.

GAO (General Accounting Office). 1990. *Global Warming: Administration Approach Cautious Pending Validation of Threat*. Washington, DC: GAO.

GAO (Government Accountability Office). 2007. *Climate Change: Financial Risks to Federal and Private Insurers in Coming Decades are Potentially Significant*. Report to the Committee on Homeland Security and Governmental Affairs, U.S. Senate. Washington, DC: GAO.

Gardiner, D., M. Anderson, R. Schlesinger, J. Fox Gorte, and D. Zeller. 2007. *Climate Risk Disclosure by the S&P 500*. Boston, MA: Ceres, Inc.

Gardner, G. T., and P. C. Stern. 2002. *Environmental Problems and Human Behavior*, 2nd edition. Boston, MA: Pearson Custom Publishing.

Gardner, G. T., and P. C. Stern. 2008. The short list: The most effective actions U.S. households can take to curb climate change. *Environment* 12-24.

Garg, A., P. R. Shukla, and M. Kapshe. 2006. The sectoral trends of multigas emissions inventory of India. *Atmospheric Environment* 40(24):4608-4620.

Gentner, D., K. J. Holyoak, and B. N. Kokinov. 2001. *The Analogical Mind: Perspectives from Cognitive Science*. Cambridge, MA: MIT Press.

Georghiou, L., J. H. Cassingena, M. Keenan, I. Miles, and R. Popper, eds. 2008. *The Handbook of Technology Foresight*. Northhampton, MA: Edward Elgar.

Geurts, J. L. A., R. D. Duke, and P. A. M. Vermeulen. 2007. Policy gaming for strategy and change. *Long Range Planning* 40(6):535-558.

Gilgoff, D. 2009. Religious groups push for climate change legislation. *U.S. News and World Report, http://politics.usnews. com*. Accessed September 25, 2009.

Gillenwater, M., D. Broekhoff, M. Trexler, J. Hyman, and R. Fowler. 2007. Policing the voluntary carbon market. *Nature Reports Climate Change*. Published online October 2007. *http://www.nature.com/2007/0711/full/climate.2007.58.html*.

Gilovich, T., and D. Griffin. 2002. *Heuristics and Biases: The Psychology of Intuitive Judgement*. Cambridge: Cambridge University Press.

Glantz, M., and D. Jamieson. 2000. Societal response to Hurricane Mitch and intra- versus intergenerational equity issues: Whose norms should apply? *Risk Analysis* 20(6):869-882.

Glick, P., A. Staudt, and B. Stein. 2009. *A New Era for Conservation: Review of Climate Change Adaptation Literature*. Washington, DC: National Wildlife Federation.

GLWQ (Great Lakes Water Quality Board). 2003. *Climate Change and Water Quality in the Great Lakes Basin*. Report to the International Joint Commission. GLWQ, Chicago, Illinois.

GM (General Motors). 2009. *GM—Environment—Partnerships Overview*. Available at *http://www.gm.com/corporate/ responsibility/partners/environment/index.jsp*. Accessed August 25, 2009.

Goodrich, C. 1960. *Government Promotion of American Canals and Railroads, 1800-1890*. New York: Columbia University Press.

Goodstein, L. 2005. Evangelical leaders swing influence behind effort to combat global warming. *New York Times,* 10 March, Page 16.

Gowda, R, and J. C. Fox. 2002. *Judgements, Decisions, and Public Policy*. Cambridge: Cambridge University Press.

Gram-Hanssen, K., F. Bartiaux, O. M. Jensen, and M. Cantaert. 2007. Do homeowners use energy labels? A comparison between Denmark and Belgium. *Energy Policy* 35(5):2879-2888.

Granade, H. C., J. Creyts, A. Derkack, P. Farese, S. Nyquist, and K. Ostrowski,. 2009. *Unlocking Energy Efficiency in the U.S. Economy*. New York: McKinsey Global Energy and Materials.

Granberg, M., and I. Elander. 2007. Local governance and climate change: Reflections on the Swedish experience. *Local Environment* 12(5):537-548.

GreeenStar. 2009. *GreeenStar—homepage.* Available at *http://www.greeenstar.com.* Accessed Oct. 15, 2010.

Greene, R. W. 2002. *Confronting Catastrophe: A GIS Handbook.* Redlands, CA: Esri Press.

Greiner, S., and A. Michaelowa. 2003. Defining investment additionality for CDM projects—practical approaches. *Energy Policy* 31(10):1007-1015.

Gropman, A. L. 1996. *Mobilizing US industry in World War II, @McNair papers; 50.* Washington, DC: Institute for National Strategic Studies

Gross, C., and K. A. Sepkowitz. 1998. The myth of the medical breakthrough: Smallpox, vaccination, and Jenner reconsidered. *International Journal of Infectious Diseases* 3(1).

Groves, D. G., and R. J. Lempert. 2007. A new analytic method for finding policy-relevant scenarios. *Global Environmental Change* 17(1):73-85.

Groves, D. G., D. Yates, and C. Tebaldi. 2008. Developing and applying uncertain global climate change projections for regional water management planning. *Water Resources Research* 44(12).

Grundig, F. 2006. Patterns of international cooperation and the explanatory power of relative gains: An analysis of co-operation on global climate change, ozone depletion, and international trade. *International Studies Quarterly* 50(4):781-801.

Gustavsson, E., I. Elander, and M. Lundmark. 2009. Multilevel governance, networking cities, and the geography of climate-change mitigation: Two Swedish examples. *Environment and Planning C: Government and Policy* 27(1):59-74.

Hacker, J. S. 2001. Learning from Defeat? Political Analysis and the Failure of Health Care Reform in the United States. *British Journal of Political Science* 31(1):61-94.

Haddow, G., J. A. Bullock, and K. Haddow. 2008. *Global Warming, Natural Hazards, and Emergency Management.* Boca Raton, FL: CRC Press.

Haites, E., and X. Wang. 2009. Ensuring the environmental effectiveness of linked emissions trading schemes over time. Mitigation and Adaptation Strategies for Global Change 14 (5):465-476.

Hall, N. L., and R. Taplin. 2007. Solar festivals and climate bills: Comparing NGO climate change campaigns in the UK and Australia. *Voluntas: International Journal of Voluntary and Nonprofit Organizations* 18(4):317-338.

Hanak, E., L. Bedsworth, S. Swanbeck, and J. Malaczynski. 2008. *Climate Policy at the Local Level: A Survey of California's Cities and Counties.* San Francisco, CA: Public Policy Institute of California.

Hassol, S. J. 2008. Improving how scientists communicate about climate change. *EOS* 89(11).

Heinz Center. 2008. *Strategies for Managing the Effects of Climate Change on Wildlife and Ecosystems.* Washington, DC: Heinz Center.

Helm, D., and C. Hepburn, eds. 2010. *The Economics and Politics of Climate Change.* Oxford: Oxford University Press.

Hirst, E., R. Goeltz, and J. Carney. 1981. *Residential Energy Use and Conservation Actions: Analysis of Disaggregate Household Data.* Report ORNL/CON-68, Oak Ridge, TN: Oak Ridge National Laboratory.

Hoffman, A. J. 2005. Climate change strategy: The business logic behind voluntary greenhouse gas reductions. *California Management Review* 47(3):21-46.

Hoffman, A. J. 2006. *Getting Ahead of the Curve: Corporate Strategies That Address Climate Change.* Arlington, VA: Pew Center on Global Climate Change.

Hoffman, M., and D. Barstow. 2007. *Revolutionizing Earth System Science Education for the 21st Century.* Report and Recommendations from a 50-State Analysis of Earth Science Education Standards. Cambridge, MA: Technical Education Research Centers.

Holling, C. S., ed. 1978. *Adaptive Environmental Assessment and Management.* London: John Wiley & Sons.

Holling, C. S., and G. K. Meffe. 1996. Command and control and the pathology of natural resource management. *Conservation Biology* 10(2):328-337.

Horne, R. E. 2009. Limits to labels: The role of eco-labels in the assessment of product sustainability and routes to sustainable consumption. *International Journal of Consumer Studies* 33(2):175-182.

Houghton, D. P. 1996. The role of analogical reasoning in novel foreign-policy situations. *British Journal of Political Science* 26(4):523-552.

House, J. I., C. Huntingford, W. Knorr, S. E. Cornell, P. M. Cox, G. R. Harris, C. D. Jones, J. A. Lowe, and I. C. Prentice. 2008. What do recent advances in quantifying climate and carbon cycle uncertainties mean for climate policy? *Environmental Research Letters* 3(4), doi: 10.1088/1748-9326/1083/1084/044002.

ICLEI. 2007. *World Mayors & Local Governments Climate Protection Agreement.* Available at *http://www.globalclimate agreement.org/.*

IETA (International Emissions Trading Association) and PWC (PricewaterhouseCoopers). 2007. *Trouble-Entry Accounting: Uncertainty in Accounting for the EU Emissions Trading Scheme and Certified Emission Reductions.* London: PWC.

IPCC (Intergovernmental Panel on Climate Change). 2001. *Climate Change 2001: Synthesis Report. A Contribution of Working Groups I, II and III to the Third Assessment Report of the Intergovernmental Panel on Climate Change.* New York: Cambridge University Press.

———. 2005. *Guidance Notes for Lead Authors of the IPCC Fourth Assessment Report on Addressing Uncertainties.* Available at *http://www.ipcc.ch/pdf/supporting-material/uncertainty-guidance-note.pdf.*

———. 2006. *IPCC Guidelines for National Greenhouse Gas Inventories—General Guidance and Reporting.* IPCC.

———. 2007a. *Climate Change 2007: Impacts, Adaptation and Vulnerability. Contribution of Working Group II to the Fourth Assessment Report of the Intergovernmental Panel on Climate Change.* M. L. Parry, O. F. Canziani, J. P. Palutikof, P. J. van der Linden and C. E. Hanson, eds. Cambridge: Cambridge University Press.

———. 2007b. *Climate Change 2007: Synthesis Report. Contribution of Working Groups I, II, and III to the Fourth Assessment Report of the Intergovernmental Panel on Climate Change, Core Writing Team.* R. K. Pachauri and A. Reisinger, eds. Geneva, Switzerland, pp. 104.

———. 2007c. Summary for Policymakers. In *Climate Change 2007: Impacts, Adaptation and Vulnerability. Contribution of Working Group II to the Fourth Assessment Report of the Intergovernmental Panel on Climate Change,* M. L. Parry, O. F. Canziani, J. P. Palutikof, P. J. van der Linden, and C. E. Hanson, eds. Cambridge, UK: Cambridge University Press, 7-22.

———. 2007d. *Climate Change 2007: Contribution of Working Group III to the Fourth Assessment Report of the IPCC on Climate Change.* B. Metz, O.R. Davidson, P.R. Bosch, R. Dave, L.A. Meyer, eds. Cambridge University Press, Cambriedge, UK and New York, New York, USA.

Jacobs, K., G. Garfin, and J. Buizer. 2010. New techniques at the science-policy interface: Climate change adaptation in the water sector. *Science and Policy* 36(10):791-798.

Jacobson, J. W., R. M . Foxx, and J. A. Mulick. 2005. *Controversial Therapies for Developmental Disabilities: Fad, Fashion, and Science in Professional Practice.* Mahwah, NJ: Lawrence Erlbaum Associates.

Jänicke, M. 2008. Ecological modernisation: New perspectives. *Journal of Cleaner Production* 16(5):557-565.

Jacques, P. J., R. E. Dunlap, and M. Freeman. 2008. The organization of denial: The link between conservative think thanks and environmental skepticism. *Environmental Politics* 17:349-385.

Jepson, P. 2005. Governance and accountability of environmental NGOs. *Environmental Science and Policy* 8(5):515-524.

Johnston, H., and J. A. Noakes. 2005. *Frames of Protest: Social Movements and the Framing Perspective.* Lanham: Rowman & Littlefield.

Jones, A. S. 2009. *Losing the News: The Future of the News That Feeds Democracy.* New York: Oxford University Press.

Jones, C. A., and D. L. Levy. 2007. North American business strategies towards climate change. *European Management Journal* 25(6):428-440.

Jones, J. 2005. *Botanical Gardens Summative Evaluation Conservatory for Botanical Science.* San Marina, CA: Huntington Botanical Gardens.

Kahneman, D., P. Slovic, and A. Tversky. 1982. *Judgment Under Uncertainty: Heuristics and Biases,* reprint edition. Cambridge: Cambridge University Press.

Kaissi, A. A., and J. W. Begun. 2008. Fads, fasions, and bandwagons in health care strategy. *Health Care Management Review* 33(2):94-102.

Kammen, D. M., and S. Pacca. 2004. Assessing the costs of electricity. *Annual Review of Energy and the Environment* (29):1-44.

Karl, T. R., J. M. Melillo, T. C. Peterson, and S. J. Hassol. 2009. *Global Climate Change Impacts in the United States.* Cambridge, UK: Cambridge University Press.

Kasperson, R. E., and J. X. Kasperson. 1991. Hidden hazards. In *Acceptable Evidence: Science and Values in Risk Management*, D. G. Mayo and R. D. Hollander, eds.. New York: Oxford University Press.

Kates, R. W., T. M. Parris, and A. A. Leiserowitz. 2005. What is sustainable development? Goals, indicators, values, and practice. *Environment* 47(3):8-21.

Keiser, J., J. Utzinger, M. Caldas De Castro, T. A. Smith, M. Tanner, and B. H. Singer. 2004. Urbanization in Sub-Saharan Africa and implication for malaria control. *American Journal of Tropical Medicine and Hygiene* 71(2):118-127.

Kempton, W. 1997. How the public views climate change. *Environment* 39(9).

Kempton, W., J. S. Boster, and J. A. Hartley. 1995. *Environmental Values in American Culture*. Cambridge, MA: MIT Press.

Kennel, C. F. 2009. Climate change: Think globally, assess regionally, act locally. *Issues in Science and Technology* 25(2).

Kerschner, S. 2009. Power companies agree to expand disclosure of climate change risk in landmark settlements wtih New York Attorney General. *American Bar Association: Environmental Disclosure Committee Newsletter* 6(1):2-5.

Keteltas, G. S., J. Lichtman, D. Lisi, and J. M. Horan. 2008. Climate change: Emerging litigation challenges. Presented at the Litigation Annual Conference in Washington D.C., April 16-18.

King County. 2007. *King County Climate Plan 2007*. Available at *http://your.kingcounty.gov/exec/news/2007/pdf/climateplan.pdf*.

Kingdon, J. W. 1995. *Agendas, Alternatives, and Public Policies*, 2nd ed. New York: Longman.

Klein, R., J. Sathaye, and T. Wilbanks. 2007. Challenges in integrating mitigation and adaptation as responses to climate change. *Mitigation and Adaptation Strategies for Global Change* (Special Issue) 12(5).

Knight, F. H. 1921. *Risk, Uncertainty, and Profit*. Boston: Hart, Schaffner & Marx; Houghton Mifflin.

Koistinen, P. A. C. 2004. *Arsenal of World War II: The Political Economy of American Warfare, 1940-1945, Modern War Studies. Lawrence:* University Press of Kansas.

Kolk, A., D. Levy, and J. Pinkse. 2008. Corporate responses in an emerging climate regime: The institutionalization and commensuration of carbon disclosure. *European Accounting Review* 17(4):719-745.

Kollmuss, A., H. Zink, and C. Polycarp. 2008. *Making Sense of the Voluntary Carbon Market: A Comparison of Carbon Offset Standards*. WWF Germany.

KPMG Carbon Advisory Group. 2009. *Accounting for Carbon: The Impact of Carbon Trading Financial Statements*. 5th in series of White Papers. KPMG.

Krier, J. E., E. Ursin, et al. 1977. *Pollution and Policy: A Case Essay on California and Federal Experience with Motor Vehicle Air Pollution, 1940-1975*. Berkeley: University of California Press.

Lazarus, R. J. 2009. Super wicked problems and climate change: Restraining the present to liberate the future. *Cornell Law Review* 94:1153-1234.

Le Quéré, C., M. R. Raupach, J. G. Canadell, G. Marland, L. Bopp, P. Ciais, T. J. Conway, S. C. Doney, R. A. Feely, P. Foster, P. Friedlingstein, K. Gurney, R. A. Houghton, J. I. House, C. Huntingford, P. E. Levy, M. R. Lomas, J. Majkut, N. Metzl, J. P. Ometto, G. P. Peters, I. C. Prentice, J. T. Randerson, S. W. Running, J. L. Sarmiento, U. Schuster, S. Sitch, T. Takahashi, N. Viovy, G. R. van der Werf, and F. Ian Woodward. 2009. Trends in the sources and sinks of carbon dioxide. *Nature Geoscience* 2:831-836.

Lee, K. N. 1993. *Compass and Gyroscope: Integrating Science and Politics for the Environment*. Washington, DC: Island Press.

———. 1999. Appraising adaptive management. *Conservation Ecology* 3(2).

Leiserowitz, A. 2006. Climate change risk perception and policy preferences: The role of affect, imagery, and values. *Climatic Change* 77(1-2):45-72.

Leiserowitz, A. 2007. Communicating the risks of global warming: American risk perceptions, affective images and interpretive communities. In *Creating a Climate for Change: Communicating Climate Change and Facilitating Social Change*, S. C. Moser and L. Dilling, eds. Cambridge: Cambridge University Press.

Leiserowitz, A., and K. Akerlof. 2010. *Race, Ethnicity and Public Responses to Climate Change*, Yale University and George Mason University, eds. New Haven, CT: Yale Project on Climate Change.

Leiserowitz, A., and G. Feinberg. 2005. American risk perceptions: Is climate change dangerous? *Risk Analysis* 25(6):1433-1442.

———. 2007. *American Support for Local Action on Global Warming. The GfK Roper/Yale Survey on Environmental Issues*. Yale School of Forestry & Environmental Studies. Available at *http://environment.yale.edu/news/5323/*.

Leiserowitz, A., E. Maibach, and C. Roser-Renouf. 2009. *Climate Change in the American Mind: Americans' Climate Change Beliefs, Attitudes, Policy Preferences, and Actions.* George Mason University, Center for Climate Change Communication.

Leiserowitz, A., E. Maibach, and C. Rosner-Renouf. 2010. *Global Warming's Six Americas: January 2010.* Yale University and George Mason University, eds. New Haven, CT: Yale Project on Climate Change. Available at *http://environment.yale.edu/uploads/SixAmericasJan2010.pdf.*

Lempert, R. J., ed. 2007. Can scenarios help policymakers be both bold and careful? In *Blindside: How to Anticipate Forcing Events and Wild Cards in Global Politics,* F. Fukuyama, ed. Washington, DC: Brookings Institute Press.

Lempert, R. J., and M. T. Collins. 2007. Managing the risk of uncertain threshold responses: Comparison of robust, optimum, and precautionary approaches. *Risk Analysis* 27(4):1009-1026.

Lempert, R. J., S. W. Popper, and S. C. Bankes. 2003. *Shaping the Next One Hundred Years: New Methods for Quantitative Long Term Policy Analysis.* pp. 187. Santa Montica, CA: Rand Corporation.

Lempert, R. J., M. E. Scheffan, and D. Sprinz. 2009. Methods for long term environmental policy challenges. *Global Environmental Politics* (3).

Levy, D. L., and P. Newell. 2005. *The Business of Global Environmental Governance.* Cambridge, MA: MIT Press.

Lindblom, C. E. 1959. The science of "muddling through." *Public Administration Quarterly* 19:79-88.

Litz, F. T. 2008. *Toward a Constructive Dialogue on Federal and State Roles in US Climate Change Policy,* Solutions White Paper Series. Pew Centre on Global Climate Change.

Lovell, H. 2009. Governing the carbon offset market. *Wiley Interdisciplinary Reviews: Climate Change,* 1: 352-353

Lovell, H., and D. M. Liverman. 2010. Understanding Carbon Offset Technologies. *New Political Economy* 15(2):255-273.

Lowe, H., G. Seufert, and F. Raes. 2000. Comparison of methods used within Member States for estimating CO_2 emissions and sinks according to UNFCCC and EU Monitoring Mechanism: Forest and other wooded land. *Biotechnologie Agronomie Societe et Environnement* 4(4):315-319.

Lutsey, N., and D. Sperling. 2008. America's bottom-up climate change mitigation policy. *Energy Policy* 36:673-685.

MacDonald-Smith, A. 2007. Climate change to boost insured losses, Allianz says. Bloomberg, September 18.

Maibach, E. W., C. Rosner-Renouf, and A. Leiserowitz. 2008. Communication and marketing as climate change- intervention assets. *American Journal of Preventive Medicine* 35(5):488-500.

Maibach, E, C. Roser-Renouf, and A. Leiserowitz. 2009. *Global Warming's Six Americas: An Audience Segmentation.* New Haven, CT: Yale Project on Climate Change, Yale University.

Manne, A. S., and R. G. Richels. 1992. *Buying Greenhouse Insurance: The Economic Costs of CO_2 Emission Limits.* Cambridge, MA: MIT Press.

Mantua, N. J., S. R. Hare, Y. Zhang, J. M. Wallace, and R. C. Francis. 1997. A Pacific decadal climate oscillation with impacts on salmon. *Bulletin of the American Meteorological Society* 78:1069-1079.

March, J. G. 1994. *A Primer on Decision Making: How Decisions Happen.* New York: Free Press.

March, J. G., and J. P. Olsen. 2004. *The Logic of Appropriateness.* ARENA Working Paper 04/09. Oslo: University of Norway, ARENA Center for Europen Studies.

Marland, G., T. Boden, and R. Andres. 2003. *Ranking of the World's Countries by 2000 Total CO_2 Emissions form Fossil-Fuel Burning, Cement Production and Gas Flaring.* Available at *http://cdiac.esd.ornl.gov/trends/emis/top2000.tot.*

Matthews, H. S., C. T. Hendrickson, and C. L. Weber. 2008a. The importance of carbon footprint estimation boundaries. *Environmental Science and Technology* 42(16):5839-5842.

Matthews, H. S., C. L. Weber, and C. T. Hendrickson. 2008b. *Estimating Carbon Footprints with Input-Output Models.* Paper read and presented at the International Input-Output Meeting on Managing the Environment, 9-11 July, in Seville, Spain.

Mayer, I. 2009. The gaming of policy and the politics of gaming: A review. *Simulation & Gaming December 2009.* 40: 825-862

Mazmanian, D., J. Jurewitz, and H. Nelson. 2008. California's climate change policy: The case of a subnational state actor tackling a global challenge. *Journal of Environment and Development* (17):401.

McCalley, L. T., and C. J. H. Midden. 2002. Energy conservation through product-integrated feedback: The roles of goal-setting and social orientation. *Journal of Economic Psychology* 23(5):589-603.

McCright, A. M., and R. E. Dunlap. 2000. Challenging global warming as a social problem: An analysis of the conservative movement's counter-claims. *Social Problems* 47(4):499-522.

———. 2003. Defeating Kyoto: The conservative movement's impact on U.S. climate change policy. *Social Problems* 50(3):348-373.

McDougall, G. H. G., J. D. Claxton, and J. R. B. Ritchie. 1983. Residential home audits: An empirical analysis of the ENEVERSAVE program. *Journal of Environmental Systems* 12(3):267-278.

MEA (Millennium Ecosystem Assessment). 2005. *Ecosystems & Human Well-Being: Volume 2: Scenarios.* Washington, DC: Island Press.

Menestrel, M., S. van den Hove, and H. C. de Bettignies. 2002. Processes and consequences in business ethical dilemmas: The oil industry and climate change. *Journal of Business Ethics* 41:251-266.

Meo, M., B. Ziebro, and A. Patton. 2004. Tulsa turnaround: From disaster to sustainability. *Natural Hazards Review* 5:1-9.

Meuleman, L., and R. Veld. 2009. *Sustainable Development and the Governance of Long Term Decisions.* Ruimtelijk, milieu en natuuronderzoek and European Environment and Sustainable Development Advisory Council Netherlands.

Meyer, J. W., and B. Rowan. 1977. Institutionalized organizations: Formal structure as myth and ceremony. *American Journal of Sociology* 83:340-363.

Meyer, P. 2004. *The Vanishing Newspaper: Saving Journalism in the Information Age.* Columbia: University of Missouri Press.

Michaelowa, A. 2005. *CDM: Current Status and Possibilities for Reform.* Paper 3 by the HWWI Research Programme International Climate Policy. Hamburg, Germany: Hamburg Institute of International Economics.

Michalak, A. M. 2008. Technical note: A geostatistical fixed-lag Kalman smoother for atmospheric inversions. *Atmospheric Chemistry and Physics Discussions* 8(2):7755-7779.

Miles, E. L., A. K. Snover, L. C. Whitely Binder, E. S. Sarachik, P. W. Mote, and N. Mantua. 2006. An approach to designing a national climate service. *Proceedings of the National Academy of Sciences of the United States of America* 103(52):19613-19615.

Mills, E. 2009. *From Risk to Opportunity: 2008—Insurer Responses to Climage Change: A Report from CERES.* Boston, MA: CERES.

Molina, M. J., and F. S. Rowland. 1974. Stratospheric sink for chlorofluoromethanes: Chlorine atom-catalysed destruction of ozone. *Nature* 249:810-812.

Morello-Frosch, R., M. Pastor, J. Sadd, and S. Shonkoff. 2009. *The Climate Gap: Inequalities in How Climate Change Hurts Americans and How to Close the Gap.* Los Angeles: University of Southern California, Program for Environmental and Regional Equity. Available at *http://college.usc.edu/geography/ESPE/documents/The_Climate_Gap_Full_Report_FINAL.pdf.*

Moser, S. C. 2005. *Stakeholder Involvement in the First U.S. National Assessment of the Potential Consequences of Climate Variability and Change: An Evolution, Finally.* Washington, DC: Committee on the Human Dimensions of Global Change, National Research Council.

Moser, S. C. 2009. Communicating climate change: History, challenges, process and future directions. *Wiley Interdisciplinary Reviews: Climate Change* 1(1):31-53.

Moser, S. C., and L. Dilling, eds. 2007. Information is not enough. In *Creating a Climate for Change: Communicating Climate Change and Facilitating Social Change.* S. C. Moser and L. Dilling, eds. New York: Cambridge University Press.

Müller, B. 2009. *Additionality in the Clean Development Mechanism: Why and What?* Oxford: Oxford Institute for Energy Studies.

NAST (National Assessment Synthesis Team). 2001. *National Assessment of the Potential Consequences of Climate Variability and Change.* Washington, DC: U.S. Global Change Research Program.

Nerlich, B., N. Koteyko, and B. Brown. 2010. Theory and language of climate change communication (advanced review). *WIREs Climate Change* 1(Jan/Feb).

Neustadt, R. E., and E. R. May. 1986. *Thinking in Time: The Uses of History for Decision-Makers.* New York: Free Press.

Newell, P. 2000. *Climate for Change: Non-state Actors and the Global Politics of the Greenhouse.* Cambridge, UK: Cambridge University Press.

———. 2008. The political economy of global environmental governance. *Review of International Studies* 34:507-529.

Newport, F. 2008. *Little Increase in Americans' Global Warming Worries. Gallup*, available at *http://www.gallup.com/poll/106660/little-increase-americans-global-warming-worries.aspx*.

Nicholls, N. 1999. Cognitive illusions, heuristics and climate prediction. *Bulletin of the American Meteorological Society* 80:1385-1397.

Nicholson-Cole, S. A. 2005. Representing climate change futures: A critique on the use of images for visual communication. *Computers, Environment and Urban Systems* 29:255-273.

Nisbet, M. 2009. Communicating climate change: Why frames matter for public engagement. *Environment* 51(2):12-25.

Nisbet, M. C., and C. Mooney. 2007. Science and society: Framing science. *Science* 316(5821):56.

Nisbet, M. C., and T. Myers. 2007. Twenty years of public opinion about global warming. *Public Opinion Quarterly* 71(3):444-470.

NOAA SAB (National Oceanic and Atmospheric Administration Science Advisory Board). 2009. *Options for Developing a National Climate Service*. Silver Spring, MD: NOAA.

Noar, S. M. 2006. A 10-year retrospective of research in health mass media campaigns: Where do we go from here? *Journal of Health Communication* 11(1):21-42.

Nordhaus, T., and M. Shellenberger. 2007. *Break Through: From the Death of Environmentalism to the Politics of Possibility*. New York: Houghton Mifflin.

Nordhaus, W. 2007. Critical assumptions in the Stern Review on Climate Change. *Science* 317(5835):203-204.

Nordhaus, W. D. 1977. Economic growth and climate: The carbon dioxide problem. *American Economic Review* 67:341-346.

Nordhaus, W. D. 1994. *Managing the Global Commons: The Economics of Climate Change*. Cambridge, MA: MIT Press.

NRC (National Research Council). 1984. *Energy Use: The Human Dimension*. Washington, DC: National Academy Press.

———. 1985. *Energy Efficiency in Buildings: Behavioral Issues*. Washington, DC: National Academy Press.

———. 1986a. *The National Climate Program: Early Achievements and Future Directions. Report of the Woods Hole Workshop, July 15-19, 1985*. Washington, DC: National Academy Press.

———. 1986b. *U.S. Participation in the TOGA Program, A Research Strategy*. Washington, DC: National Academy Press.

———. 1990. *The U.S. Global Change Research Program: An Assessment of the FY 1991 Plans*. Washington, DC: National Academy Press.

———. 1996. *Understanding Risk: Informing Decisions in a Democratic Society*. Washington, DC: National Academy Press.

———. 1999. *Making Climate Forecasts Matter*. Washington, DC: National Academy Press.

———. 2001. *A Climate Services Vision: First Steps Toward the Future*. Washington, DC: National Academy Press.

———. 2002a. *Communicating Uncertainties in Weather and Climate Information: A Workshop Summary*. Washington, DC: National Academy Press.

———. 2002b. *New Tools for Environmental Protection: Education, Information, and Voluntary Measures*. Washington, DC: National Academy Press.

———. 2005. *Decision Making for the Environment: Social and Behavioral Science Research Priorities*. Washington, DC: The National Academies Press.

———. 2007a. *Analysis of Global Change Assessments: Lessons Learned*. Washington, DC: The National Academies Press.

———. 2007b. *Contributions of Land Remote Sensing for Decisions about Food Security and Human Health*. Washington, DC: The National Academies Press.

———. 2007c. *Earth Science and Application from Space: National Imperatives for the Next Decade and Beyond*. Washington, DC: The National Academies Press.

———. 2007d. *Evaluating Progress of the U.S. Climate Change Science Program: Methods and Preliminary Results*. Washington, DC: The National Academies Press.

———. 2008a. *Potential Impacts of Climate Change on U.S. Transportation*. Washington, DC: The National Academies Press.

———. 2008b. *Public Participation in Environmental Decision Making*. Washington, DC: The National Academies Press.

———. 2008c. *Research and Networks for Decision Support in the NOAA Sectoral Applications Research Program*. Washington, DC: The National Academies Press.

———. 2009a. *Informing Decisions in a Changing Climate*. Washington, DC: The National Academies Press.

————. 2009b. *Learning Science in Informal Environments: Places, People, and Pursuits.* Washington, DC: The National Academies Press.

————. 2009c. *Letter Report on the Orbiting Carbon Observatory.* Committee on Methods for Estimating Greenhouse Gas Emissions. Washington, DC: The National Academies Press.

————. 2009d. *Restructuring Federal Climate Research to Meet the Challenges of Climate Change.* Washington, DC: The National Academies Press.

————. 2009e. *Uncertainty Management in Statistics and Remote Sensing: Summary of a Workshop.* Washington, DC: The National Academies Press.

————. 2010a. *Adapting to the Impacts of Climate Change.* Washington, DC: The National Academies Press.

————. 2010b. *Advancing the Science of Climate Change.* Washington, DC: The National Academies Press.

————. 2010c. *Climate Stabilization Targets: Emissions, Concentrations, and Impacts over Decades to Millennia.* Washington, DC: The National Academies Press.

————. 2010d. *Limiting the Magnitude of Future Climate Change.* Washington, DC: The National Academies Press.

————. 2010e. *Verifying Greenhouse Gas Emissions for a Climate Treaty.* Washington, DC: The National Academies Press.

NYCPCC (New York City Panel on Climate Change). 2009. *Climate Risk Information.* New York: NYCPCC.

O'Connor, R. E., R. J. Bord, and A. Fisher. 1998. The curious impact of knowledge about climate change on risk perceptions and willingness to sacrifice. *Risk Decision Policy* 3:145-155.

Okereke, C. 2007. An exploration of motivations, drivers and barriers to carbon management: The UK FTSE 100. *European Management Journal* 25(6):475-486.

Oldstone, M. B. A. 1998. *Viruses, Plagues, and History.* New York: Oxford University Press.

Olsson, P., L. H. Gunderson, S. R. Carpenter, P. Ryan, L. Lebel, C. Folke, and C. S. Holling. 2006. Shooting the

O'Neill, S., and S. Nicholson-Cole. 2009. "Fear won't do it": Promoting positive engagement with climate change rapids: Navigating transitions to adaptive governance of social-ecological systems. *Ecology and Society* 11(1):1-18. through visual and iconic representations. *Science Communication* 30:355.

O'Neill, S. J., and M. Hulme. 2009. An iconic approach for representing climate change. *Global Environmental Change* 19:402-410.

The Oregon Climate Trust. 2009. *The Climate Trust Homepage.* Available at *http://www.climatetrust.org/*.

Oreskes, N. 2004. The Scientific Consensus on Climate Change. *Science* 306(5702):1686.

O'Riordan, T. 2000. *Climate Change and Business Response.* Norwich: University of East Anglia.

Overpeck, J., C. Anderson, D. Cayan, M. Dettinger, K. Dow, H. Hartmann, J. Jones, E. Miles, P. Mote, M. Shafer, B. Udall, and D. White. 2009. *Climate Services: The RISA Experience.* Boulder: Western Water Assessment. Available at *http://wwa.colorado.edu/about_us/docs/RISA_NCS_May09_finalV3b.pdf*.

Padgett, J., A. Steinemann, J. Clarke, and M. Vandenbergh. 2007. A comparison of carbon calculators. *Environmental Impact Assessment Review* 28(2-3):106-115.

PAHO (Pan American Health Organization). 2009. *Dengue.* Washington, DC: PAHO. Available at *http://new.paho.org/hq/index.php?option=com_content&task=view&id=264&Itemid=363*.

Paisley, W. J. 2001. Public communication campaigns: The American experience. In *Public Communication Campaigns*, R. E. Rice and C. K. Atkin, eds. Thousand Oaks, CA: Sage.

Parformak, P. W., F. Sissine, and E. A. Fischer. 2009. *Energy Efficiency in Buildings: Critical Barriers and Congressional Policy.* Congressional Research Service, Washington D.C.

Patashnik, E. 2003. After the public interest prevails: The political sustainability of policy reform. *Governance* 16(6):203-234.

Patton, A., ed. 1994. *From Rooftop to River: Tulsa's Approach to Floodplain and Stormwater Management.* Tulsa, OK: Department of Public Works.

Pearson, T. R. H., S. Brown, and K. Andrasko. 2008. Comparison of registry methodologies for reporting carbon benefits for afforestation projects in the United States. *Environmental Science and Policy* 11(6):490-504.

Penman, J., D. Kruger, I. Galbally, T. Hiraishi, B. Nyenzi, S. Emmanul, L. Buendia, R. Hoppaus, T. Martinsen, J. Meijer, K. Miwa, and K. Tanabe. 2000. *Good Practice Guidance and Uncertainty Management in National Greenhouse Gas Inventories.* Japan: IPCC National Greenhouse Gas Inventories Programme.

Peterson, T. D., and A. Z. Rose. 2006. Reducing conflicts between climate policy and energy policy in the US: The important role of the states. *Energy Policy* 34(5):619-631.

Pew Center. 2008. *Business Environmental Leadership Council (BELC). Pew Center on Global Climate Change: The Pew Center on Global Climate Change.* Available at *http://www.pewclimate.org/business/belc.*

———. 2009. *Climate Change 101: State Action.* Pew Center on Global Climate Change. Available at *http://www.pewclimate. org/what_s_being_done/in_the_states/regional_initiatives.cfm.*

Pew Research Center for the People and the Press. 2009. *Economy, Jobs Trump All Other Policy Priorities in 2009.* Washington, DC: Pew Research Center for the People and the Press.

Pielke, R. A. 2000a. Policy history of the U.S. Global Change Research Program. Part I. Administrative development. *Global Environmental Change* (10):9-25.

———. 2000b. Policy history of the U.S. Global Change Research Program. Part II. Legislative process. *Global Environmental Change* (10):133-144.

Princen, T. 2009. Long-term decision making: Biological and psychological evidence. *Global Environmental Politics* 9(3):9-19.

PRNewsWire. 2009. *Investors Achieve Major Company Commitments on Climate Change.* Available at *http://news.prnewswire. com/DisplayReleaseContent.aspx?ACCT=104&STORY=/www/story/08-24-2009/0005082141&EDATE=.*

Pumphrey, C. 2008. *Global Climate Change National Security Implications.* Carlisle, PA: Strategic Studies Institute, U.S. Army War College.

Quadra Planning Consultants Ltd. and Galiano Institute for Environmental and Social Research. 2004. *Seafood Watch Evaluation: Summary Report.* Saltspring Island, BC, Canada.

Quality Offset Initiative. 2009. *Maintaining Carbon Market Integrity: Why Renewable Energy Certificates Are Not Carbon Offsets.* June 2009 *http://offsetqualityinitiative.org/OQI%20REC%20Brief%20Web.pdf.*

Rabe, B. G. 2008. States on steroids: The intergovernmental odyssey of American climate policy. *Review of Policy Research* 25(2):105-128.

———. 2009. Second-generation climate policies in the states: Proliferation, diffusion, and regionalization. In *Changing Climates in North American Politics: Institutions Policy Making, and Multilevel Governance,* H. Selin and S. D. VanDeveer, eds. Cambridge, MA: MIT Press.

Rabe, B. and C. Borick. 2009. "To Believe or Not to Believe: An Examination of the Factors that Determine Acceptance of Global Warming" *Paper presented at the annual meeting of the WPSA ANNUAL MEETING "Ideas, Interests and Institutions" on March 19, 2009, Hyatt Regency Vancouver, BC Canada, Vancouver, BC, Canada Online.* Available at *http://www. allacademic.com/meta/p316790_index.html.* Accessed October 13, 2010.

Read, D., A. Bostrom, M. G. Morgan, B. Fischhoff, B., and T. Smuts. 1994. What do people know about global climate change? Survey results of educated laypeople. *Risk Analysis* 14:971-982.

Reap, J., F. Roman, S. Duncan, and B. Bras. 2008a. A survey of unresolved problems in life cycle assessment. Part 1: Goal & scope and inventory analysis. *International Journal of Life Cycle Assessment* 13(4):290-300, doi: 210.1007/s11367-11008-10008-x.

———. 2008b. A survey of unresolved problems in life cycle assessment. Part 2: Impact assessment and interpretation. *International Journal of Life Cycle Assessment* 13(5):374-388, doi: 310.1007/s11367-11008-10009-11369.

Rittel, H. W. J., and M. M. Webber. 1973. Dilemmas in a general theory of planning. *Policy Sciences* 4:155-169.

Roberts, J. T., and B. Parks. 2006. *A Climate of Injustice: Global Inequality, North-South Politics and Climate Policy.* Boston: MIT Press.

Roberts, N. C., and P. J. King. 1991. Policy entrepreneurs: Their activity structure and function in the policy process. *Journal of Public Administration Research and Theory* 1(2):147-175.

Rogers, D. J., S. E. Randolph, R. W. Snow, and S. I. Hay. 2002. Satellite imagery in the study and forecast of malaria. *Nature* 415:710-715.

Rosegrant, M. W., C. Ringler, and T. Zhu. 2009. Water for agriculture: Maintaining food security under growing scarcity. *Annual Review of Environment and Resources* 34:205-222

Rosenzweig, C., and W. D. Solecki. 2001. *Climate Change and a Global City: The Potential Consquences of Climate Variability and Change, Metro East Coast* (MEC). Report for the U.S. Global Change Research Program, National Assessment of Potential Consequences of Climate Variability and Change for the United States. New York: Columbia Earth Institute, Columbia University.

Roser, D. 2009. The discount rate: A small number with a big impact. *Applied Ethics: Life, Environment and Society* 10-25.

Rowland, I. H. 2000. Beauty and the beast?: BP's and Exxon's positions on global climate change. *Environment and Planning: Government and Policy* 18:339-354.

Rutz, D., R. Janssen, P. Helm, R. Geraghty, M. Black, B. Wason, J. Neeft, E. van Thuijl, S. Borg, S. Bürkner, et al. 2007. The EU Carbon Labelling Initiative. Paper read at the 15th European Biomass Conference and Exhibition. May 7th-11th, 2007, ICC Berlin, Germany.

Rypdal, K., and W. Winiwarter. 2001. Uncertainties in greenhouse gas emission inventories—evaluation, comparability and implications. *Environmental Science and Policy* 4(2-3):107-116-107–116.

Saad, L.-Gallup Poll. Increased Number Think Global Warming is "Exaggerated" 2009. Available from http://www.gallup.com/poll/116590/increased-number-think-global-warming-exaggerated.aspx.

Sabin, P. 2010. The climate crisis and energy transition: Lessons from history. *Environmental History.* 15(1) 76-93.

Saltelli, A., S. Tarantola, and F. Campolongo. 2000. Sensitivity analysis as an ingredient of modeling. *Statistical Science* 377-395.

Sanchez, M. C., R. E. Brown, C. Webber, and G. K. Homan. 2008. Savings estimates for the United States Environmental Protection Agency's ENERGY STAR voluntary product labeling program. *Energy Policy* 36(6):2098-2108.

Sarewitz, D., and R. A. Pielke. 2007. The neglected heart of science policy: Reconciling supply and demand for science. *Environmental Science and Policy* (10):5-16.

Schmall, D., and M. Sanders. 2008. California Air Resources Board Issues Draft Plan For Reducing Greenhouse Gas Emission. Available at *http://209.85.229.132/search?q=cache:DPwK-PWAk-UJ:www.paulhastings.com/assets/publications/937.pdf%3Fwt.mc_ID%3D937.pdf+CARB+linked+to+CCAR&cd=6&hl=en&ct=clnk&gl=ca.*

Schmidt, H. J. 2009. Carbon footprinting, labelling and life cycle assessment. *International Journal of Life Cycle Assessment* 14:6-9.

Schneider, S. H., B. L. Turner II, and H. Morehouse Garriga. 1998. Imaginable suprise in global change science. *Journal of Risk Research* 1(2):165-185.

Schneider, S. H., S. Semenov, A. Patwardhan, I. Burton, C. H. D. Magadza, M. Oppenheimer, A. B. Pittock, A. Rahman, J. B. Smith, A. Suarez, and F. Yamin. 2007. Assessing key vulnerabilities and the risk from climate change. In *Climate Change 2007: Impacts, Adaptation and Vulnerability. Contribution of Working Group II to the Fourth Assessment Report of the Intergovernmental Panel on Climate Change*, M. L. Parry, O. F. Canziani, J. P. Palutikof, P. J. van der Linden, and C. E. Hanson, eds. Cambridge, UK: Cambridge University Press.

Scholz, J. T., and B. Stiftel. 2005. *Adaptive Governance and Water Conflict: New Institutions for Collaborative Planning.* Washington, DC: Resources for the Future.

Schroeder, H., and H. Bulkeley. 2009. Global cities and the governamce of climate change: What is the role of law in cities? *Fordham Urban Law Journal* 36:313-537.

Scott, W. R. 2001. *Institutions and Organizations.* Thousand Oaks, CA: Sage.

SEC (Securities and Exchange Commission), ed. 2010. Commission guidance regarding disclosure related to climate change: Final rule. *Federal Register.* 75(25) 6290-6297

SEG (Scientific Expert Group on Climate Change) 2007. *Confronting Climate Change: Avoiding the Unmanageable and Managing the Unavoidable,* R. Bierbaum, J. Holdren, M. MacCracken, R. Moss, and P. Raven, eds. Prepared for the United Nations Commission on Sustainable Development by Sigma Xi, Research Triangle Park, NC and the United Nations Foundation, Washington, DC.

Shaw, A., et al. 2009. Making local futures tangible—Synthesizing, downscaling, and visualizing climate change scenarios for participatory capacity building. *Global Environmental Change* 19(4):447-463.

Slocum, T., D. C. Cliburn, J. J. Feddema, and J. R. Miller. 2003. Evaluating the usability of a tool for visualizing the uncertainty of the future global water balance. *Cartography and Geographic Information Science* 34(4):299-317.

Slovic, P. 1999. Trust, emotion, sex, politics, and science: Surveying the risk-assessment battlefield. *Risk Analysis* 19(4):444-470.

Slovic, P. 2000. *The Perception of Risk.* London: Earthscan.

Snow, D. A., E. B. Rochford, Jr., S. K. Worden, and R. D. Benford. 1986. Frame alignment processes, micromobilization, and movement participation. *American Sociological Review* 51(4):464-481.

Snyder, L. 2007. Health communication campaigns and their impact on behavior. *Journal of Nutrition Education and Behavior* 39(2):S32-40.

Snyder, L. B., and M. A. Hamilton. 2002. Meta-analysis of U.S. health campaign effects on behavior: Emphasize enforcement, exposure, and new information, and beware the secular trend. Pp. 357-383 in *Public Health Communication: Evidence for Behavior Change.* R. Hornik, ed. Hillsdale, NJ: Lawrence Erlbaum Associates.

Socolow, R. H., and S. H. Lam. 2007. Good enough tools for global warming policy making. *Philosophical Transactions of the Royal Society A* (365):897-934.

Solomon, S., G. K. Plattner, R. Knutti, and P. Friedlingstein. 2009. Irreversible climate change due to carbon dioxide emissions. *Proceedings of the National Academy of Sciences of the United States of America* 106(6):1704-1709.

Staudt, A., G. Barnhardt, P. Glick, and K. Theoharides. 2009. Climate Change Adaption Across the Landscape: A Survey of Federal and State Agencies, Conservation Organizations and Academic Institutions in the United States. Washington, DC: The Associate of Fish and Wildlife Agencies, Defenders of Wildlife, The Nature Conservancy and The National Wildlife Federation.

Steenhof, P. 2009. Estimating greenhouse gas emission reduction potential: An assessment of models and methods for the power sector. *Environmental Modeling and Assessment* 14(1):17-28.

Steg, L. 2008. Promoting household energy conservation. *Energy Policy* 36(12):4449-4453.

Stern, N. 2007. *The Economics of Climate Change: The Stern Review.* Cambridge, UK: Cambridge University Press.

Stern, N., S. Peters, V. Bakhshi, A. Bowen, C. Cameron, S. Catovsky, D. Crane, S. Cruickshank, S. Dietz, and N. Edmonson. 2006. *Stern Review: The Economics of Climate Change.* HM Treasury, London HM Treasury, London.

Stern, P. C. 1986. Blind spots in policy analysis: What economics doesn't say about energy use. *Journal of Policy Analysis and Management* 5(2):200-227.

Stern, P. C., and G. T. Gardner. 1981. Psychological research and energy policy. *American Psychologist* 36(4):329-342.

Stolaroff, J. K., C. L. Weber, and H. S. Matthews. 2009. Design issues in a mandatory greenhouse gas emissions registry for the United States. *Energy Policy* 37(9) 3463-3466

Suddaby, R. and R. Greenwood. 2005. Rhetorical strategies of legitimacy. *Administrative Science Quaterly* 50:35-67.

Summers, L., and R. Zeckhauser. 2009. Policymaking for posterity. *Journal of Risk and Uncertainty* 37(2-3):115-140.

Sunstein, C. R. 2005. *Laws of Fear: Beyond the Precautionary Principle.* New York: Cambridge University Press.

———. 2007. Of Montreal and Kyoto: A tale of two protocols. *Harvard Environmental Law Review* 31(1):1-65.

Swiss Re. 1994. *Global Warming, Elements of Risk.* Zurich: Swiss Re.

———. 2004. Sustainability Report 2004. Available at *http://www.swissre.com.*

TFCC (Task Force Climate Change)/Oceanographer of the Navy. 2009. *U.S. Navy Arctic Road Map.* Washington, DC: U.S. Department of Defense.

Tellus Institute. 1992. *Tellus Packaging Study.* Boston: Tellus Institute of Resource and Environmental Strategies.

TNC (The Nature Conservancy). 2010. *Where Is the San Pedro River? Map and Watershed Graphic.* Available from *http://www. nature.org/wherewework/northamerica/states/arizona/features/art26621.html.* Accessed June 6, 2010.

TNC New Mexico Conservation Science Program. 2008. *Implications of Recent Climate Change on Conservation Priorities in New Mexico.* Santa Fe, NM: TNC.

Tonn, B., A. Hemrick, and F. Conrad. 2006. Cognitive representations of the future: survey results. *Futures* 38:810-829.

Tversky, A., and D. Kahneman. 1981. The framing of decisions and the psychology of choice. *Science* 211(4481):453-458.

UNDP (United Nations Development Programme). 2002. A climate risk management approach to disaster reduction and adaption to climate change. Havana: UNDP and Harvard Medical School Center for Human Health and the Global Environment.

UNEP FI (United Nations Environment Programme Finance Initiative). 2002. *Climate Change and the Financial Services Industry: Threats and Opportunities.* Nairobi, Kenya: UNEP.

Union of Concerned Scientists. 2006. *The Changing Northest Climate: Our Choices, Our Legacy.* Cambridge, MA: Union of Concerned Scientists.

USCAP (U.S. Climate Action Partnership). 2009. *A Blueprint for Legislative Action: Consensus Recommendations for U.S. Climate Protection Legislation.* Washington D.C.: USCAP.

U.S. Climate Action Report. 2010. *Fifth National Communication of the United States of America Under the United Nations Framework Convention on Climate Change.* Washington DC: United States Department of State.

USGBC (U.S. Green Building Council). 2010. *LEED Core Concepts and Strategies Online Course.* Available at *http://www.usgbc.org/DisplayPage.aspx?CMSPageID=1720.* Accessed June 14, 2010.

USGCRP (U.S. Global Change Research Program). 2001. *The Potential Consequences of Climate Variability and Change,* National Assessment Synthesis Team, ed. Cambridge: Cambridge University Press.

———. 2009. Global Climate Change Impacts in the United States: A State of Knowledge Report from the U.S. Global Change Research Program, T. R. Karl, J. M. Melillo, and T. C. Peterson, eds. New York: Cambridge University Press.

U.S. Office of Global Change. 2007. *Fourth Climate Action Report to the UN Framework Convention on Climate Change.* Washington, DC: U.S. Department of State.

Van Asselt, M., and E. Vos. 2006. The precautionary principle and the uncertainty paradox. *Journal of Risk Research* 9:313-336.

Vandenbergh, M., J. Barkenbus, and J. Gilligan. 2008. Individual carbon emissions: The low-hanging fruit. *UCLA Law Review* 55:1701-1758.

van Houwelingen, J. H., and W. F. van Raaij. 1989. The effect of goal-setting and daily electronic feedback on in-home energy use. *Journal of Consumer Research* 16(1):98-105.

Victor, D. G., J. C. House, and S. Joy. 2005. Climate: Enhanced: A Madisonian approach to climate policy. *Science* 309:1820-1821.

Vosniadou, S., and A. Ortony, eds. 1989. *Similarity and Analogical Reasoning.* New York: Cambridge University Press.

Walters, C. J. 1986. *Adaptive Management of Renewable Resources.* New York: Macmillan.

Wara, M. W., and D. G. Victor. 2008. *A Realistic Policy on International Carbon Offsets.* Program on Energy and Sustainable Development, Working Paper 74.

Waxman, H. A., and E. J. Markey. 2009. *American Clean Energy and Security Act of 2009 Discussion Draft Full Text.* Available at *http://energycommerce.house.gov/Press_111/20090331/acesa_discussiondraft.pdf.*

WBCSD (World Business Council for Sustainable Development) and WRI (World Resources Institute). 2005. *The Greenhouse Gas Protocol: The GHG Protocol for Project Accounting.* Conches-Geneva: World Business Council for Sustainable Development and Washington D.C.: World Resources Institute

———. 2007. *The Greenhouse Gas Protocol.* Washington, DC: WRI.

WCI (Western Climate Initiative). 2008. *The WCI Cap & Trade Program.* Available at *http://www.westernclimateinitiative.org/the-wci-cap-and-trade-program.*

———. 2009. *WCI Offsets Committee White Paper.* Boulder: WCI.

WCRP (World Climate Research Programme). 1985. Scientific plan for the tropical ocean global atmosphere programme. In *WCRP Publication.* Geneva: World Meteorological Organization.

———. 1986a. International implementation plan for TOGA. In *ITPO Document 2.* Geneva: World Meteorological Organization.

———. 1986b. *Scientific Plan for the World Ocean Circulation Experiment.* WCRP Publication Series, number 6. Geneva: World Meteorological Organization.

Weber, E. U. 2006. Experience-based and description-based perceptions of long-term risk: Why global warming does not scare us (yet). *Climatic Change* 77(1-2):103-120.

Weidema, B. P., M. Thrane, P. Christensen, J. Schmidt, and S. Lokke. 2008. A catalyst for life cycle assessment? *Journal of Industrial Ecology* 12(1):3-6.

Weitzman, M. L. 2001. Gamma discounting. *American Economic Review* 91(1):260-271.

Westerling, A. L., H. G. Hidalgo, D. R. Cayan, and T. W. Swetnam. 2006. Warming and earlier spring increase western U.S. forest wildfire activity. *Science* 313(5789):940-943.

White, R. 2007. *Carbon Labelling: Evidence, Issues and Questions.* Oxford: Environmental Change Insitute.

WHO (World Health Organization). 2008. *Protecting Health from Climate Change: World Health Day 2008*. Geneva: WHO.

Wiener, J. B. 2009. Engaging China on climate change. *RFF Resources Magazine* 29-33.

Wiener, J. B., R. B. Stewart, J. K. Hammitt, J.-C. Hourcade, D. G. Victor, J. C. House, and S. Joy. 2006. Madison and climate change policy. *Science* 311:335-336, doi: 310.1126/science.1311.5759.1335c.

Wilby, R.L., J. Troni, Y. Biot, L. Tedd, B.C. Hewitson, D.G. Smith, and R.T. Sutton. 2009. A Review of Climate Risk Information for Adaptation and Development Planning. *International Journal of Climatology* 29:1193-1215.

Wilhite, D. A., L. Botterill, and K. Monnik. 2005. National drought policy: Lessons learned from Australia, South Africa, and the United States. In *Drought and Water Crises: Science, Technology, and Management Issues*, D. A. Wilhite, ed. Boca Raton, FL: CRC Press.

Willows, R.I., and R.K. Connell. 2003. Climate adaptation: Risk, uncertainty and decision-making. UKCIP Technical Report. Oxford: UKCIP.

Winkler, H. 2008. Measurable, reportable and verifiable: the keys to mitigation in the Copenhagen deal. *Climate Policy* 8(7).

Witte, K., and M. Allen. 2000. A meta-analysis of fear appeals: Implications for effective public health campaigns. *Health Education and Behavior* 27(5).

WMO (World Meteorological Organization). 2007. *Scientific Assessment of Ozone Depletion: 2006*. Nairobi, Kenya: United Nations Environment Programme.

———. 2009. *Summary of the Expert Segment, version 8*. Paper read at World Climate Conference 3, Geneva, August 31-September 4.

Woolsey, G. 2008. The fifth column: The case of the confounded water model or all streams do not flow north. *Interfaces* 38(3):210-211.

World Bank. 2006. Managing climate risk. In *Integrating Adaption into World Bank Group Operations*. Washington, DC: World Bank Group.

———. 2009. *State and Trends of the Carbon Market 2009*. Available at *http://siteresources.worldbank.org/INTCARBONFI-NANCE/Resources/State_of_the_Market_press_release_The_FINAL_2009.pdf*.

———. 2010. *World Development Report 2010: Development and Climate Change*. Washington, DC: International Bank for Reconstruction and Development/World Bank.

WRI (World Resources Institute. 2008. *GHG Emission Registries, Bottom Line*. Washington, DC: World Resources Institute.

———. 2009. *Greenhouse Gas Protocol: Users of the Corporate Standard*. Available at *http://www.ghgprotocol.org/standards/corporate-standard/users-of-the-corporate-standard#hide*.

Young, B., C. Suarez, and K. Gladman. 2009. *Climate Risk Disclosure in SEC Filings: An Analysis of 10-K Reporting by Oil and Gas, Insurance, Coal, Transportation and Electric Companies Commissioned by CERES and Environmental Defense Fund*. Boston, MA: CERES.

Panel on Informing Effective Decisions and Actions Related to Climate Change Statement of Task

The Panel on Informing Effective Decisions and Actions Related to Climate Change will describe and assess different activities, products, strategies, and tools for informing decision makers about climate change and helping them plan and execute effective, integrated responses. The panel will describe the different types of climate change–related decisions and actions being made at various levels and in different sectors and regions, develop a framework for analyzing and informing these actions and decisions, and evaluate the activities, products, and tools that could help ensure these actions and decisions are informed by the best available technical knowledge. The tools, products, and activities considered by the panel will include, but are not necessarily limited to, observing systems, climate models, monitoring and early warning systems, assessments, integrated assessment models, outreach activities, and communication networks between information providers and users. The panel will also discuss steps to better educate and train future generations of scientists, decision makers, and citizens to meet the challenges associated with climate change.

Ultimately, the goal of this panel is to answer the fourth question in the Statement of Task for the study: "What can be done to inform effective decisions and actions related to climate change?"

This question will be expanded over the course of the study to include more specific subquestions,[1] such as the following:

- What climate-relevant information and other support do different kinds of decision makers need to respond effectively to climate change (including mitigation and adaptation), and what approaches and tools are most effective at providing this information and support?

---

[1] These subquestions are only examples of the types of questions to be addressed by the panel, to indicate the level of specificity intended. Some of these illustrative questions may be revised or dropped, and other questions may be added, at the discretion of the panel and the Committee on America's Climate Choices.

- What roles can federal, state, and local governments and other groups (e.g, the academic community) play in providing effective "climate services"—the timely production and delivery of information, data, and knowledge to decision makers affected by climate?
- What information and tools (e.g., monitoring, metrics, integrated assessment models, etc.) do we need to evaluate the progress of different responses to climate change?
- How can decisions and actions related to climate change be made more flexible and responsive to changing conditions and new information?
- What can current efforts and past experiences (both failures and successes) teach us about responding effectively to climate change?

The panel will be challenged to produce a report that is broad and authoritative, yet concise and useful to decision makers. The costs, benefits, limitations, tradeoffs, and uncertainties associated with different options and strategies should be assessed qualitatively and, to the extent practicable, quantitatively, using the scenarios of future climate change and vulnerability developed in coordination with the Committee on America's Climate Choices and the other study panels. The panel should also provide policy-relevant (but not policy-prescriptive) input to the committee on the following overarching questions:

- What short-term actions can be taken to better inform decisions and actions related to climate change?
- What promising long-term strategies, investments, and opportunities could be pursued to better inform decisions and actions related to climate change?
- What are the major scientific and technological advances (e.g., new observations, improved models, research programs, etc.) needed to better inform decisions and actions related to climate change?
- What are the major impediments (e.g., practical, institutional, economic, ethical, intergenerational, etc.) to effectively informing decisions and actions related to climate change, and what can be done to overcome these impediments?
- What can be done to more effectively inform decisions and actions related to climate change at different levels (e.g., local, state, regional, national, and in collaboration with the international community) and in different sectors (e.g., nongovernmental organizations, the business community, the research and academic communities, individuals and households, etc.)?

The American Experience with Complex Decisions: Past Examples

Throughout its history, the United States has acted decisively to confront major challenges, even in the absence of complete information or national consensus. National action has been initiated by a variety of individuals and institutions, including citizens, the private sector, and government. Sometimes new scientific knowledge was instrumental in prompting action, but in other cases political, corporate, or moral leadership responded decisively despite uncertain or incomplete scientific knowledge, potential costs, or conflicting public opinion.

This appendix describes five historical examples where the United States successfully confronted and overcame major national and international challenges. There are two reasons to pay attention to such case studies. First, analogies often play an important role in human decision making (e.g., Gentner et al., 2001; Vosniadou and Ortony, 1989). Second, historians and political scientists have identified a number of examples in which key leaders drew upon historical analogies to make decisions about national and foreign policy (Hacker, 2001; Houghton, 1996; Neustadt and May, 1986). Likewise, scientists, journalists, environmentalists, and labor, religious, political, and business leaders have often drawn upon historical analogies to help articulate and explain the climate change problem and its potential solutions (Sabin, 2010). Reasoning by historical analogy can be both useful and challenging: useful because analogies can at times help to identify and illuminate important problem features and potential solutions, and challenging because at other times analogies can misdirect attention and lead to the misapplication of the "lessons of history." While climate change is a relatively new and highly complex problem—and in many ways remains unique—it too shares a number of similar features with prior national and international challenges.

Historical analogies also remind us that the United States has successfully overcome complex and difficult problems in the past. Each example below shares important similarities with the challenge of climate change; however, each also differs in important ways. Many other historical analogues have been used to think about climate change (e.g., the Manhattan Project, the Apollo Space Program, the New Deal), but they are not included here.

THE MONTREAL PROTOCOL

The Montreal Protocol on Substances that Deplete the Ozone Layer is often recognized as a potential model for climate change. Like climate change, ozone depletion is a global environmental threat. In this case, emissions of human-produced chlorofluorocarbons (CFCs) are destroying the ozone layer that protects the Earth's surface from harmful solar ultraviolet (UV) light. Like the greenhouse gas (GHG) emissions that cause climate change, these emissions come from a variety of industrial processes taking place in both developed and developing countries, with the bulk of such emissions originating historically from the industrialized world. Also as in the case of climate change, scientific research discovered an unintended consequence of modern industrial activities that is largely invisible to the eye yet has potentially very serious global consequences. Furthermore, the early years of ozone layer research were filled with scientific uncertainties. For example, in 1974, based on laboratory research, chemists Mario Molina and Sherwood Rowland first hypothesized that CFCs were stable enough to rise to the stratosphere, where they would break down the Earth's protective ozone layer (Molina and Rowland, 1974). Their research was roundly criticized by a number of companies that produced or relied upon CFCs. Nonetheless, the news media reported their hypothesis and identified common household products (such as aerosol spray cans) as one of the sources of CFCs. The public quickly responded, with many choosing to avoid CFC-based products. It was not until 1985 that British Antarctic Survey scientists finally discovered the formation of an ozone "hole" in the stratosphere over Antarctica (Farman et al., 1985). That same year, the Vienna Convention for the Protection of the Ozone Layer was negotiated and signed by many of the world's largest emitters; this was quickly followed by the Montreal Protocol, which entered into force in 1989.

Technological innovation and market position played critical roles in the policy-making process, because the same companies that had produced CFCs invented more benign substitutes. The structure of the Vienna Convention was also important, as it included a periodic review of the evolving science, a structure by which the treaty could be revised and updated over time, and a special fund to assist developing countries in complying with the treaty. Over the years, as the science has progressed, the treaty has been progressively tightened to achieve a further and faster phase-out of ozone-destroying compounds. As a result, the Montreal Protocol has been hailed internationally as one of the most successful international agreements ever (DeSombre, 2000).

While there are some similarities between the problems of climate change and ozone depletion, there are also some very important differences (Bodansky, 2001; Grundig, 2006). For example, the problematic substances (CFCs) for ozone were produced by a relatively small number of companies; substitutes were developed by these same companies (who stood to profit from the transition); and while CFCs were important to certain products and industrial processes, they were not fundamental to the operation of modern society. Climate change, however, is driven primarily by the burning of fossil fuels, for which there are currently few comparable alternatives; they are produced by some of the world's largest companies; they provide the primary source of income for a number of key nations; and they supply the primary source of energy for the world. Furthermore, ozone depletion threatened significant personal health consequences because UV-B light is associated with increased rates of skin cancer. By contrast, while climate change is projected to have significant health consequences, the impacts will be neither universal nor as personally relevant to most Americans. People around the world could see themselves as more or less vulnerable to the risk of increased UV light due to ozone depletion, while the health risks of climate change are likely to be much more heterogeneous. In fact, studies have found that a majority of Americans currently believe climate change will have a small or no impact on human health, or they simply have no idea (Leiserowitz et al., 2009). Finally, while CFCs were used in some consumer products such as aerosol spray cans and refrigerators, fossil fuels power much of the world's transportation system and electrical grid and provide key inputs into countless goods and foodstuffs (Sunstein, 2007).

THE ERADICATION OF SMALLPOX

Limiting the severity and adapting to the impacts of climate change will require the participation of individuals, organizations, and governments in every nation. The daunting scope of this task also has precedent, however. In 1979, the United Nations World Health Organization (WHO) formally declared victory in its 20-year campaign to eradicate smallpox worldwide.

Smallpox was one of the most deadly and contagious diseases known to humankind. It originated about 10,000 years ago and became endemic across Europe and Asia. Before widespread vaccinations became available during the 19th century, the disease killed about half a million people annually (0.5 percent of the population) in Europe alone. By the 20th century, smallpox still killed about 2 million people each year worldwide. In 1959 the United Nations began—and in 1967 greatly intensified—a campaign to eradicate the disease worldwide, a task made possible because smallpox exists only in humans and has no other carriers. Using extensive networks to reach every village on Earth, particularly in Africa and the Indian subcontinent, WHO teams identified each outbreak, isolated the victims, and vaccinated the surrounding population. Advertising campaigns and financial incentives encouraged even illiterate villagers to quickly report any smallpox outbreaks. Near-universal participation was necessary because any unreported carriers could harbor the disease. After years of hard and coordinated work costing hundreds of millions of dollars, the campaign achieved a final success. The last naturally occurring case of smallpox was diagnosed and contained in Somalia in 1977 (Fenner, 1993; Oldstone, 1998).

Despite similarities in scope, however, the eradication of smallpox differs in important ways from efforts to reduce climate change. For example, the disease's impacts were immediate and personal—the disease horribly disfigured and often killed its victims. Thus, individuals, communities, and entire nations could readily see and personalize the threat. Compared to climate change, the required responses—quarantine and vaccination—were relatively quick, were inexpensive, and did not fundamentally challenge existing social and economic patterns. Nonetheless, there are parallels to some of the risks associated with climate change, including increases in vector-borne or diarrheal infections that often afflict the poor; effective responses can reduce overall vulnerability to these impacts of climate change just as it did to smallpox. The eradication of smallpox also stands as an example of how even adversaries, such as the United States and the Soviet Union, could work together through the United Nations to eliminate a common threat to humanity.

THE CLEAN AIR ACT

Over the past 50 years, environmental protection has proven one of the great public policy success stories in the United States. In particular, the 1970 Clean Air Act and its major 1990 amendments have dramatically reduced unsightly and unhealthy air pollution at a cost representing a tiny fraction of the benefits produced.

In the 1950s and 1960s, the air above many American towns and cities had become deadly with increasing industrialization and automobile use. In 1966, an air pollution inversion killed 168 people in New York City. At times, ozone levels in Los Angeles' air rose to five times above safe levels and visibility dropped to mere blocks. Noxious smog engulfed steel towns in the industrial heartland. During these two decades, the federal government established research programs to develop air pollution monitoring and abatement technology. California and other states began to regulate their emissions. In 1970, Congress passed the landmark Clean Air Act, authorizing the federal government and the states to regulate industrial and automotive emissions to meet national air quality standards (Krier, 1977). In 1990, Congress significantly enhanced the original act, in particular establishing a cap-and-trade system for sulfur dioxide (the source of acid rain), a forerunner of the system some have proposed for limiting GHGs.

The 1970 clean air legislation occurred as one element in a social transformation—the rise of environmental consciousness in the United States. The first Earth Day was held that year. The Environmental Protection Agency was established in 1971. The Clean Air Act was a dramatic success. Since 1970, the economy has more than doubled in size, yet pollutants such as sulfur have dropped by a third, lead has dropped by 98 percent, and the air across the nation is visibly cleaner and meets health standards far more often. Over its first 20 years, the Clean Air Act is estimated to have cost the United States about $500 billion, while saving over $20 trillion, a benefit-cost savings of over 40 to 1 (EPA, 1997).

Cleaning America's air, however, also represents a different challenge than climate change. The Clean Air Act was largely a national project, achievable without the co-operation of other nations (although many other nations were undertaking parallel efforts). The impacts of dirty air were also far more visible and immediate to citizens than the impacts of climate change. In addition, the job of cleaning America's air is still not complete. Many cities still consistently violate health standards, and, with a growing economy and traffic, continued improvements in air quality may require new technology transitions. Nonetheless, the Clean Air Act provides an example of a hotly contested environmental policy that transformed technology across many sectors of the U.S. economy, significantly reduced pollution at a fraction of the cost initially estimated by many, and has made a dramatic, observable difference in people's lives.

THE TRANSCONTINENTAL RAILROAD

To address climate change, the government must help catalyze a technology and infrastructure revolution that will transform the way Americans produce and consume energy. This would not be the first time the U.S. government has facilitated such a transformation. The first transcontinental railroad, completed in 1869, is widely considered one of the great engineering triumphs of the 19th century. Over the following decades, the massive project successfully achieved the main goals of the policy makers who helped to finance it—linking the far-flung pieces of a country recently shattered by civil war and enabling the world's first, and still strongest, continental economy (Ambrose, 2000; Bain, 1999; Goodrich, 1960).

The trip west to California by ship or wagon had previously taken months. The railroad reduced it to days. Visionaries had dreamed of a Pacific railroad since the 1830s, but the project's risk and expense put it outside the reach of any private entrepreneurs.

Sectional disagreements over a northern or southern route prevented government action until, in the midst of the Civil War, Congress approved the Pacific Railroad Act, which incentivized private firms with large subsidies, paying them in cash and land for each mile of track laid. The government dictated the basic route (roughly following what is now Interstate 80) but left the details to the railroads. The legislation launched a process rife with amazing determination, thievery, heroism, cruelty, and corruption as multiple lines raced their way east and west, eventually meeting at Promontory Summit, Utah.

Addressing climate change will also require widespread deployment of new technology and infrastructure. Just as in the building of the transcontinental railroad, the government will need to chart a broad plan, provide incentives, and take some of the risk, while leaving the private sector to make most of the specific engineering and investment decisions. Building the transcontinental railroad, however, is only a partially useful analogue to climate change. The United States accomplished the endeavor alone, without need for cooperation with other nations. The project disregarded environmental concerns, the rights of indigenous peoples, the treatment of immigrant workers, and proper oversight of public funds that would prove rightly intolerable today. The benefits (and dangers) of the new railroad were immediate and largely obvious to all concerned. Yet the Pacific Railroad does stand as a towering example of policy makers successfully pursuing a long-term, transformational goal without a detailed long-term plan. Instead, the federal government provided strong financial incentives to the private sector that catalyzed the widespread deployment and use of a new technology that transformed the world.

WORLD WAR II

The massive national mobilization required to fight and win World War II may have some useful lessons for the prevention of catastrophic climate change (Bartels, 2001; Brown, 2009). Like WWII, reducing global emissions of GHGs will require a national focus, sense of urgency, dedication to success, cooperation with other countries, and national mobilization, including individuals and all levels of government, diverse economic sectors, and broad civil society. In response to the threat of Nazi Germany and Imperial Japan, and after the tragedy of the attack on Pearl Harbor, the United States literally reinvented and reorganized itself. Within months, many factories had been retooled from commercial to military purposes. During the war, millions of American men and women were drafted or volunteered for military service, while millions more worked on the home front in support of the war effort, including in factories, on farms, and in construction (Gropman, 1996; Koistinen, 2004). A wide variety of commodities were rationed, including cars, fuel, food, and clothing, and many Americans planted "Victory Gardens" to feed their families during the war. Moreover, the outcome of the war itself was deeply uncertain, and, as it proceeded, Americans endured and surmounted a number of major military setbacks and losses. Nonetheless, the country and its leaders were willing to act despite these enormous uncertainties and large costs in both human lives and national treasure. Moreover, the United States partnered with the other Allies, including ideological foes like the Soviet Union, to defeat their common enemy. Winning WWII thus required an extraordinary level of coordination both within the United States and internationally. And in the process, the United States reinvented itself, emerging from the war as a global military, economic, and cultural superpower.

Preventing dangerous levels of climate change will also require changes in the way American society produces and consumes energy and significant changes across economic and political sectors, both within the United States and internationally. While WWII reminds us what the United States can achieve when it is motivated, it is also important to recognize that climate change presents a different set of challenges. In WWII, the United States faced an existential threat from other human beings—namely the Axis powers, led by Hitler, Mussolini, and Hirohito—an enemy that was easy to understand, vilify, and mobilize the nation to fight. By contrast, climate change does not have an easily identifiable villain. In the words of the cartoon character Pogo, "we have met the enemy and he is us." Most human activities in the modern world result in the release of GHG emissions. While fingers of blame are often pointed at particular leaders, industries, and entire nations, the truth is that almost all human beings are complicit, albeit to widely differing degrees, in the problem. Risk-perception researchers have also found that human beings are generally more sensitive to and concerned about threats from other human beings or human technologies than from natural hazards, which are often viewed more fatalistically as uncontrollable acts of nature or God (Slovic, 2000). Unlike the bombing of Pearl Harbor or the terrorist attacks of September 11, 2001, climate change will manifest primarily as more frequent or severe natural hazards (e.g., heat waves, droughts, floods, disease outbreaks, etc.)—harm by a thousand (seemingly natural) cuts rather than a single catastrophic event. Furthermore, while fascism was easily understood as a direct threat to the nation's security (and one's own liberty), climate change is currently perceived by many as a threat to unseen others (future generations, people, and species far away), although it is increasingly raised as a new threat to national security (Fingar, 2009; Leiserowitz et al., 2009). Finally, Americans' response to WWII was deeply rooted in the values of self-defense, patriotism, and national pride. The fight against climate change, however, has not yet tapped into these core values. Nonetheless, WWII stands as a powerful reminder that when the United States is sufficiently motivated and mobilized, it can literally reinvent and transform itself and the world with speed and innovation.

Climate change presents a technical, social, and political challenge that is in some ways similar to—although in other ways quite unique from—many challenges the United States has faced before. The United States has the proven ability to revolutionize technology and the nation's infrastructure, mobilize around a common purpose, work with other nations to combat common threats, and solve major environmental problems at far less cost than originally expected. Previous generations have successfully addressed problems of similarly daunting complexity, uncertainty, and scale.

Comparison of CO_2 Emissions for States Versus National, United States, in 1999 and 2000

Rank	National or Subnational Jurisdication	MMTCE
1	United States	1528.7
2	China (Mainland)	761.59
3	Russian Federation	391.66
4	Japan	323.28
5	India	292.27
6	Germany	214.39
7	**Texas**	**181.11**
8	United Kingdon	154.98
9	Canada	118.96
10	Italy (Including San Marino)	116.86
11	Republic of Korea	116.54
12	Mexico	115.71
13	Saudi Arabia	102.17
14	**California**	**99.52**
15	France (Including Monaco)	98.92
16	Australia	94.09
17	Ukraine	93.55
18	South Africa	89.32
19	Islamic Republic of Iran	84.69
20	Brazil	83.93
21	Poland	82.25
22	Spain	77.22
23	**Pennsylvania**	**74.28**

Rank	National or Subnational Jurisdiction	MMTCE
24	Indonesia	73.57
25	**Ohio**	**71.49**
26	**Florida**	**64.28**
27	**Indiana**	**63.95**
28	**Illinois**	**63.13**
29	Turkey	60.47
30	Taiwan	57.99
31	**New York**	**57.86**
32	**Lousiana**	**56.83**
33	Thailand	54.22
34	Democratic People's Republic of Korea	51.54
35	**Michigan**	**51.46**
36	**Georgia**	**44.93**
37	Venezuela	43.05
38	**Kentucky**	**39.97**
39	**North Carolina**	**39.86**
40	Malaysia	39.41
41	Egypt	38.82
42	Netherlands	37.9
43	Argentina	37.72
44	**Alabama**	**37.62**
45	**Tennessee**	**34.34**
46	**New Jersey**	**34.14**
47	**Missouri**	**33.95**
48	Kazakhstan	33.1
49	Czech Republic	32.42
50	Uzbekistan	32.38
51	**West Virginia**	**30.81**

NOTE: MMTCE, million metric tons of carbon equivalent.

SOURCES: World data from Marland et al. (2003); U.S. state data from EPA (2004).

State Greenhouse Gas Emissions Targets and Baselines

State	Target Reductions	Baseline Year
WA	1990 levels by 2020, 25% below 1990 levels by 2035, and 50% below 1990 levels by 2050	1990
MT	1990 emission levels by 2020	1990
OR	10% below 1990 levels by 2020 and 75% below 1990 levels by 2050	1990
CA	1990 levels by 2020, and 80% below 1990 levels by 2050	1990
UT	2005 levels by 2020	2005
CO	20% below 2005 levels by 2020 and 80% below 2005 levels by 2050	2005
AZ	2000 levels by 2020, and 50% below 2000 levels by 2040	2000
NM	2000 emission levels by 2012, 10% below 2000 levels by 2020, and 75% below 2000 emission levels by 2050	2000
HI	Reduced to 1990 levels by 2020	1990
MN	15% by 2015, 30% by 2025, and 80% by 2050, based on 2005 levels	2005
MI	20% below 2005 levels by 2025 and 80% below 2005 by 2050	2005
IL	1990 levels by 2020 and 60% below 1990 levels by 2050	1990
FL	2000 levels by 2017, 1990 levels by 2025, and 80% below 1990 levels by 2050	2000, 1990
VA	30% below business as usual by 2025	Below BAU
MD	25% below 2006 levels by 2020	2006
NJ	1990 levels by 2020 and to 80% below 2006 levels by 2050	1990, 2006
NY	5% below 1990 levels by 2010, and 10% below 1990 levels by 2020	1990
CT	10% below 1990 levels by 2020	1990
RI	1990 levels by 2010, 10% below 1990 levels by 2020, and 75-85% below 2001 levels in the long term	1990, 2001
MA	10% and 25% below 1990 levels by 2020, as well as targets for 2030 and 2040	1990

State	Target Reductions	Baseline Year
MH	1990 levels by 2010, 10% below 1990 levels by 2020, and 75%-85% below 2001 levels in the long term	1990, 2001
VT	1990 levels by 2010, 10% below 1990 levels by 2020, and 75%-85% below 2001 levels in the long term	1990, 2001
ME	1990 levels by 2010, 10% below 1990 levels by 2020, and 75%-80% below 2003 levels in the long term	1990, 2003

APPENDIX E

America's Climate Choices: Membership Lists

COMMITTEE ON AMERICA'S CLIMATE CHOICES

ALBERT CARNESALE (Chair), University of California, Los Angeles
WILLIAM CHAMEIDES (Vice Chair), Duke University, Durham, North Carolina
DONALD F. BOESCH, University of Maryland Center for Environmental Science, Cambridge
MARILYN A. BROWN, Georgia Institute of Technology, Atlanta
JONATHAN CANNON, University of Virginia, Charlottesville
THOMAS DIETZ, Michigan State University, East Lansing
GEORGE C. EADS, Charles River Associates, Washington, D.C.
ROBERT W. FRI, Resources for the Future, Washington, D.C.
JAMES E. GERINGER, Environmental Systems Research Institute, Cheyenne, Wyoming
DENNIS L. HARTMANN, University of Washington, Seattle
CHARLES O. HOLLIDAY, JR., DuPont, Wilmington, Delaware
KATHARINE L. JACOBS,* Arizona Water Institute, Tucson
THOMAS KARL,* National Oceanic and Atmospheric Administration, Asheville, North Carolina
DIANA M. LIVERMAN, University of Arizona, Tuscon and University of Oxford, United Kingdom
PAMELA A. MATSON, Stanford University, California
PETER H. RAVEN, Missouri Botanical Garden, St. Louis
RICHARD SCHMALENSEE, Massachusetts Institute of Technology, Cambridge
PHILIP R. SHARP, Resources for the Future, Washington, D.C.
PEGGY M. SHEPARD, WE ACT for Environmental Justice, New York, New York
ROBERT H. SOCOLOW, Princeton University, New Jersey
SUSAN SOLOMON, National Oceanic and Atmospheric Administration, Boulder, Colorado
BJORN STIGSON, World Business Council for Sustainable Development, Geneva, Switzerland

THOMAS J. WILBANKS, Oak Ridge National Laboratory, Tennessee
PETER ZANDAN, Public Strategies, Inc., Austin, Texas

PANEL ON LIMITING THE MAGNITUDE OF FUTURE CLIMATE CHANGE

ROBERT W. FRI (Chair), Resources for the Future, Washington, D.C.
MARILYN A. BROWN (Vice Chair), Georgia Institute of Technology, Atlanta
DOUG ARENT, National Renewable Energy Laboratory, Golden, Colorado
ANN CARLSON, University of California, Los Angeles
MAJORA CARTER, Majora Carter Group, LLC, Bronx, New York
LEON CLARKE, Joint Global Change Research Institute (Pacific Northwest National Laboratory/University of Marland), College Park, Maryland
FRANCISCO DE LA CHESNAYE, Electric Power Research Institute, Washington, D.C.
GEORGE C. EADS, Charles River Associates, Washington, D.C.
GENEVIEVE GIULIANO, University of Southern California, Los Angeles
ANDREW J. HOFFMAN, University of Michigan, Ann Arbor
ROBERT O. KEOHANE, Princeton University, New Jersey
LOREN LUTZENHISER, Portland State University, Oregon
BRUCE MCCARL, Texas A&M University, College Station
MACK MCFARLAND, DuPont, Wilmington, Delaware
MARY D. NICHOLS, California Air Resources Board, Sacramento
EDWARD S. RUBIN, Carnegie Mellon University, Pittsburgh, Pennsylvania
THOMAS H. TIETENBERG, Colby College (retired), Waterville, Maine
JAMES A. TRAINHAM, RTI International, Research Triangle Park, North Carolina

PANEL ON ADAPTING TO THE IMPACTS OF CLIMATE CHANGE

KATHARINE L. JACOBS* (Chair, through January 3, 2010), University of Arizona, Tucson
THOMAS J. WILBANKS (Chair), Oak Ridge National Laboratory, Tennessee
BRUCE P. BAUGHMAN, IEM, Inc., Alabaster, Alabama
ROBERT BEACHY,* Donald Danforth Plant Sciences Center, Saint Louis, Missouri
GEORGES C. BENJAMIN, American Public Health Association, Washington, D.C.
JAMES L. BUIZER, Arizona State University, Tempe
F. STUART CHAPIN III, University of Alaska, Fairbanks
W. PETER CHERRY, Science Applications International Corporation, Ann Arbor, Michigan
BRAXTON DAVIS, South Carolina Department of Health and Environmental Control, Charleston
KRISTIE L. EBI, IPCC Technical Support Unit WGII, Stanford, California

JEREMY HARRIS, Sustainable Cities Institute, Honolulu, Hawaii
ROBERT W. KATES, Independent Scholar, Bangor, Maine
HOWARD C. KUNREUTHER, University of Pennsylvania Wharton School of Business, Philadelphia
LINDA O. MEARNS, National Center for Atmospheric Research, Boulder
PHILIP MOTE, Oregon State University, Corvallis
ANDREW A. ROSENBERG, Conservation International, Arlington, Virginia
HENRY G. SCHWARTZ, JR., Jacobs Civil (retired), Saint Louis, Missouri
JOEL B. SMITH, Stratus Consulting, Inc., Boulder, Colorado
GARY W. YOHE, Wesleyan University, Middletown, Connecticut

PANEL ON ADVANCING THE SCIENCE OF CLIMATE CHANGE

PAMELA A. MATSON (Chair), Stanford University, California
THOMAS DIETZ (Vice Chair), Michigan State University, East Lansing
WALEED ABDALATI, University of Colorado at Boulder
ANTONIO J. BUSALACCHI, JR., University of Maryland, College Park
KEN CALDEIRA, Carnegie Institution of Washington, Stanford, California
ROBERT W. CORELL, H. John Heinz III Center for Science, Economics and the Environment, Washington, D.C.
RUTH S. DEFRIES, Columbia University, New York, New York
INEZ Y. FUNG, University of California, Berkeley
STEVEN GAINES, University of California, Santa Barbara
GEORGE M. HORNBERGER, Vanderbilt University, Nashville, Tennessee
MARIA CARMEN LEMOS, University of Michigan, Ann Arbor
SUSANNE C. MOSER, Susanne Moser Research & Consulting, Santa Cruz, California
RICHARD H. MOSS, Joint Global Change Research Institute (Pacific Northwest National Laboratory/University of Marland), College Park, Maryland
EDWARD A. PARSON, University of Michigan, Ann Arbor
A. R. RAVISHANKARA, National Oceanic and Atmospheric Administration, Boulder, Colorado
RAYMOND W. SCHMITT, Woods Hole Oceanographic Institution, Massachusetts
B. L. TURNER II, Arizona State University, Tempe
WARREN M. WASHINGTON, National Center for Atmospheric Research, Boulder, Colorado
JOHN P. WEYANT, Stanford University, California
DAVID A. WHELAN, The Boeing Company, Seal Beach, California

PANEL ON INFORMING EFFECTIVE DECISIONS AND
ACTIONS RELATED TO CLIMATE CHANGE

DIANA LIVERMAN (Co-chair), University of Arizona, Tucson and University of Oxford, United Kingdom
PETER RAVEN (Co-chair), Missouri Botanical Garden, Saint Louis
DANIEL BARSTOW, Challenger Center for Space Science Education, Alexandria, Virginia
ROSINA M. BIERBAUM, University of Michigan, Ann Arbor
DANIEL W. BROMLEY, University of Wisconsin-Madison
ANTHONY LEISEROWITZ, Yale University
ROBERT J. LEMPERT, The RAND Corporation, Santa Monica, California
JIM LOPEZ,* King County, Washington
EDWARD L. MILES, University of Washington, Seattle
BERRIEN MOORE III, Climate Central, Princeton, New Jersey
MARK D. NEWTON, Dell, Inc., Round Rock, Texas
VENKATACHALAM RAMASWAMY, National Oceanic and Atmospheric Administration, Princeton, New Jersey
RICHARD RICHELS, Electric Power Research Institute, Inc., Washington, DC.
DOUGLAS P. SCOTT, Illinois Environmental Protection Agency, Springfield
KATHLEEN J. TIERNEY, University of Colorado at Boulder
CHRIS WALKER, The Carbon Trust LLC, New York, New York
SHARI T. WILSON, Maryland Department of the Environment, Baltimore

Asterisks (*) denote members who resigned during the study process.

Panel on Informing Effective Decisions and Actions Related to Climate Change Biographical Sketches

Diana Liverman holds joint appointments between Oxford University (as Senior Research Fellow in the Environmental Change Institute - ECI) and the University of Arizona (where she co-directs the Institute of the Environment). Her research has focused on the human dimensions of global environmental change, including climate impacts, governance, and policy; climate and development; and the political ecology of environment, land use, and development in Latin America. She has current projects on climate vulnerability and adaptation, climate impacts on food systems, and carbon offsets, and has interest in connecting research to stakeholders and climate science to the arts and creative sector. She has led or coordinated major research programs for the Tyndall Center for Climate Change, the James Martin 21st Century School at Oxford, the Global Environmental Change and Food Systems project, the U.K. Climate Impacts Program, and the Climate Assessment for the Southwest. Her advisory roles have included the National Research Council (NRC) Committee on the Human Dimensions of Global Environmental Change (chair) and the scientific advisory committees for the InterAmerican Institute for Global Change (co-chair). She has a B.A. in geography from University College London, an M.A. from the University of Toronto, and a Ph.D. from the University of California, Los Angeles (UCLA).

Peter H. Raven (NAS) is President of the Missouri Botanical Garden; George Engelmann Professor of Botany, Washington University in St. Louis; and a National Academy of Sciences (NAS) member. Dr. Raven's primary interests are in conservation, global sustainability, plant systematics, biogeography, and evolution. Dr. Raven, a recipient of the National Medal of Science, was a member of President Bill Clinton's Committee of Advisors on Science and Technology. He also served for 12 years as home secretary of the NAS and is a member of the academies of science in Argentina, Brazil, Chile, China, Denmark, India, Italy, Mexico, Russia, Sweden, the United Kingdom, and several other countries. He is the author of numerous books and reports, both popular and scientific. He earned his Ph.D. at UCLA.

Daniel Barstow is President of the Challenger Center for Space Science Education. Formed in the wake of the Challenger Space Shuttle tragedy, Challenger now has 46 Learning Centers throughout the United States, each providing simulated space missions that engage and inspire students and connect them with modern tools of Earth and space exploration. Barstow leads the Challenger Center in strengthening its core operations, growing the network, and expanding its impact on science, technology, engineering, and math education. Over the past two decades, his work has focused on revolutionizing Earth and space science education by emphasizing Earth as a dynamic integrated system, Earth visualization technology, and inquiry-based learning. He was the founding chair of the national Climate Literacy Network and is actively involved in climate education through a variety of partnerships, programs, and policy reform initiatives.

Rosina M. Bierbaum is dean of the University of Michigan School of Natural Resources and Environment. Previously, she served as acting director of the Office of Science and Technology Policy (OSTP) in the Executive Office of the President. Before her appointment as acting director, she was the associate director for environment at OSTP, serving as the administration's senior scientific adviser on environmental research and development on a wide range of issues, including global change, air and water quality, ecosystem management, and energy research and development. Dr. Bierbaum worked closely with the President's National Science and Technology Council and co-chaired its Committee on Environmental and Natural Resources, which coordinated the $5 billion federal research and development efforts in this area, including the (then) $2 billion U.S. Global Change Research Program. She is a fellow of the American Association for the Advancement of Science and the American Academy of Arts and Sciences and is currently a member of the President's Council of Advisors on Science and Technology. She has served on numerous scientific advisory committees and is a board member for several foundations. Dr. Bierbaum received her Ph.D. in ecology and evolution from the State University of New York at Stony Brook.

Daniel W. Bromley is Anderson-Bascom Professor of applied economics at the University of Wisconsin-Madison, and Visiting Professor of Resource Economics at Humboldt University-Berlin. Professor Bromley has published extensively on the institutional foundations of the economy; legal and philosophical dimensions of property rights; economics of natural resources and the environment; and economic development. He has been editor of the journal Land Economics since 1974. He is a Fellow of the Association of Environmental and Resource Economists, the Agricultural and Applied Economics Association, and is listed in Who's Who in Economics. He recently completed a 3-year term as chair of the U.S. Federal Advisory Committee on Marine Protected Areas. He has been a consultant to the Global Environment Facility, the World Bank, the

Ford Foundation, the State of Alaska, the U.S. Agency for International Development, the Asian Development Bank, the Organization for Economic Cooperation and Development, and the Ministry for the Environment in New Zealand. Dr. Bromley's research interests concern the existing institutional arrangements in an economy, and the process of institutional change. He also served as a member of the Committee on the Alaska Groundfish Fishery and Steller Sea Lions, and as a member of the Ocean Studies Board of the NAS. Dr. Bromley received his Ph.D. in natural resource economics from Oregon State University in 1969.

Anthony Leiserowitz, Ph.D., is director of the Yale Project on Climate Change at the School of Forestry and Environmental Studies at Yale University. He is also a principal investigator at the Center for Research on Environmental Decisions at Columbia University. He is an expert on American and international public opinion on global warming, including public perception of climate change risks, support and opposition for climate policies, and willingness to make individual behavioral change. His research investigates the psychological, cultural, political, and geographic factors that drive public environmental perception and behavior. He has conducted survey, experimental, and field research at scales ranging from the global (140+ countries) to the national (United States), municipal (New York City), and local levels (among the Inupiaq Eskimo). He also recently conducted the first empirical assessment of worldwide public values, attitudes, and behaviors regarding global sustainability, including environmental protection, economic growth, and human development.

Robert Lempert is a senior scientist at the RAND Corporation and director of the Frederick S. Pardee Center for Longer Range Global Policy and the Future Human Condition. His research focuses decision making under uncertainty, with an emphasis on climate change, energy, and the environment. Currently, Dr. Lempert's research team assists a number of natural resource agencies in their efforts to include climate change in their long-range plans. Dr. Lempert is a Fellow of the American Physical Society, a member of the Council on Foreign Relations, and a member of the NAS Climate Research Committee. A professor of policy analysis in the Pardee RAND Graduate School, Dr. Lempert is an author of the book Shaping the Next One Hundred Years: New Methods for Quantitative, Longer-Term Policy Analysis.

Edward L. Miles is the Virginia and Prentice Bloedel Professor of Marine and Public Affairs in the School of Marine Affairs at the University of Washington and senior fellow at the Joint Institute for the Study of Atmosphere and Oceans. Since 1965, Dr. Miles has worked at the interface of the natural and social sciences and law with a focus on outer space, the oceans, and the global and regional climate systems. Trained originally in political science and international relations, he has invested more than 30 years in learning about oceanography and fisheries science/management and 20 years in learning about the planetary climate system. His research and teaching interests have encompassed international science and technology policy; the design, creation, and management of international environmental regimes; a wide variety of problems in national and international ocean policy; and the impacts of climate variability and climate change at global and regional spatial scales. Dr. Miles is a member of the NAS and the NRC's Committee on the Human Dimensions of Global Change and Policy and Global Affairs Committee. He is also a member of the American Academy of Arts and Sciences. He received his Ph.D. in international relations/comparative politics from the Graduate School of International Studies and the University of Denver.

Berrien Moore III joined the University of New Hampshire (UNH) faculty in 1969, soon after receiving his Ph.D. in mathematics from the University of Virginia, and became a tenured professor in 1976. He was named University Distinguished Professor in 1997. He has published extensively on the global carbon cycle, biogeochemistry, remote sensing, and environmental policy. He led the Institute for the Study of Earth, Oceans, and Space at UNH as Director from 1987 to early 2008. Upon stepping down from his position at UNH, he became the Executive Director of Climate Central, an emerging, nonprofit, nonpartisan think tank dedicated to producing and providing the public, business and civic leaders, and policy makers with objective and understandable information about climate change and potential solutions. The group is based in Princeton, New Jersey, and Palo Alto, California. Professor Moore continues as Director Emeritus, Institute for the Study of Earth, Oceans, and Space, UNH.

Mark Newton is responsible for Dell corporate environmental sustainability policy and strategy. In this role he leads global policy development, manages stakeholder engagements, and informs corporate strategy on environmental issues including material use, energy efficiency, product recycling, and climate strategy. Mr. Newton joined Dell in 2003 as Manager of Worldwide Environmental Affairs. Under his leadership, Dell integrated global environmental design requirements into the business as part of its ongoing commitment to environmental responsibility. His team established product compliance assurance processes, introduced stakeholder concerns into the business and led policy and process development activities. Prior to joining Dell, Mr. Newton also led product-focused environmental technology programs at Apple and Motorola, and applied chemistry at DEKA R&D. He received a doctorate in chemistry in 1993 from the University of Texas at Dallas.

Venkatachalam Ramaswamy is the director of NOAA's Geophysical Fluid Dynamics Laboratory and is a lecturer with the rank of Professor in the Atmospheric and Oceanic Sciences Program (Department of Geosciences) and the Princeton Environmental

Institute at Princeton University. His primary research is on numerical modeling of the global climate system and investigating the radiative and climatic influences of greenhouse gases, aerosols, and clouds. He has led key chapters in several international and national scientific assessments (e.g., Intergovernmental Panel on Climate Change [IPCC], U.S. Global Change Research Program, and reports [e.g., NRC] on ozone depletion, aerosol climate forcing, climate modeling, and climate change). He was coordinating lead author of chapters in the IPCC Third and Fourth Assessment Reports (2001, 2007) and was also a co-author of the Summary for Policymakers in both reports. He is a member of the World Climate Research Program's Joint Scientific Committee, which provides advice on cutting-edge worldwide research in climate and climate change. He was a member of the organizing committee and a participant in the World Climate Conference-3 (2009). Besides modeling of atmospheric processes such as radiation, aerosols, clouds, the stratosphere, and the hydrologic cycle, he has made use of observations from various platforms, combining them with appropriate model simulations to yield critical information on changes in the climate system, including knowledge of the climate feedbacks. His recent investigations include studies on understanding the roles of different species and processes in the global and regional climate change of the 20th century, and using the IPCC emissions scenarios to determine the projections of climate change in the 21st century.

Richard Richels directs global climate change research at the Electric Power Research Institute (EPRI). His current research focus is the economics of mitigating greenhouse gas emissions. In previous assignments, he directed EPRI's energy analysis, environmental risk, and utility planning research activities. Dr. Richels has served as a lead author for the IPCC Second, Third, and Fourth Scientific Assessments and served on the Synthesis Team for the U.S. National Assessment of Climate Change Impacts on the United States. He also served on the Scientific Steering Committee for the U.S. Carbon Cycle Program. He currently serves on the Advisory Committee for Carnegie-Mellon University's Center for Integrated Study of the Human Dimensions of Global Change, the U.S. Government's Climate Change Science Program Product Development Advisory Committee, and the NAS Climate Research Committee. Dr. Richels received a B.S. in physics from the College of William and Mary in 1968. He was awarded an M.S. in 1973 and Ph.D. in 1976 from Harvard University's Division of Applied Sciences, where he concentrated in decision sciences. While at Harvard he was a member of the Energy and Environmental Policy Center.

Doug Scott was appointed director of the Illinois Environmental Protection Agency effective July 1, 2005, and continues to serve under the leadership of Governor Pat Quinn. Doug Scott was born in Rockford in 1960 and graduated from Rockford East High School in 1978. Doug served as Assistant City Attorney and City Attorney for

Rockford from 1985 to 1995. From 1995 to 2001, he served as an Illinois State Representative for the 67th District. He served on the Energy and Environment Committee and was a member of the committee that rewrote the states' electric utility laws. Doug Scott was elected to the Office of the Mayor of Rockford in April 2001 and served a 4-year term. In addition to being elected to leadership positions in the Illinois Municipal League, United States Conference of Mayors, and national League of Cities, Scott has served as President of the Illinois Chapter of the National Brownfields Association. Director Scott took over leadership of the nation's oldest state environmental agency on the 35th anniversary date of the Illinois EPA's start on July 1, 1970. He is committed to maintaining and enhancing the Agency's key role in protecting our air, land, and water, making government more accountable and accessible to citizens and the regulated community, including local governments and business. He returned home after receiving his bachelor's degree with honors from the University of Tulsa in 1982, and graduating with a law degree with honors from Marquette University in 1985.

Kathleen Tierney is a professor of sociology and director of the Natural Hazards Center at the University of Colorado at Boulder. The Hazards Center is housed in the Institute of Behavioral Science, where Prof. Tierney holds a joint appointment. Dr. Tierney's research focuses on the social dimensions of hazards and disasters, including natural, technological, and human-induced extreme events. She is the author, with Michael Lindell and Ronald Perry, of Facing the Unexpected: Disaster Preparedness and Response in the United States (Joseph Henry Press, 2001). With William Waugh, she recently co-edited Emergency Management: Principles and Practice for Local Government (International City and County Management Association, 2007). She is co-editor of the journal Natural Hazards Review and a former member of the NRC Committee on Disaster Research in the Social Sciences. Her current projects focus on theory and research on disaster resilience; warning systems for extreme weather events; and factors affecting the vulnerability of interdependent critical infrastructure systems.

Chris Walker was the Director (Chief Executive) for North America for The Climate Group–North America and recently accepted the position as head of The Carbon Trust LLC in New York. Prior to these appointments, he was head of Swiss Re's Sustainability Business Development. Here he ran the unit responsible for developing commercial applications to Swiss Re Sustainability commitments and, in particular, business opportunities in sustainability, ecosystem markets, emissions reductions, and renewables. Mr. Walker also served as the government affairs liaison on climate change and GHG emissions issues. While based at Swiss Re's Zurich headquarters, he created and advanced from concept to initiation the company's Greenhouse Gas Risk Solutions unit, specializing in greenhouse gas risk mitigation and opportunity innovation. In 2000, he created and led Swiss Re Group's worldwide GHG emissions market feasibility study determining the market facilitation role for Swiss Re. Mr. Walker received his B.A. in government from St. John's University, attended the Institute on Comparative Political and Economic Systems at Georgetown, and is also a graduate of the St. John's School of Law. Prior to joining Swiss Re in 1996, he practiced law in New York and New Jersey.

Shari T. Wilson was sworn in by Governor Martin O'Malley as Maryland's Secretary of the Environment on March 15, 2007. As Secretary of the Environment, Ms. Wilson directs regulatory, enforcement, and voluntary programs for air quality control of stationary and mobile sources, hazardous and solid waste management and cleanup, oil control, lead paint risk reduction, wastewater treatment, public drinking water supply, wetlands protection, surface and groundwater quality, mining, dam safety, risk assessment, and loan and grant programs for wastewater, water supply, and environmental restoration projects. Secretary Wilson administers a combined operating and capital budget of $460 million and leads a diverse staff of 950 scientists, engineers, and professionals with other technical and administrative expertise. Ms. Wilson is a member of the Governor's Cabinet, BRAC Sub-Cabinet, Chesapeake Bay Sub-Cabinet, and Smart Growth Sub-Cabinet. Prior to becoming Secretary, Ms. Wilson served as a chief solicitor in the Baltimore City Law Department in Land Use and as a manager in the City's Planning Department from 2004 to 2007. Ms. Wilson holds three degrees: a juris doctorate from the University of Baltimore School of Law, a master's degree from the University of Virginia, and a bachelor's degree from the University of Richmond.